Mössbauer Spectroscopy

Yutaka Yoshida · Guido Langouche
Editors

Mössbauer Spectroscopy

Tutorial Book

 Springer

Editors
Yutaka Yoshida
Shizuoka Institute of Science
  and Technology
Fukuroi
Japan

Guido Langouche
Physics Department
Institute for Nuclear and Radiation Physics
Leuven
Belgium

ISBN 978-3-642-44178-3   ISBN 978-3-642-32220-4 (eBook)
DOI 10.1007/978-3-642-32220-4
Springer Heidelberg New York Dordrecht London

Preface

Since the discovery of the Mössbauer Effect many excellent books have been published for researchers and for doctoral and master level students. However, there appears to be no textbook available for final year bachelor students, nor for people working in industry who have received only basic courses in classical mechanics, electromagnetism, quantum mechanics, chemistry and materials science. The challenge of this book is to give an introduction to Mössbauer Spectroscopy for this level.

The ultimate goal of this book is to give the reader not only a scientific introduction to the technique, but also to demonstrate in an attractive way the power of Mössbauer spectroscopy in many fields of science, in order to create interest among the readers in joining the community of Mössbauer spectroscopists. This is particularly important at times where in many Mössbauer laboratories succession is at stake. This book is based on tutorial lectures, organized at the occasion of the 2011 International Conference on the Application of Mössbauer Spectroscopy (ICAME2011) in Kobe.

In Chap. 1 is written by Saburo Nasu, the reader will find a general introduction to Mössbauer Spectroscopy. What is the Mössbauer effect and What is the characteristic feature of Mössbauer spectroscopy? These questions are answered briefly in this chapter. Mössbauer spectroscopy is based on recoilless emission and resonant absorption of gamma radiation by atomic nuclei. Since the electric and magnetic hyperfine interactions of Mössbauer probe atom in solids can be described from the Mössbauer spectra, the essence of experiments, the hyperfine interactions and the spectral line shape are discussed. A few typical examples are also given for laboratory experiments and new nuclear resonance techniques with synchrotron radiation.

Chapter 2 is devoted to chemical applications of Mössbauer spectroscopy, and the authors are Philipp Gütlich and Yann Garcia. They begin with a brief recapitulation of the hyperfine interactions and the relevant parameters observable in a Mössbauer spectrum. The main chapter with selected examples of chemical applications of Mössbauer spectroscopy follows and is subdivided into sections on: Basic information on structure and bonding, switchable molecules, mixed-valence

compounds, molecule-based magnetism, industrial chemical problems like corrosion, and application of a portable miniaturized Mössbauer spectrometer for
applications outside the laboratory and in space. This lecture ends with an outlook
to future developments.

In Chap. 3, Robert E. Vandenberghe describes the applications of Mössbauer
spectroscopy in earth science. With iron as the fourth most abundant element in the
earth crust, ^{57}Fe Mössbauer spectroscopy has become a suitable additional technique for the characterization of all kind of soil materials and minerals. In this
chapter a review of the most important soil materials and minerals is presented. It
starts with a description of the Mössbauer spectroscopic features of the iron oxides
and hydroxides, which are essentially present in soils and sediments. Further, the
Mössbauer spectra from sulfides and carbonates are briefly considered. Finally, the
Mössbauer features of the typical and most common silicate minerals are represented. Because the spectral analysis is not always a straightforward procedure,
some typical examples are given showing the power of Mössbauer spectroscopy in
the characterization of minerals.

Chapter 4 concerns with Fe-based nanostructures investigated by Mössbauer
Spectrometry, and the author is Jean-Marc Greneche. For the last two decades,
numerous projects and studies were devoted to nanoscience and nanotechnology.
The understanding of physical properties requires to correlate the structural,
chemical and physical properties of nanostructures. Numerous new characterization techniques appear for the same period to investigate structural properties at
local nanometer scale. But further local spectroscopic techniques including zero-
field and in-field ^{57}Fe Mössbauer spectrometry have to be used in order to probe
surface and bulk structures, to determine the role of surface or grain boundaries in
the case of nanoparticles and nanostructured powders and static magnetic properties or superparamagnetic relaxation phenomena in the case of magnetic nanostructures. He illustrates thus how in situ ^{57}Fe Mössbauer spectrometry can be
extremely relevant when it is used to investigate local properties of nanocrystalline
alloys, nanostructured powders, nanoparticles and assemblies of particles and
functionalized nanostructures and mesoporous hybrids.

In Chap. 5, subsequently, Mössbauer studies on magnetic multilayers and
interfaces are explained by Teruya Shinjo and Ko Mibu. Mössbauer spectroscopy
has been used as a fundamental analytical tool to characterize various magnetic
materials. It is described in this chapter that a magnetic hyperfine interaction
observed in a Mössbauer spectrum is information particularly useful to investigate
magnetic materials. Examples of studies on various magnetic materials utilizing
^{57}Fe and also ^{119}Sn Mössbauer effect measurements are introduced. Modern
devices including magnetic materials often consist of multilayered structures and
therefore interface properties of ultrathin magnetic layers are of great importance.
Mössbauer spectroscopic studies focusing on magnetic properties of multilayers
and interfaces are described and unique information obtained from Mössbauer
spectra is explained.

Finally, in Chap. 6, Guido Langouche and Yutaka Yoshida provide "implantation Mössbauer spectroscopy" between 1983 and 2011, where three accelerator

facilities, i.e., Hahn-Meitner Institute Berlin, ISOLDE-CERN and RIKEN, have been used for materials research. The techniques developed are "Coulomb-excitation and recoil-implantation", "on-line isotope mass separation and implantation" and "projectile fragmentation and implantation", respectively. The physics on dilute atoms in materials, the final lattice sites and their chemical states as well as diffusion phenomena can be studied immediately after each implantation of the nuclear probes. In order to get such atomistic information, however, it is quite important to select a proper system investigated in a research project, and in addition, to consider both the defects produced near the nuclear probes and their motions during the measurements.

In the last few years many leading scientists in our Mössbauer community passed away in different countries: Hendrik de Waard, Uli Gonser, F. E. Fujita, and recently also Rudolf L. Mössbauer, the discoverer of the effect called after him. We all are strongly indebted to their research work which has been challenging during their whole life. We are extremely grateful to them, and we will do our best in transferring their heritage to the next generations of the Mössbauer family.

Fukuro, September 2012   Yutaka Yoshida
Leuven   Guido Langouche

Contents

Contents

Chapter 1
General Introduction to Mössbauer Spectroscopy

Saburo Nasu

Abstract This chapter describes a general introduction of the Mössbauer spectroscopy. What is the Mössbauer effect and what is the characteristic feature of the Mössbauer spectroscopy? These questions are answered briefly in this chapter. Mössbauer spectroscopy is based on recoilless emission and resonant absorption of gamma radiation by atomic nuclei. Since the electric and magnetic hyperfine interactions of Mössbauer probe atom in solids can be described from the Mössbauer spectra, the essence of experiments, the hyperfine interactions and the spectral line shape are discussed. In addition, the experiments and the new resonance technique with synchrotron radiation have been also briefly described.

In 1956 and 1957, young physicist R. Mössbauer has performed the experiments concerning the scattering of the 129 keV γ-ray of ^{191}Ir by Ir and discovered an increase in scattering at low temperatures. Results obtained and his interpretations were published in 1958 [1–3], which is the beginning of the Mössbauer effects study and its development as the Mössbauer spectroscopy. The Nobel Prize for physics 1961 was awarded to him [4]. Mössbauer spectroscopy is the recoilless emission and the recoilless resonant absorption of the γ-ray by the nucleus. After the ^{57}Fe is found as the most suitable nucleus for the recoilless resonance, the ^{57}Fe Mössbauer spectroscopy is recognized as one of the powerful analytical tools for the study in the fields of solid states and material science. Up to now the Mössbauer spectroscopy is contributing to the progress in physics, chemistry, biology, metallurgy, mineralogy and so on.

In this chapter, we present a general introduction to the Mössbauer spectroscopy, although the many excellent textbooks for the Mössbauer effect and its spectroscopy have been already published [5–15]. First we discuss the nuclear

---

S. Nasu (✉)
Osaka University, Toyonaka, Osaka 560-8531, Japan
e-mail: sabunasu@yahoo.co.jp

Y. Yoshida and G. Langouche (eds.), *Mössbauer Spectroscopy*,
DOI: 10.1007/978-3-642-32220-4_1, © Springer-Verlag Berlin Heidelberg 2013

levels and their decays and then discuss the radiation of γ quanta. Nuclear levels are influenced by the magnetic fields and the electronic charge through the Coulomb interactions, which is so called hyperfine interactions of the nucleus, and discussed in this chapter. Recent development of the synchrotron radiation facilities made possible to use this modern X-ray source as a source for the nuclear resonance instead of the radioisotope. Brief description for the nuclear scattering with the synchrotron radiation is presented in the final section of this chapter.

1.1 What is the Mössbauer Effect?

Mössbauer effect is, as already mentioned, the recoilless emission and the recoilless resonant absorption of γ photons by the nucleus. Two nuclei, which are in principal identical nucleus, are necessary to observe the Mössbauer effect. Source is the nucleus, which emits γ photon, and the absorber is the nucleus that resonantly absorbs γ photon. Since the source emits γ photons, source nuclei have unstable nuclear levels and prepared in principally by the nuclear reactions induced for the stable nuclei by the bombardments of particles like deuteron, proton and neutron so on. Excited nucleus is a parent nucleus having a moderate lifetime and decays to the first excited state of Mössbauer source nucleus, which is followed, by the decay to the stable ground state of the Mössbauer nucleus. Decay to the ground state creates the emission of γ-ray photons. The γ-ray energy emitted from nucleus is usually very high compared to the photons emitted by the atomic process. During the emission and absorption of γ-ray, the momentum and energy conservation law needs usually the recoil of the nucleus. However, when the γ-ray has relatively low energy and the source nuclei embedded in solid, there is a large probability to have zero recoil energy (recoilless) in the γ-ray emission process. In the same way for the absorption process, there is a large probability to have recoilless resonant absorption. Figure 1.1 shows schematic illustration of the source and absorber in a Mössbauer experiment. Creation of the source and the de-excitation of the absorber after resonant absorption are also indicated. Doppler velocity is used to modulate the γ-ray energy by the first order Doppler effect, for example, by the mechanical motion of the source. What is the Mössbauer effect? The answer for this question is, by one word, as follows; that the Mössbauer effect is recoilless emission and absorption of γ photons by atomic nucleus.

1.1.1 Nuclear Levels

As shown in Fig. 1.1, the parent nuclei for Mössbauer experiment should have a moderate lifetime (for ^{57}Fe Mössbauer experiment, ^{57}Co has a half-life 270 days). After the decay of the parent nucleus by α or β and so on to the excited state of Mössbauer nucleus ^{57}Co decays by E. C. (electron capture) to ^{57}Fe excited states.

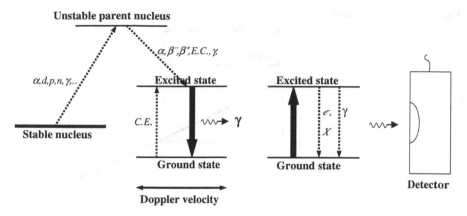

Fig. 1.1 Schematic illustration of the source and the absorber in a Mössbauer experiment

Finally this excited states decay to ground state of the Mössbauer nucleus. For ^{57}Fe, ^{57}Co decays to two excited states of ^{57}Fe, but decays 99.8 % to the lowest energy level 14.4 keV that is called as a first excited state from ground state of ^{57}Fe. Ground state of ^{57}Fe is a stable nucleus and shows no decay. Each nuclear level has a unique energy, spin, parity and decay constant λ (probability of decaying per unit time). Ground state is, of course, zero energy and decay constant is zero.

1.1.2 Decay[1]

From the definition of λ, the number decay in a time dt is given by

$$dN = -\lambda N(t)dt \qquad (1.1)$$

where $N(t)$ is the number of particles present at time t. Integration yields the exponential decay law,

$$N(t) = N(0)e^{-\lambda t}. \qquad (1.2)$$

Figure 1.2 shows $\log N(t)$ versus t. Half-life and mean life (lifetime) are indicated. The mean life is the average time a particle exists before it decays; it is connected to λ and $t_{1/2}$ by

$$\tau = \frac{1}{\lambda} = \frac{t_{1/2}}{\ln 2} = 1.44 t_{1/2}. \qquad (1.3)$$

[1] Frauenfelder H. and. Henly M. E: Subatomic Physics (Prentice-Hall, Inc., Englewood Cliffs, New Jersey 1974).

Fig. 1.2 Exponential decay

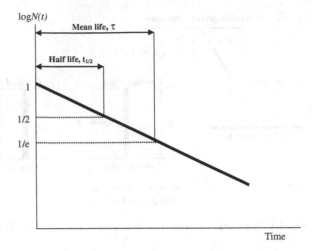

To relate the exponential decay to properties of the decaying state, the time dependence of the wave functions of a particle at rest is shown explicitly as

$$\psi(t) = \psi(0)e^{-iEt/\hbar}. \tag{1.4}$$

If the energy E of this state is real, the probability of finding the particle not a function of time because

$$|\psi(t)|^2 = |\psi(0)|^2. \tag{1.5}$$

A particle described as a wave function as shown in (1.4) with real E does not decay. To introduce an exponential decay of a state described by $\psi(t)$, a small imaginary part is added to the energy,

$$E = E_0 - \frac{1}{2}i\Gamma, \tag{1.6}$$

where E_0 and Γ are real and where the factor $\frac{1}{2}$ is chosen for convenience. With (1.6), the probability becomes

$$|\psi(t)|^2 = |\psi(0)|^2 e^{-\Gamma t/\hbar}. \tag{1.7}$$

It agrees with the exponential law (1.2) if

$$\Gamma = \lambda\hbar. \tag{1.8}$$

With (1.4) and (1.6) the wave function of a decaying state is

$$\psi(t) = \psi(0)e^{-iE_0 t/\hbar}e^{-\Gamma t/2\hbar}. \tag{1.9}$$

By the Fourier inversion of $\psi(t)$ to $g(\omega)$, $g(\omega)$ is given by

$$g(\omega) = \frac{1}{\sqrt{2}}\psi(0)\int\limits_{0}^{\infty} dt\, e^{+i(\omega - E_0/\hbar)t}e^{-\Gamma t/2\hbar}, \tag{1.10}$$

or

$$g(\omega) = \frac{\psi(0)}{\sqrt{2}}\frac{i\hbar}{(\hbar\omega - E_0) + i\Gamma/2}, \tag{1.11}$$

where the decay starts at time $t = 0$ and then the lower limit on the integral can be set equal to zero. Since $E = \hbar\omega$, the probability density $P(E)$ of finding energy E is proportional to $|g(\omega)|^2 = g^*(\omega)g(\omega)$:

$$P(E) = const.g^*(\omega)g(\omega) = const.\frac{\hbar^2}{2\pi}\frac{|\psi(0)|^2}{(\hbar\omega - E_0)^2 + \Gamma^2/4}. \tag{1.12}$$

The condition

$$\int\limits_{-\infty}^{+\infty} P(E)dE = 1 \tag{1.13}$$

yields

$$const. = \frac{\Gamma}{\hbar^2|\psi(0)|^2}, $$

and $P(E)$ becomes

$$P(E) = \frac{\Gamma}{2\pi}\frac{1}{(E - E_0)^2 + (\Gamma/2)^2}. \tag{1.14}$$

This equation gives the very important fact that the energy of decaying state is not a constant and is distributed over a region with a width determined by the decay constant. The width is called natural line width. The shape of the distribution is called a Lorentzian or Breit–Wigner curve as shown in Fig. 1.3.

Fig. 1.3 Lorentzian distribution curve of decaying state. Γ is the full width of half maximum

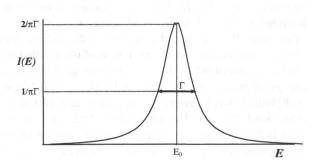

From (1.8), the product of lifetime and width is

$$\tau \cdot \Gamma = \hbar \qquad (1.15)$$

This relation can be interpreted as a Heisenberg uncertainty relation, $\Delta t \cdot \Delta E \geq \hbar$. To measure the energy of the state within an uncertainty $\Delta E = \Gamma$, a time $\Delta t = \tau$ is needed. For ^{57}Fe, the lifetime of the first excited state is 141 ns and the width of this state given by $\Gamma = \hbar/\tau = 4.67 \times 10^{-9}$ eV. The ratio of decay energy 14.4 keV of the first excited state to width is $E/\Gamma = 3.1 \times 10^{12}$ which is extremely large, that can make the physical measurements in very high precession.

1.1.3 Recoilless Emission and Recoilless Resonant Absorption of γ Photons

Figure 1.1 shows the essence of the Mössbauer effect and the widths of the excited states of source and absorber are extremely small which is the order of 10^{-9} eV for ^{57}Fe. The width of the emitted γ-ray is the order of 10^{-9} eV. When the γ-ray emission and absorption processes do not need any energy and momentum transfer to nuclei, the resonant absorption of γ photons can be expected. However, the energy and momentum conservation law is always important for the emission and absorption process of γ photons. Consider the γ-ray emission process by the decay from first excite state to ground state of ^{57}Fe, momentum conservation gives $p_{nucleus} = -p_{photon}$. The magnitude of the photon momentum is connected with the photon energy by $p_{photon} = E_{photon}/c$. There will be a recoil energy loss R in this process. To calculate R, we assume that the photon is emitted by a nucleus of mass M that is at rest before the decay. Since the nuclei are vey heavy, we can use the nonrelativistic approximation to connect the magnitude of the momentum with the recoil energy R that is the recoil energy of the free nucleus and given by

$$R = p_{nucleus}^2/2M = E_0^2/2Mc^2. \qquad (1.16)$$

For the absorption process, the same conservation law should be satisfied. For the transition $E_0 = 14.4$ keV in ^{57}Fe, R is 1.9×10^{-3} eV which is 10^6 times large compared to the natural line width of the excited state and no resonance between source and absorber for the free nucleus can be expected. When the nucleus is bounded into the solid, the recoil energy can be dispersed by the excitation of the solid. When the source and absorber are the nuclei embedded into the solid, recoil energy R may be used for the excitation of phonon that is the vibration state of solid. Phonon is quantized as discrete value in solid and in usual metals the excitation energy of phonon states is the order of 10^{-1}–10^{-2} eV and there is rather large probability to have a zero phonon excitation in emission and absorption process; in other word, the recoilless emission and recoilless resonant absorption of photon. This is the most important characteristic feature of the Mössbauer effect. As a consequence, the γ photon emitted by the decay from the first excited state that has a natural line width Γ is resonantly absorbed by the absorber nucleus to excite to the first excited state apart 14.4 keV for ^{57}Fe from ground state that also has same natural line width Γ. Spectrum obtained by the Mössbauer experiment as shown in Fig. 1.1 consists of the γ photon counts passing through the absorber as a function of Doppler velocity v of the source. By the Doppler velocity v the photon energy E_0 is changed by the first order Doppler effect $\Delta E = \frac{v}{c} E_0$. Finally Mössbauer spectrum consists of γ-ray count rate versus Doppler velocity v. Since source line shape is a Lorentzian with width Γ and absorber line shape is also Lorentzian with width Γ, the observed spectrum by transmission of the γ-ray should have a Lorentzian with width of 2Γ because of the convolution of source and absorber Lorentz functions. This is an ideal case for the ideal source and very thin absorber. However, the real source shows a little self-absorption of γ photons and the absorber has a finite thickness showing a thickness effect, although the resultant spectrum is very close to Lorentz function [17, 18].

A general expression for the fraction of zero phonon or recoilless process is

$$f = \exp\left(-\frac{4\pi^2\langle x^2\rangle}{\lambda^2}\right) = \exp\left[-\kappa^2\langle x^2\rangle\right], \qquad (1.17)$$

where λ is the wavelength of the γ-ray, $\kappa = 2\pi/\lambda = E/\hbar c$, and $\langle x^2\rangle$ is the component of the mean square vibrational amplitude of the emitting nucleus in the direction of the γ-ray. In order to obtain a value of f close to unity, we require $\kappa^2\langle x^2\rangle \ll 1$, which requires that the root mean square displacement of the nucleus is small compared to the wavelength of the γ-ray.

Emission of γ-ray by the decay from excited state $|i\rangle$ to ground state $|f\rangle$ is accompanied the conservation of energy and the angular momentum. Using the quantities denoted in Fig. 1.4, following conservation should be satisfied. For energy conservation,

$$E_i = E_f + \hbar\omega$$

Angular momentum and parity conservations give

$$|j_e - j_g| \le L \le |j_e + j_g| \text{ and } L \ne 0 \qquad (1.18)$$

$$|M| = |m_e - m_g| \le L \text{ and } \pi_e = \pi_g\pi, \qquad (1.19)$$

Fig. 1.4 Initial and final states of nuclear levels, photon energy, their angular momentums and parities

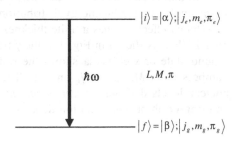

$|i\rangle = |\alpha\rangle; |j_e, m_e, \pi_e\rangle$

$\hbar\omega \qquad L, M, \pi$

$|f\rangle = |\beta\rangle; |j_g, m_g, \pi_g\rangle$

Fig. 1.5 Decay of ^{57}Co to ^{57}Fe

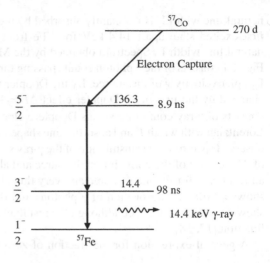

where suffixes e and g denote the excited state and ground state, respectively. It is known that any 2^L pole multipole radiation field carries angular momentum L and z-component of angular momentum M. The parity π of a multipole field is $(-1)^L$ or $(-1)^{L-1}$ for EL or ML radiation, respectively [16].

For the ^{57}Fe Mössbauer level whose parent nucleus is ^{57}Co, the decay scheme is shown in Fig. 1.5. Spin and parity of the first excited state are 3/2 and $-$. For the ground state, spin and parity are 1/2 and $-$. Because of no parity change the multipole radiation field of the γ-ray emission decaying from excited state to ground state is M1 and E2 radiation. Higher order multipolarity is less expected and known as negligible small like 0.0006 % for the 14.4 keV γ radiation from the first excited state of ^{57}Fe [17]. For γ-ray, the distribution function $F_L^M(\theta)$ can be found by calculating the energy flow (Poynting vector) as a function of θ for multipole radiation characterized by the quantum numbers L and M. For dipole radiation, one obtains

$$F_1^0(\theta) = 3 \sin^2 \theta,$$
$$F_1^{\pm 1}(\theta) = \frac{3}{2}(1 + \cos^2 \theta).$$

(1.20)

Usually ^{57}Co source for the ^{57}Fe Mössbauer experiment is doped into the metal like Rh matrix and distributed uniformly without self-absorption showing a rather sharp single Lorentzian energy distribution. Absorber is a specimen to study by the Mössbauer effect and has a finite thickness containing multiple phases of resonant nuclei ^{57}Fe. As shown in Fig. 1.4, the γ transition of nucleus from excited state to ground state or vice versa shows the radiation field depending on the quantum numbers of L, M, j_e, m_e, j_g and m_g. That is, the transition probability between nuclear levels depends on $\Delta m = m_e - m_g$ and the angular θ dependence of the emitting or absorbing radiation depends on the L and M given by (1.20).

For a thick absorber specimen containing n multiple phases, the following expression is frequently used for the analysis of the transmission spectra [18–20].

$$p(v) = N_0(1 - f_s) + N_b + N_0 f_s \int_{-\infty}^{+\infty} s(E - v)e^{-\sigma(E)}dE, \qquad (1.21)$$

where N_0 is counts from resonant radiation, N_b counts from other radiation and f_s recoilless fraction of source. Source has Lorentzian line shape as mentioned before and is given by

$$s(E - v) = \frac{\Gamma_s}{2\pi} \frac{1}{(E - v)^2 + (\Gamma_s/2)^2}.$$

Cross section $\sigma(E)$ for the resonance absorption is usually expected to have the form

$$\sigma(E) = \sum_n \frac{T_n(\Gamma_n/2)^2}{(E - E_n)^2 + (\Gamma_n/2)^2}. \qquad (1.22)$$

where $T = \sum_n T_n = n_a a f_a \sigma_0 t$ is a total effective thickness of the absorber. Other notations are listed in Table 1.1.

When the absorber thickness is thin, the (1.21) can be rewritten as follows;

$$c(v) \equiv \frac{N_0 + N_b - p(v)}{N_0 f_s} = \int_{-\infty}^{+\infty} s(E - v)(1 - e^{-\sigma(E)})dE$$

$$\simeq \sum_n \frac{\Gamma_n}{\Gamma_s + \Gamma_n} \frac{T_n((\Gamma_s + \Gamma_n)/2)^2}{(v - E_n)^2 + ((\Gamma_s + \Gamma_n)/2)^2} \qquad (1.23)$$

When the absorber is thin, the observed transmission Mössbauer spectrum is a superposition of several Lorentz functions.

Table 1.1 List of the parameters used in Eq. (1.22) and T

Γ_s	Full width at half maximum (FWHM) of the source
Γ_n	FWHM of the absorption
E_n	Absorption line center
T_n	Effective thickness of nth atomic site
n_a	Number of atoms per cubic centimeter of absorber volume
f_a	Recoilless fraction of absorber
a	Fractional abundance of atoms which can absorb resonantly
σ_0^a	Absorption cross section at resonance in square centimeter
t	Thickness of absorber in centimeter

[a] $\sigma_0 = \frac{\lambda^2}{2\pi} \times \frac{2j_e+1}{2j_g+1} \times \frac{1}{1+\alpha}$ and α is the internal conversion coefficient of γ transition

1.2 Application to the Material Research

Previous discussion on the nuclear levels is for the point nucleus or for the bare nucleus. But the nucleus has a finite charge radius and surrounded by the atomic electron clouds. The interaction with the electron clouds leads the shift and/or split of the nuclear levels, which is called as the hyperfine interactions. When the nucleus has a magnetic dipole moment, the magnetic interactions with the magnetic field due to the atom's own electrons lifts all of their $(2j + 1)$-fold degeneracy of the nuclear levels which is called as the nuclear Zeeman effect. Coulomb interactions between electron charge clouds and the proton charge distribution within the nucleus make a shift and/or split of the nuclear levels. From the observation of these shift and split, it is possible to determine the electronic states of the Mössbauer active atom, using nuclear parameters like a charge radius and electromagnetic moments of the nucleus.

Soon after its discovery the application of the Mössbauer effect has been made by the fact that the energy of electromagnetic radiation can be measured with very great precision. Using this characteristic feature, study on gravitational red shift has been performed [21], but details concerning this type of the experiment are not described here. Atomic motion and lattice vibration of solid are also important research subjects using Mössbauer effect. However, the Mössbauer effect is fundamentally concerned only with processes in which the quantum state of the lattice remains unchanged. The information concerning the motion of the lattice atoms is less obtained in an experiment where only recoilless γ-rays are observed.

A development of synchrotron radiation facility made possible to perform the nuclear resonant scattering with synchrotron radiation. Elastic scattering is identical, in principle, to the Mössbauer resonance by γ photons from radioactive nuclei. From the inelastic scattering one can observe the scattering involved phonon annihilation and creation in solid. Nuclear resonant scattering with synchrotron radiation will briefly described in final part of this chapter.

1.2.1 Hyperfine Interactions

Coulomb interactions between electron charge clouds and the proton charge distribution within the nucleus is given by

$$E_{el} = \iint \frac{\rho_e(r_e)\rho_N(r_N)}{|r_e - r_N|} dr_e dr_N,$$

and

$$\frac{1}{|r_e - r_N|} = 4\pi \sum_{k=0}^{\infty} \sum_{m=-k}^{k} \frac{1}{2k+1} \frac{r_<^k}{r_>^{k+1}} Y_{km}^*(\hat{r}_e) Y_{km}(\hat{r}_N). \tag{1.24}$$

where the suffixes N and e denote the nucleus and the electron. From (1.24), the monopole interaction for $k = 0$ is given by

$$E_0 = \iint \frac{1}{r_N} \rho_e(r_e)\rho_N(r_N)dr_e dr_N.$$

This interaction energy can be computed classically by considering a uniformly charged spherical nucleus imbedded in its s-electron charge cloud and makes a shift of the nuclear levels. Observed shift of the resonance spectrum is called as "isomer shift" because the shift depends on the difference in the nuclear radii of the isomeric (ex.) and ground (gd.) states. A change in the s-electron density that might be due to the change in valence will cause the change in isomer shift. It implies that for ^{57}Fe the isomer shift values depend on the valence state of Fe atom.

The change in γ-ray energy due to the monopole interaction is therefore the difference of two terms written for the nucleus in isomeric (ex.) and ground (gd.) states,

$$\Delta E_{ex} - \Delta E_{gd} = \frac{2\pi}{5} Ze^2 |\psi(0)|^2 (R_{ex}^2 - R_{gd}^2).$$

ΔE is the difference from the point nucleus. Observed isomer shift is the difference between source and absorber and given by

$$\text{Isomer shift (IS)} = \frac{2\pi}{5} Ze^2 \left[|\psi_a(0)|^2 - |\psi_s(0)|^2 \right] (R_{ex}^2 - R_{gd}^2).$$

Figure 1.6a shows the isomer shift and the expected Mössbauer absorption spectrum.

Since the nucleus has no electronic dipole moment from the parity, the electronic dipole interaction for $k = 1$ does not exist. Next interaction is the electronic quadrupole interaction for $k = 2$ which is given by

$$E_2 = \frac{4}{5} \pi \sum_{m=-2}^{2} \int \rho_N(r_N) r_N^2 Y_{2m}(\hat{r}_N)dr_N \int \rho_e(r_e) \frac{1}{r_e^3} Y_{2m}^*(\hat{r}_e)dr_e.$$

This is the result of the interaction of the nuclear quadrupole moment Q with the electric field gradient EFG that is due to other charges in the solid. Nuclear quadrupole moment Q is expressed by the left-side integral and the electric field gradient is expressed by the right-side integral in the above equation. The nuclear quadrupole moment reflects the deviation of the nucleus from spherical symmetry.

A flattened nucleus has negative Q while an elongated nucleus has a positive Q. Nuclei whose spin is 0 or 1/2 are spherically symmetric and have a zero Q; thus the ground state of ^{57}Fe, with $j_g = 1/2$, cannot exhibit quadrupole splitting. The electric field gradient is obtained by applying the gradient operator to the three components of the electric field that is a vector. Consequently the electric field gradient EFG is a 3×3 tensor. However, this tensor is reduced to diagonal form

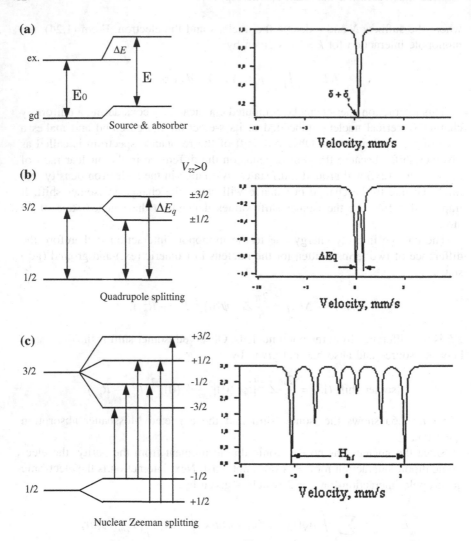

Fig. 1.6 Shift and splitting of nuclear levels of ^{57}Fe nucleus by the hyperfine interactions and expected Mössbauer spectra. **a** Center shift of the observed absorption line. In this case the observed shift is the sum of isomer shift and second order Doppler shift. **b** Electric quadrupole splitting ΔEq and **c** magnetic hyperfine splitting that is a nuclear Zeeman splitting. H_{hf} is the magnitude of the hyperfine magnetic field at nucleus

using a proper coordinate system and can be completely specified by three components $\partial^2 V/\partial x^2$, $\partial^2 V/\partial y^2$ and $\partial^2 V/\partial z^2$ those are generally denoted as V_{xx}, V_{yy} and V_{zz}. These three components must obey the following Laplace equation,

$$V_{xx} + V_{yy} + V_{zz} = 0.$$

Consequently, there remain only two independent components, usually chosen as V_{zz} and η which is the asymmetry parameter defined by

$$\eta = \frac{V_{xx} - V_{yy}}{V_{zz}}.$$

Usually the components are chosen as $|V_{zz}| > |V_{xx}| \geq |V_{yy}|$ and then $0 \leq \eta \leq 1$.

The excited state of ^{57}Fe has spin 3/2 and has a positive Q value showing a quadrupole splitting. From the quadrupole interaction Hamiltonian, the eigenvalues are

$$E_Q = \frac{V_{zz}Q}{4j(2j-1)} \left[3m_j^2 - j(j+1) \right] (1 + \eta^2/3)^{\frac{1}{2}}, m_j = j, j-1, \,,\,,\, -j.$$

Figure 1.6b shows the quadrupole splitting of ^{57}Fe excited state $j = 3/2$ and the resultant Mössbauer spectrum which is a doublet.

As described in the first part of this section, when the nucleus has a nuclear magnetic dipole moment, the magnetic hyperfine interaction with magnetic state of its own electrons lifts all of their $(2j + 1)$-fold degeneracy of the nuclear levels as the nuclear Zeeman effect. For ^{57}Fe, both of excited and ground states have nuclear magnetic dipole moments and interact with the magnetic fields being created by the electronic states like electron spin-polarization, electron orbital current and dipole field by other electron spin or external magnetic field. Nuclear levels of excited and ground states are divided into six different levels. Excited state $(j_e = 3/2)$ splits into four sublevels and ground state $(j_g = 1/2)$ splits into two sublevels and eight γ transitions between excited and ground states can be expected. However, the radiation field is $M1$ and its selection rule excludes the transition of $\Delta m = \pm 2$ and resultant Mössbauer absorption spectrum consists of 6 lines as shown in Fig. 1.6c. The hyperfine magnetic field from a single electron is

$$H = -2\mu_B \left[\frac{8\pi}{3} S|\psi(0)|^2 + \left(\frac{L}{r^3} \right) + \left(\frac{3r(S \cdot r) - r^2 S}{r^5} \right) \right].$$

The first term is the Fermi contact interaction and is only operable for s electrons. The second term is due to the orbital current. The third term represents the dipole field due to the electron spin. These two latter terms are generally smaller than the contact term and vanish for s-state ions. For ^{57}Fe in Fe^{3+} ($S = 5/2$, $L = 0$), the contact interaction gives about -60 T. For ^{57}Fe in Fe^{2+} ($S = 2$, $L = 2$), the field is somewhat smaller because of smaller spin and also appreciable positive orbital contribution. At room temperature the hyperfine magnetic field at ^{57}Fe in metallic iron is -33 T and this is the reference value to determined the hyperfine magnetic field in magnetic materials using ^{57}Fe Mössbauer spectroscopy. Nuclear levels of ^{57}Fe under magnetic field and the expected Mössbauer spectrum are shown in Fig. 1.6c.

Figure 1.6c shows the ^{57}Fe spectrum obtained from magnetically ordered materials like ferromagnet or antiferromagnet that have a cubic symmetry to vanish the quadrupole interaction. Generally the magnetic substance has non-cubic symmetry having the electric field gradient and shows the quadrupole interaction. In this case the ^{57}Fe Mössbauer spectrum shows a combined quadrupole and magnetic hyperfine interactions. For the analysis of the combined quadrupole and magnetic interaction, the several methods and computer programs for the numerical analysis have been published from rather old time [22, 23].

1.2.2 Dynamical Effects on the Spectral Shape

Previous section, the static hyperfine interactions have been concerned about the application to the material research. One might imagine that the Mössbauer spectroscopy can apply to study of the dynamics of the lattice in which the Mössbauer active atoms is embedded. Atomic jump associated with the atomic diffusion of the Mössbauer active atoms can be also studies by the Mössbauer spectroscopy [24].

For the study of lattice dynamics, recoilless fraction f (1.17) indicates the vibrational dynamics of the lattice that concerns the mean square displacements of Mössbauer atoms $<x^2>$. Similarly the second order Doppler shift (SOD) that is given by

$$\text{SOD (Second order Doppler shift)} = -\frac{\langle v^2 \rangle}{2c},$$

where $<v^2>$ is the mean square velocity of Mössbauer active atoms in lattice. These $<x^2>$ and $<v^2>$ can be evaluated using various models for lattice dynamics, for example, using Debye model.

Mössbauer spectroscopy can be used to observe the effects of diffusive motion in solids under three different conditions; (1) when the diffusing atom itself is a Mössbauer active atom, (2) when a Mössbauer active atom experiences a changing hyperfine interaction as a result of the motion of a more rapidly diffusing species in the material, and (3) when the γ-ray is scattered by some material and the Mössbauer effect is used for energy analysis of the scattered γ-ray. In the first case, if the atom with resonant nucleus moves rapidly enough to travel a distance that is large relative to γ-ray wavelength in a time that is short relative to the nuclear lifetime, one can expect a line broadening since the motion destroys the coherence of the γ-ray.

If the direction of the main component V_{zz} of the electric field gradient (EFG) fluctuates rapidly as a result of diffusion, the Mössbauer spectrum may be changed in shape. Two atomic sites have very different isomer shift values and the jump between these two states also may show the time-dependent Mössbauer spectra as a function of the jump frequency. The time scales over which the Mössbauer effect can be used to observe the dynamical effect is determined by the characteristic times associated with the resonance; the natural lifetime and the Larmor precession times of the hyperfine interactions.

Application to the study on the magnetic materials is frequently performed and recognized as the most important subject in the field of material science. Recently nano-meter size magnet has been interested in a magnetic device technology. It is well known that the nano-meter size fine magnetic particles (ferromagnet and antiferromagnet) show superparamagnetism. From the early stage of the development in the Mössbauer spectroscopy, the ^{57}Fe Mössbauer effect has been used to study on the behavior of superparamagnetic particles that contain ^{57}Fe Mossbauer active atoms [25]. The Zeeman splitting of the nuclear excited state is resolved if the spacing between the levels is larger than the width of the levels, that is, $\omega_L = |g_e|\mu_N H_{hf}/\hbar > \Gamma_N$ or $\tau_L < \tau_N$. This condition is generally satisfied for ^{57}Fe in magnetically ordered materials because of the large value of the hyperfine field. Hyperfine field $H_{hf} = 50$ T gives $\tau_L = 4 \times 10^{-9}$ s, whereas $\tau_N = 1.4 \times 10^{-7}$ s. It implies there is always sufficient time for several complete Larmor precessions to take place before the nuclear decays and τ_N is not time scale to determine the relaxation behavior. τ_L is considered as the measuring time for the observation of a hyperfine interaction. If the relaxation time τ is such that $\tau \gg \tau_L$, the orientation of H_{hf} hardly changes during one Larmor precession time and one can observe the six lines ^{57}Fe Mössbauer spectrum. If $\tau \ll \tau_L$, the orientation of H_{hf} changes many times before the completion of one Larmor precession and one can observe the singlet or doublet Mössbauer spectrum like paramagnetic substance. However, when $\tau \sim \tau_L$, the complicated spectra as a function of τ have been observed as shown in Fig. 1.7.

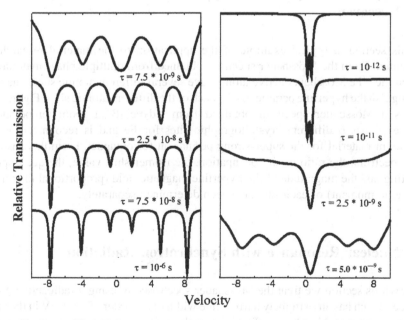

Fig. 1.7 Calculated ^{57}Fe Mossbauer spectra as a function of relaxation time τ from 10^{-12} s to 10^{-6} s. Hyperfine parameters like H_{hf} and ΔE_q are used for the values of α-FeOOH

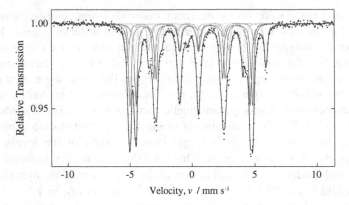

Fig. 1.8 ^{57}Fe Mossbauer spectrum obtained from $Nd_2Fe_{14}B$ at room temperature. Spectrum consists of six different sextets that correspond to ^{57}Fe at six different atomic sites. Hyperfine interaction parameters like H_{hf}, ΔE_q, IS and relative aria contribution can be determined from the analysis of the spectrum

Figure 1.7 shows the calculated Mössbauer spectra as a function of relaxation time τ assuming the magnitude of saturated hyperfine field and the quadrupole splitting are same as observed for α-FeOOH.

1.3 Typical Example of the Application to the Material Research

In this section, a typical example of the application to the material research is presented. From the Mössbauer spectrum obtained from multiple phase material in which the ^{57}Fe Mössbauer active atoms have several different atomic sites, one can distinguish the hyperfine parameters for each of the different atomic sites. Figure 1.8 shows a Mössbauer spectrum obtained from $Nd_2Fe_{14}B$ at room temperature. $Nd_2Fe_{14}B$ has 6 different crystallographic sites for Fe and is recognized as an important material for the super-strong permanent ferromagnet. From this absorption spectrum, the ratio of site occupation, the isomer shift value, the quadrupole splitting and the magnitude of the hyperfine magnetic field (proportional to atomic magnetic moment) of each site have been determined separately.

1.4 Nuclear Resonance with Synchrotron Radiation

In previous section we treat the Mössbauer spectroscopy using a radioactive γ-ray source, which has an extremely narrow line width of the order of 10^{-9} eV in the case of ^{57}Fe resonance. Mössbauer effect is a recoilless emission and a recoilless resonant absorption of low energy γ photons in a combination with source and absorber nuclei.

The recent development of the synchrotron radiation facilities (especially the development of the insertion devices like undulator) has made possible to perform the nuclear resonance scattering with the synchrotron radiation because of its extremely high Brilliance (photons/sec/mrad/0.1 % bandwidth) like $10^{+17} \sim 10^{+19}$ for the standard undulator in SPring8 [26]. On the other hand, the bandwidth of the synchrotron X-ray is known to be about a few eV using double single crystal monochrometer and is extremely large to excite the nuclear levels. Development of high-resolution monochrometer using 2 nested channel cuts Si crystals reduced the bandwidth of undulator X-ray to 1.6–3.5 meV with 5×10^8–4×10^9 photons/sec/100 mA, but its bandwidth is still 10^6 times larger bandwidth compared to γ photons in Mössbauer nucleus [27]. In order to eliminate the difficulty to observe the photons by the nuclear resonant scattering with synchrotron radiation, it is now a standard technique to use the pulsed operation of the synchrotron storage ring and to use the characteristic time difference in the scattering processes. Thomson scattering occurs instantaneously, but the emission of photons as a de-excitation of nuclear levels after the nuclear resonant scattering, roughly speaking, was delayed in time by the lifetime of the excited state. Schematic illustration for the scattering with the synchrotron radiation is shown in Fig. 1.9. Characteristic features of the synchrotron radiation are as follows; (1) non- radioactive source, (2) high brilliance, (3) white X-ray in principle, (2) small beam size with well-collimated beam, (3) linear polarization, and (4) pulsed operation to form a time domain. In 1974, Ruby proposed to use the synchrotron radiation for the source in Mössbauer spectroscopy [28] and now we have many facilities for the synchrotron radiation at many places in the world [26].

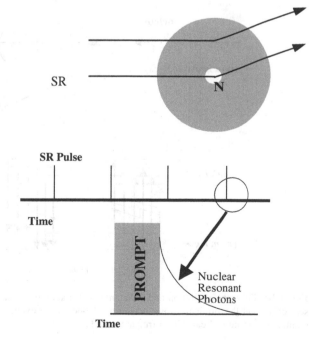

Fig. 1.9 Schematic illustration of the scattering process with synchrotron radiation. *N* nucleus. *SR pulse duration* single banch mode operation of the storage ring

1.4.1 Elastic Scattering

Nuclear excitation and nuclear resonant scattering with synchrotron radiation have opened new fields in Mössbauer spectroscopy and have quite different aspects with the spectroscopy using a radioactive source. For example, as shown in Fig. 1.10, when the high brilliant radiation pulse passed through the resonant material and excite collectively the assemblies of the resonance nuclei in time shorter than the lifetime of the nuclear excited state, the nuclear excitons are formed and their coherent radiation decay occurs within much shorter period compared with an usual spontaneous emission with natural lifetime. This is called as speed-up of the nuclear de-excitation. The other de-excitations of the nuclei through the incoherent channels like electron emission by internal conversion process are suppressed. Synchrotron radiation is linearly polarized and the excitation and the de-excitation of the nuclear levels obey to the selection rule of magnetic dipole (M1) transition for the ^{57}Fe resonance. As shown in Fig. 1.10, the coherent de-excitation of nuclear levels creates a quantum beat Ω given by

$$\Omega = \omega_{n,m}(n, m, r) - \omega_{n,m'}(n, m', r').$$

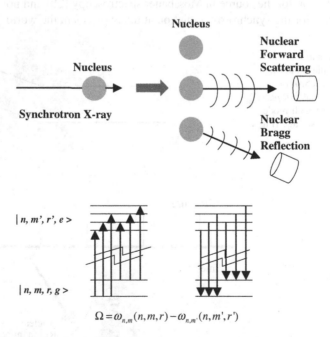

Fig. 1.10 Coherent excitation and coherent de-excitation of the nucleus for the nuclear forward scattering and the nuclear Bragg scattering. Coherent de-excitation of the resonant scattering creates the quantum beat with a frequency of Ω

Fig. 1.11 Energy spectrum of nuclear resonant scattering from α-Fe foil at room temperature, reported by Seto et al. [29]. *Dotted line* is a contribution of multi-phonon excitations [29]

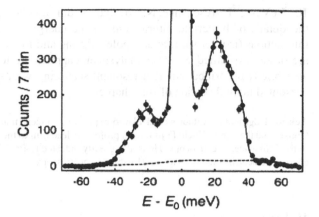

By the observation of quantum beat Ω, one can determine the details on the nuclear levels, from which the hyperfine interaction parameters are evaluated precisely.

1.4.2 Inelastic Scattering

High-resolution monochrometor reduces the bandwidth of the synchrotron radiation and also can scan the radiation energy in the region of meV that is the energy range of the phonon in the scattering material. Using the synchrotron radiation, it is possible to observe the nuclear inelastic scattering process that is the creation of phonon and the annihilation of phonon by the interaction between synchrotron radiation and scattering material. Figure 1.11 shows the energy profile of the inelastic scattering photons obtained from α-Fe at 300 K, which was a first successful observation of a nuclear inelastic resonance spectrum associated with the excitation of the lattice phonons [29]. From the inelastic scattering, the localized partial density of phonon state of scattering material can be determined [29, 30].

1.5 Summary

In this chapter, a general introduction to the Mössbauer spectroscopy has been described. First the Mössbauer effect with recoilless emission and absorption of nuclear γ-ray using radioactive source is described and emphasized the extremely narrow bandwidth that is the most important feature of the Mössbauer spectroscopy;

10^{-9} eV for ^{57}Fe resonance. Narrow bandwidth can make possible easily to study on the details of hyperfine interactions in the energy range of 10^{-7} eV. Hyperfine interactions (isomer shift, quadrupole splitting and magnetic hyperfine interaction) are briefly explained. Spectral analysis and application to the material research are also briefly described. Nuclear resonant scattering with synchrotron radiation was presented in final section of this chapter.

Acknowledgments Author would like to express his sincere thanks for the late professors Uli Gonser and the late F. Eiichi Fujita. Both professors lead him to the field of science and especially to the Mössbauer spectroscopy. He is also greatly indebted to Prof. T. Shinjo for his guidance and encouragement in the study using Mössbauer spectroscopy.

References

1. R.L. Mössbauer, Z. Physik **151**, 124 (1958)
2. R.L. Mössbauer, Z. Naturwissenshaften **45**, 538 (1958)
3. R. L. Mössbauer, Z. Naturforsch **14a**, 211 (1959)
4. R. L. Mössbauer, *Nobel Lecture* December 11 (1961)
5. H. Frauenfelder, *The Mössbauer Effect*, Frontiers in Physics (W. A. Benjamin, Inc., New York 1963)
6. G. K. Wertheim, *Mössbauer Effect: Principles and Applications* (Academic Press, New York, 1964)
7. L. May (ed.), *An Introduction to Mössbauer Spectroscopy* (Plenum Press, New York, 1971)
8. N.N. Greenwood, T.C. Gibb, *Mössbauer Spectroscopy* (Chapman and Hall Ltd, London, 1971)
9. U. Gonser (ed.), *Mössbauer Spectroscopy*, Topics in Applied Physics, vol. 5 (Heidelberg, New York, 1975)
10. R.L. Cohen (ed.), *Application of Mössbauer Spectroscopy*, vol. I (Academic Press, New York, 1976)
11. R.L. Cohen (ed.), *Application of Mössbauer Spectroscopy*, vol. II (Academic Press, New York, 1980)
12. U. Gonser (ed.): *Mössbauer Spectroscopy II. The Exotic Side of the Method*, Topics in Current Physics (Springer, Heidelberg, New York, 1981)
13. P. Gütlich, R. Link, A.X. Trautwein, *Mössbauer Spectroscopy and Transition Metal Chemistry*, Inorganic Chemistry Concepts 3 (Springer, Heidelberg, 1978)
14. T.C. Gibb, *Principles of Mössbauer Spectroscopy Studies in Chemical Physics* (Chapman and Hall, London, 1976)
15. I.J. Gruverman (ed.), *Mössbauer Effect Methodology*, vol 1–10 (Plenum Press, New York, 1965–1975)
16. K. Siegbahn (ed.), α,β,γ *Ray Spectroscopy*, vol. 2 (North-Holland Publishing Company, Amsterdam, 1965), p. 863
17. C.M. Lederer, J.M. Hollander, I. Perlman, *Table of Isotopes*, 6th edn. (Wiley, New York, 1968), p. 191
18. S. Margulies, J.R. Ehrman, Nuclear Inst. Meth. **12**, 131 (1961)
19. M. Celia Dibar-Ure, P.A. Flinn, *Mössbauer Effect Methodology*, vol 7, ed. by. J. Gruverman, C.W. Seidel, D.K. Dieterly (Plenum Press, New York, 1974), p. 245
20. R. Kato, Y. Tamada, T. Ono, S. Nasu, Jpn. J. Appl. Phys. **49**, 033003 (2010)
21. R.V. Pound, G.A. Rebka Jr, Phys. Rev. Lett. **4**, 337 (1960)
22. W. Kundig, Nucl. Inst. Methods **48**, 219 (1967)

23. G.R. Hoy, S. Chandra, J. Chem. Phys. **47**, 961 (1967)
24. S. Dattagupta, *Relaxation Phenomena in Condensed Matter Physics* (Academic Press, Orlando, 1987)
25. J. L. Dormann, D. Fiorani, E. Tronc, *Magnetic Relaxation in Fine-Particle Systems* Advances in Chemical Physics, vol XCVIII (Wiley, New York, 1997)
26. E. Gerdau, H. de Waard (eds.), *Hyperfine Interactions*, vol. 123–125 (Baltzer Science Publishers, The Netherlands, 1999/2000)
27. Y. Yoda, Presentation in JSPS Belgium-Japan Binational Seminar (2003)
28. S.L. Ruby, J. Phys. **C6**, 209 (1974)
29. M. Seto, Y. Yoda, S. Kikuta, X.W. Zhang, M. Ando, Phys. Rev. Lett. **74**, 3828 (1995)
30. W. Sturhahn, T.S. Toellner, E.E. Alp, X.W. Zhang, M. Ando, Y. Yoda, S. Kikuta, M. Seto, C.W. Kimball, B. Dabrowski, Phys. Rev. Lett. **74**, 3832 (1995)

Author Biography

Saburo Nasu
1965 Graduated Kyoto University
1967 Master of Engineering, Kyoto University
1971 Doctor of Engineering, Kyoto University
Post-doc at University of Saarland
Research associate of Welding Research Institute
of Osaka University
Research assistant of University of Saarland
Research associate, Assistant professor, Professor
of Faculty of Eng. Science, Osaka University
From 2005 Emeritus professor of Osaka University
Guest researcher of JAEA at Tokai, Institute
for Chemical research, Kyoto University
Guest researcher of iCeMS, Kyoto University

Typical publications

1. Nasu Saburo: Mössbauer Spectroscopy and its Application to Materials Research. High Temperature Materials and Processes 17 (1–2), 45–56 (1998)
2. Nasu Saburo: High Pressure Experiments with Synchrotron Radiation. Hyperfine Int. 113, 97–109 (1998)
3. Nasu Saburo: Mössbauer Study of Defects and Local Structure in Solids. Radiation Effects & Defects in Solids 148, 181–190 (1999)
4. Nasu Saburo et al.: Pressure Induced Ferromagnetism of $SrFeO_3$. Proceeding of AIRAPT-17, 763–766 (2000)
5. Nasu Saburo: High-Pressure Mössbauer spectroscopy using synchrotron radiation and radioactive sources. Hyperfine Int., 128, 101–113 (2000)
6. Nasu Saburo: High-pressure Mössbauer study using a diamond anvil cell. RIKEN Review, 27, 67–71 (2000)

7. Callens R. et al.: Stroboscopic detection of nuclear forward-scattered synchrotron radiation. Phys. Rev. B, 65, 180404(R) (2002)
8. Honmma Y et al.: Magnetostriction in the Ferromagnetic State of UGa_2. Physica B, 312-313, 904–905 (2002)
9. Baba K. et al.: 111Cd Time Differential Perturbed Angular Correlation Study of Deformed Ni. Materials Transactions, 43, No. 8, 2125–2129 (2002)
10. Kamimura T. et al.: Mössbauer Spectroscopic Study of Rust Formed on a Weathering Steel and a Mild Steel Exposed for a Long Term in an Industrial Environment. Materials Transactions, 43, No. 4, 694–703 (2002)
11. Nasu S. et al.: Mössbauer Study of ε-Fe under an External Magnetic Field. J. Physics: Condensed Matter. 14, No. 44, 11167–11171 (2002)
12. Callens R. et al.: Stroboscopic Detection of Nuclear Forward-scattered Synchrotron Radiation. Phys. Rev. B, 65, 180404-1(R)–180404-4(R) (2002)
13. Kawakami T. et al.: Pressure-induced Transition from Charge- disproportionated antiferromagnetic State to Charge-uniform Ferromagnetic State in $Sr_2/^3La_1/^3FeO_3$. Phys. Rev. Lett. 88, 037602-1–037602-4 (2002)
14. Kawakami T. et al.: High-Pressure Mössbauer and x-ray Powder Diffraction Studies of $SrFeO_3$. J. Phys. Soc. Jpn. 72, No. 1, 33–36 (2003)

Chapter 2
Chemical Applications of Mössbauer Spectroscopy

Philipp Gütlich and Yann Garcia

Abstract The Tutorial Lecture begins with a brief recapitulation of the hyperfine interactions and the relevant parameters observable in a Mössbauer spectrum. The main chapter with selected examples of chemical applications of Mössbauer spectroscopy follows and is subdivided into sections on: basic information on structure and bonding; switchable molecules (thermal spin transition in mono- and oligonuclear coordination compounds, light-induced spin transition, nuclear-decay-induced spin transition, spin transition in metallomesogens); mixed-valency in biferrocenes and other iron coordination compounds, and in an europium intermetallic compound; electron transfer in Prussian blue-analog complexes; molecule-based magnetism; industrial chemical problems like corrosion; application of a portable miniaturized Mössbauer spectrometer for applications outside the laboratory and in space. The Lecture ends with concluding remarks and an outlook to future developments.

Abbreviations

δ	Isomer shift
ΔE_Q	Quadrupole splitting
ΔE_M	Magnetic splitting
EFG	Electric field gradient
B_c	Fermi contact field
B_D	Spin dipolar field
eQ	Quadrupole moment

P. Gütlich (✉)
Institute of Inorganic and Analytical Chemistry, University of Mainz, 55099, Mainz, Germany

Y. Garcia
Institute of Condensed Matter and Nanosciences, MOST-Inorganic Chemistry, Université Catholique de Louvain, 1348, Louvain-la-Neuve, Belgium

Y. Yoshida and G. Langouche (eds.), *Mössbauer Spectroscopy*,
DOI: 10.1007/978-3-642-32220-4_2, © Springer-Verlag Berlin Heidelberg 2013

β_N	Nuclear Bohr magneton
g_N	Nuclear Landé factor
g	Landé splitting factor
H_{eff}	Effective hyperfine field
H_{ext}	External field
HS	High-spin
LS	Low-spin
SCO	Spin crossover
ST	Spin transition
LIESST	Light induced excited spin state trapping
NIESST	Nuclear decay-induced excited spin state trapping
ZFS	Zero-field splitting
AF	Antiferromagnetic
$T_{1/2}$	Transition temperature
LC	Liquid-crystal
1D	One-dimensional
EXAFS	Extended X-ray absorption fine structure
DSC	Differential scanning calorimetry
TGA	Thermo-gravimetric analysis
MAS	Mössbauer absorption spectroscopy
MES	Mössbauer emission spectroscopy
MIMOS	Miniaturized Mössbauer spectrometer
APXS	Alpha particle X-ray spectrometer
MER	Mars exploration rover
RAT	Rock abrasion tool
NFS	Nuclear forward scattering
NIS	Nuclear inelastic scattering
PDOS	Partial density of states
ptz	1-propyl-tetrazole
mtz	1-methyl-tetrazole
phen	1,10-phenanthroline
phdia	4,7-phenanthroline-5,6-diamine
bpym	Bipyrimidine
pmatrz	4-amino-3,5-bis{[(2-pyridyl-methyl)amino]methyl}-4H-1,2,4-triazole
iptrz	4-isopropyl-1,2,4-triazole
hyetrz	4-(2′-hydroxy-ethyl)-1,2,4-triazole
C_{10}-tba	3,5-bis(decyloxy)-N-(4H-1,2,4-triazol-4-yl)benzamide
C_{12}-tba	3,5-bis(dodecyloxy)-N-(4H-1,2,4-triazol-4-yl)benzamide
C_n-tba	3,5-bis(alkoxy)-N-(4H-1,2,4-triazol-4-yl)benzamide
TB-LMTO-ASA	Tight-binding linear Muffin-Tin Orbital atomic-sphere approximation

2.1 Introduction

Shortly after the discovery of *recoilless nuclear resonance fluorescence of gamma radiation* by Rudolf Mössbauer [1–4], physicists and chemists explored the possibility to use this effect as a basis for a new physical technique in materials science. One of the first publications, by Kistner and Sunyar [5] reported the magnetic hyperfine splitting of α-Fe_2O_3 (Fig. 2.1a). This was the first spectrum of a material that reflected all three types of hyperfine interactions between nuclear moments and electrons penetrating the nucleus that can be observed in a Mössbauer spectrum: the electric monopole interaction resulting in the *isomer shift*, the electric quadrupole interaction causing the *quadrupole splitting* and the magnetic dipole interaction giving rise to *magnetic splitting of degenerate nuclear levels*. These interactions are all taken into account in Fig. 2.1b (middle) in the case of Fe_2O_3, and are compared on the left side of Fig. 2.1b with metallic iron which shows a magnetic splitting only and on the right side with stainless steel that displays an isomer shift only. This report demonstrated the usefulness of the Mössbauer effect as a new spectroscopic technique—Mössbauer spectroscopy—for the non destructive characterization of materials and initiated overwhelming research activities in physics, chemistry, geo- and earth sciences, and even industrial applications.

The purpose of this tutorial lecture is to demonstrate how Mössbauer spectroscopy can help solving chemical problems. There are several thousands of excellent examples that have appeared in the literature since the discovery of the Mössbauer effect [6–17]. Because of limited space we shall present a selection of chemical applications of Mössbauer spectroscopy mainly from our own work dealing with investigations of transition metal compounds featuring:

- Bonding and structural properties
- Valence state
- Solid state reactions
- Electron transfer reactions
- Mixed valency
- Spin crossover
- Magnetic properties

The major part of the chapter will be devoted to the phenomenon of thermally induced spin crossover in iron complex compounds. This research topic has recently gained increasing interest by chemists and physicists for the promising potential of technical applications as devices. It will be demonstrated that Mössbauer spectroscopy, together with magnetic measurements, is particularly suited to follow the electronic structure dynamics of such materials under various conditions.

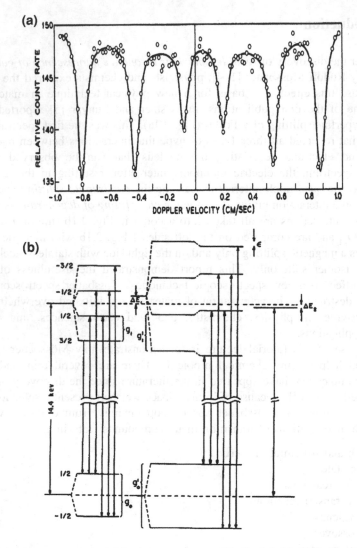

Fig. 2.1 a Room temperature Mössbauer spectrum of Fe_2O_3. **b** Schematic representation of the nuclear sub states of the ground and the 14.4 keV excited states of ^{57}Fe bound in metallic iron (*left*), Fe_2O_3 (*middle*) and stainless steel (*right*) [5]

2.2 Hyperfine Interactions and Mössbauer Parameters

Three kinds of hyperfine interactions may be observed in a Mössbauer spectrum:

Electric monopole interaction between protons of the nucleus and electrons (mainly s-electrons) penetrating the nuclear field. The observable Mössbauer parameter is the "isomer shift δ". Isomer shift values give information on the oxidation state, spin state, and bonding properties such as covalency and electronegativity.

Table 2.1 Conditions for hyperfine interactions and resulting Mössbauer parameters

Type of interaction	Nuclear condition	Electronic condition	Consequence
Electric monopole interaction	$R_e^2 \neq R_g^2$	$\|\Psi(0)\|_A^2 \neq \|\Psi(0)\|_S^2$	Different shift of nuclear levels $\Rightarrow$ *Isomer shift* δ
Electric quadrupole interaction	Electric quadrupole moment $eQ \neq 0$ $(I > 1/2)$	EFG $\neq 0$	Nuclear states split into $I + \frac{1}{2}$ substates $\|I, \pm m_I >$ (twofold degenerate) $\Rightarrow$ *Quadrupole splitting* ΔE_Q
Magnetic dipole interaction	Magnetic dipole moment $\mu \neq 0$ $(I > 0)$	$H \neq 0$	Nuclear states $\|I >$ split into $2I + 1$ substates $\|I, m_I >$ with $m_I = + I, +I\text{-}1, \ldots, -I$ $\Rightarrow$ *Magnetic dipole splitting* ΔE_M

Electric quadrupole interaction between the nuclear quadrupole moment and an inhomogeneous electric field at the nucleus. The observable Mössbauer parameter is the "quadrupole splitting ΔE_Q". The information derived from the quadrupole splitting refers to oxidation state, spin state and site symmetry.

Magnetic dipole interaction between the nuclear magnetic dipole moment and a magnetic field at the nucleus. The observable Mössbauer parameter is the "magnetic splitting ΔE_M". This quantity gives information on the magnetic properties of the material under study.

The following table summarizes the conditions, regarding the electronic and the nuclear properties, that lead to the three kinds of hyperfine interactions observable in a Mössbauer spectrum (Table 2.1).

2.2.1 Electric Monopole Interaction: Isomer Shift

Electric monopole interaction is the Coulomb interaction between protons of the nucleus and electrons (mainly s-electrons) penetrating the nuclear field. In a typical Mössbauer experiment, the source (S) material (e.g. ^{57}Co embedded in Rh metal) is generally different from the absorber (A) material under study. The nuclear radius in the excited state is different (in the case of ^{57}Fe, it is smaller) than that in the ground state: $R_e \neq R_g$. If the source and absorber materials are different, the electronic densities set up by all s-electrons (1s, 2s, 3s, etc.) of the electronic shells are different at the nuclei of the source and the absorber: $\|\Psi(0)\|_A^2 \neq \|\Psi(0)\|_S^2$. The result is that the electric monopole interactions are different in the source and the absorber and therefore affect the nuclear ground and excited state levels to a different extent. This leads to the measured isomer shift δ (see Fig. 2.2).

The isomer shift depends directly on the s-electron densities (as sum of contributions from all s-electron shells), but may be influenced indirectly via shielding

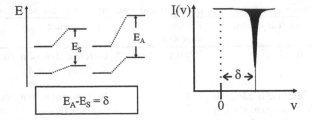

Fig. 2.2 Electric monopole interaction between electrons and protons perturbs the energy levels of the nuclear ground and excited states (with different radii). The energy changes are different in the source (S) and absorber (A) as a result of different electron densities at the source and absorber nuclei, the result is manifested as isomer shift in the Mössbauer spectrum

effects of p-, d- and f-electrons, which are not capable (if neglecting relativistic effects) of penetrating the nuclear field. Results from Hartree–Fock calculations of the contributions of s-orbitals to the total electron density at the iron nucleus as a function of oxidation state and configuration have shown that (a) nominally the largest contributions originate from the filled 1s and 2s shells (1s $\sim 10^4$ au^{-3}, 2s $\sim 10^3$ au^{-3}, 3s $\sim 10^2$ au^{-3}), and (b) significant changes in the electron densities arise from the noticeably different contributions from the 3s shell populations due to different shielding effects of $3d^n$. The reason becomes apparent on inspecting the strongly overlapping distribution functions of 3s and 3d electrons.

Chemical bonds between metal ion and ligands in coordination compounds can be viewed as the result of the balance between σ-donation (s-electrons from ligands are donated into s-orbitals of the metal) and d_π–p_π back donation (d-electrons move from d-orbitals of the metal to empty π-orbitals of the ligands). Both bonding mechanisms influence the isomer shift in the same direction, but to different extent, depending on the nature of the ligands and thus on the weight of the atomic orbitals of the metal and ligands participating in the molecular orbitals (covalency effects). This is the reason why isomer shift scales for different compounds of the same oxidation state often cover a broad range of values. The most valuable information derived from isomer shift data refers to the valence state of a Mössbauer-active atom embedded in a solid material as shown in Fig. 2.3.

2.2.2 Electric Quadrupole Interaction: Quadrupole Splitting

Electric quadrupole interaction occurs if at least one of the nuclear states involved possesses a quadrupole moment eQ (which is the case for nuclear states with spin $I > 1/2$) and if the electric field at the nucleus is inhomogeneous. In the case of ^{57}Fe the first excited state (14.4 keV state) has spin $I = 3/2$ and therefore also an electric quadrupole moment.

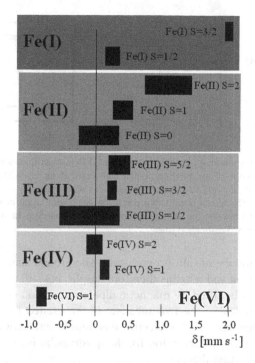

Fig. 2.3 The figure shows ranges of isomer shift values as expected for different oxidation and spin states. The most positive isomer shift occurs with formally Fe^I compounds with spin S = 3/2. In this case, the seven d-electrons exert a very strong shielding of the s-electrons towards the nuclear charge, this reduces markedly the s-electron density ρ_A giving a strongly negative quantity $|\Psi(0)|^2_A \neq |\Psi(0)|^2_S$. As the nuclear factor $(R^2_e - R^2_g)$ is negative for ^{57}Fe, the measured isomer shift becomes strongly positive. At the other end of the isomer shift scale are strongly negative values expected for formally Fe^{VI} compounds with spin S = 1. The reason is that iron Fe^{VI} compounds have only two d-electrons, the shielding effect for s-electrons is very weak in this case and the s-electron density ρ_A at the nucleus becomes relatively high which—multiplied by the negative nuclear factor $(R^2_e - R^2_g)$—pushes the isomer shift value strongly in a negative direction. It is seen from the figure that some isomer shift regions do not overlap, e.g. Fe^{II} HS compounds with S = 2 can be easily assigned from a Mössbauer spectrum. In other cases with more or less strong overlapping δ values unambiguous assignment to certain oxidation and spin states may not be possible. In such cases the quadrupole splitting parameter ΔE_Q will be included in the analysis and leads to a conclusive characterization in most cases

If the electric field gradient (EFG) is non-zero, for instance due to a non-cubic valence electron distribution and/or non-cubic lattice site symmetry, electric quadrupole interaction as visualized by the precession of the quadrupole moment vector about the field gradient axis sets in and splits the degenerate $I = 3/2$ level into two substates with magnetic spin quantum numbers $m_I = \pm 3/2$ and $\pm 1/2$ (Fig. 2.4). The energy difference between the two substates ΔE_Q is observed in the spectrum as the separation between the two resonance lines. These two resonance lines in the spectrum refer to the two transitions between the two substates of the

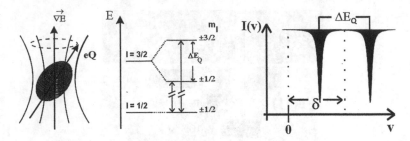

Fig. 2.4 In the case of a non-zero electric field gradient (EFG), electric quadrupole interaction as visualized by the precession of the quadrupole moment vector about the field gradient axis sets in and splits the degenerate I = 3/2 level into two substates with magnetic spin quantum numbers $m_I = \pm 3/2$ and $\pm 1/2$. This gives rise to two transition lines with equal probability (intensity). The energy difference between the two sub states ΔE_Q is observed in the spectrum as the separation between the two resonance lines

split excited state and the unsplit ground state. The ground state with $I = 1/2$ has no quadrupole moment and remains therefore unsplit, but still twofold degenerate. This degeneracy can be lifted by magnetic dipole interaction (Zeeman effect, see below). The same holds for the two substates of the excited $I = 3/2$ level, which are still twofold degenerate after electric quadrupole interaction. This becomes apparent by looking at the expression for the quadrupolar interaction energies E_Q derived from perturbation theory:

$$E_Q(I, m_I) = \left(\frac{eQV_{zz}}{4I(2I-1)} \right) [3m_I^2 - I(I+1)] \quad \text{(for axial symmetry)}.$$

For ^{57}Fe Mössbauer spectroscopy, electric quadrupole interaction in the absence of magnetic dipole interaction leads to a doublet, the separation of the two resonance lines giving the quadrupole interaction energy ΔE_Q which is proportional to the quadrupole moment eQ and the electric field gradient. The electric field E at the nucleus is the negative gradient of the potential, $-\nabla V$, and the electric field gradient EFG is given by the nine components $V_{ij} = (\partial^2 V/\partial i \partial j)$ $(i, j = x, y, z)$ of the 3×3s rank EFG tensor. Only five of these components are independent because of the symmetric form of the tensor, i.e. $V_{ij} = V_{ji}$ and because of Laplace's equation which requires that the tensor be traceless: $\nabla V_{ii} = 0$. In the principal axes system the off-diagonal elements vanish, and for axial symmetry (fourfold or threefold axis of symmetry passing through the Mössbauer nucleus yielding $V_{xx} = V_{yy}$) the EFG is then solely given by the tensor component V_{zz}. For non-axial symmetry the asymmetry parameter $\eta = (V_{xx} - V_{yy})/V_{zz}$ is required in addition. When choosing the principal axes ordering such that $V_{zz} \geq V_{xx} \geq V_{yy}$, the asymmetry parameter range becomes $0 \leq \eta \leq 1$.

In principle, there are two sources which can contribute to the total EFG: (1) charges (or dipoles) on distant ions surrounding the Mössbauer atom in non-cubic symmetry, usually termed *lattice contribution* to the EFG; (2) anisotropic (non-cubic) electron distribution in the valence shell of the Mössbauer atom, usually called *valence electron contribution* to the EFG. The latter comes about mainly in two ways: (1) Anisotropic population of the metal d-orbitals visualized in the frame of simple crystal field theory with axial distortion to molecular symmetry lower than O_h (an example is given below); (2) anisotropic covalency effects in molecular orbitals between the metal center and ligands with different σ-bonding and π-back bonding capability. It is understood that both sources of valence electron contributions are jointly operative and cannot be separated.

The electric quadrupole splitting gives information on the oxidation state, the spin state and the local symmetry of the Mössbauer atom. Note that the isomer shift parameter δ is given by the distance of the barycenter of the quadrupole doublet from zero Doppler velocity (Fig. 2.4).

2.2.3 Magnetic Dipole Interaction: Magnetic Splitting (Nuclear Zeeman Effect)

The requirements for magnetic dipole interaction (nuclear Zeeman effect) to be observed are that (1) the nuclear states involved must possess a magnetic dipole moment and (2) a magnetic field must be present at the nucleus. A nuclear state with spin $I > 1/2$ possesses a magnetic dipole moment μ. This is the case for both the ground state with $I = 1/2$ and the first excited state with $I = 3/2$ of ^{57}Fe. Magnetic dipole interaction (visualized as the precession of the magnetic dipole moment vector about the axis of the magnetic field; Fig. 2.5) leads to splitting of the states $|I, m_I >$ into $2I + 1$ substates characterized by the magnetic spin quantum numbers m_I. Thus the excited state with $I = 3/2$ is split into four, and the ground state with $I = 1/2$ into two substates. These substates are no longer degenerate. The energies of the sublevels are given from first-order perturbation theory by

$$E_M(m_I) = -\mu B m_I / I = -g_N \beta_N B m_I,$$

where g_N is the nuclear Landé factor and β_N the nuclear Bohr magneton. Note that the sign of the magnetic spin quantum numbers m_I of the sublevels have a different

Fig. 2.5 Magnetic dipole interaction (visualized as the precession of the magnetic dipole moment vector about the axis of the magnetic field) leads to splitting of the states |I, m_I > into 2I + 1 substates characterized by the magnetic spin quantum numbers m_I

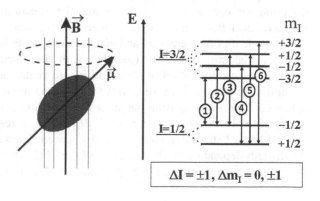

Fig. 2.6 Typical ^{57}Fe Mössbauer spectrum resulting from magnetic dipole interaction. The energies of the ground and excited state splitting can be determined as depicted in the figure and described in the text

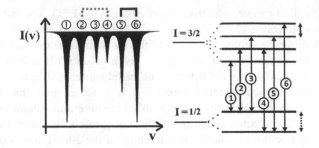

sequence in the excited state and the ground state, this being due to the different signs of the magnetic moments of the two states. The allowed gamma transitions between the sublevels of the excited state and those of the ground state are given by the selection rules for magnetic dipole transitions: $\Delta I = \pm 1$, $\Delta m_I = 0, \pm 1$. The six allowed transitions in the case of ^{57}Fe are shown in Figs. 2.5 and 2.6.

The separation between the lines 2 and 4 (also between 3 and 5) refers to the magnetic dipole splitting of the ground state. The separation between lines 5 and 6 (also between 1 and 2, 2 and 3, 4 and 5) refers to the magnetic dipole splitting of the excited $I = 3/2$ state (Fig. 2.6). The magnetic hyperfine splitting enables one to determine the effective magnetic field (size and direction) acting at the nucleus. Such a field can be externally applied. But many substances can also create a magnetic field of their own through various mechanisms, e.g.:

- The *Fermi contact field* B_C arises from a net spin-up or spin-down s-electron density at the nucleus as a consequence of spin polarization of inner filled s-shells by spin-polarized partially filled outer shells;
- a contribution B_L may arise from the *orbital motion* of valence electrons with the orbital momentum quantum number L;
- a contribution B_D, called *spin-dipolar field*, may arise from the total electron spin of the atom under consideration.

All contributions may be present and add to the total effective magnetic field $B_{eff} = B_C + B_L + B_D$. By applying an external magnetic field of known size and direction one can determine the size and the direction of the intrinsic effective magnetic field B_{eff} of the material under investigation.

Magnetic dipole interaction and electric quadrupole interaction may be present in a material simultaneously (together with the electric monopole interaction which is always present). The perturbations are treated depending on their relative strengths. In the case of relatively weak quadrupole interaction the nuclear sublevels $|I, m_I\rangle$ arising from magnetic dipole splitting are additionally shifted by the quadrupole interaction energies $E_Q(I, m_I)$; as a result, the sublevels of the excited $I = 3/2$ state are no longer equally spaced. The shifts by E_Q are upwards or downwards depending on the direction of the EFG. This enables one to determine the sign of the quadrupole splitting parameter ΔE_Q (Fig. 2.7).

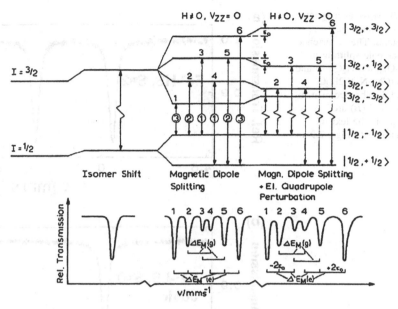

Fig. 2.7 Nuclear energy level scheme (^{57}Fe) for electric monopole interaction (causing the isomer *shift, left*), pure magnetic dipole interaction (causing magnetic *splitting, middle*), and combined magnetic dipole interaction and electric quadrupole interaction (*right*)

2.3 Selected Applications

2.3.1 Basic Information on Structure and Bonding

2.3.1.1 Quadrupole Splitting in Three Typical Fe^{II} Compounds

Figure 2.8 shows the Mössbauer spectra of three selected Fe^{II} compounds.

Ferrous sulphate, formulated as [Fe(H$_2$O)$_6$]SO$_4$·H$_2$O, is a high-spin (HS) compound with spin $S = 2$ and shows a large quadrupole splitting of ca. 3 mm s^{-1}. K$_4$[Fe(CN)$_6$] is a low-spin (LS) compound with $S = 0$ and cubic (O_h) molecular symmetry and shows no quadrupole splitting. Na$_2$[Fe(CN)$_5$NO] is also LS with $S = 0$, but strong tetragonal distortion from O_h symmetry due to the replacement of one of the six CN$^-$ ligands by NO, gives rise to a significant quadrupole splitting. The occurrence of quadrupole splitting in [Fe(H$_2$O)$_6$]-SO$_4$·H$_2$O and Na$_2$[Fe(CN)$_5$NO]·2H$_2$O and the absence of it in K$_4$[Fe(CN)$_6$] are explained in Figs. 2.9 and 2.10.

For HS Fe^{II} with 3d^6 electron configuration, the six 3d electrons are distributed under O_h symmetry as shown in Fig. 2.9 (left). The two degenerate e.$_g$ orbitals carry one electron each, and the three degenerate t$_{2g}$ orbitals are occupied by 1$^{1/3}$ electrons on average. As the t$_{2g}$ and e$_g$ orbitals are cubic subgroups, there is no valence electron contribution to the EFG independent of the number of electrons

Fig. 2.8 Mössbauer spectra of three selected Fe^{II} compounds. The occurrence of quadrupole splitting in $[Fe(H_2O)_6]SO_4 \cdot H_2O$ (*A*) and $Na_2[Fe(CN)_5NO] \cdot 2H_2O$ (*C*) and the absence of it in $K_4[Fe(CN)_6]$ (*B*) are explained in the text in connection with Figs. 2.9 and 2.10

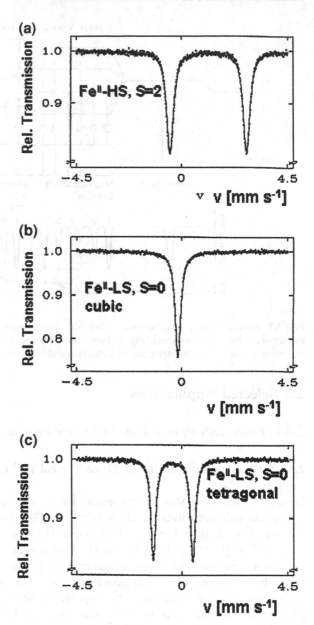

occupying them. There is also no lattice contribution to the EFG arising from the coordination sphere of six identical H_2O ligands arranged with O_h symmetry. Thus, there is no quadrupole splitting expected under O_h symmetry. $[Fe(H_2O)_6]^{2+}$, however, is a "Jahn–Teller-active" complex ion. It is unstable under O_h symmetry. It undergoes axial distortion with symmetry lowering to (e.g.) D_{4h} as schematized in Fig. 2.9 (right), either compressing or elongating the octahedron in z-direction.

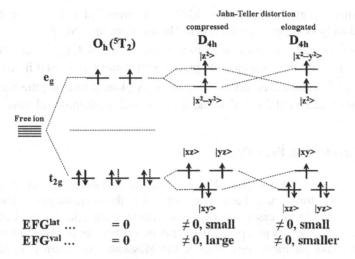

Fig. 2.9 Quadrupole splitting in the HS $[Fe(H_2O)_6]^{2+}$ complex ion arises from a non-cubic valence electron distribution due to Jahn–Teller distortion with lowering of symmetry from O_h (EFG = 0) to D_{4h} with valence electron population in a compressed octahedron as shown in the figure and described in the text

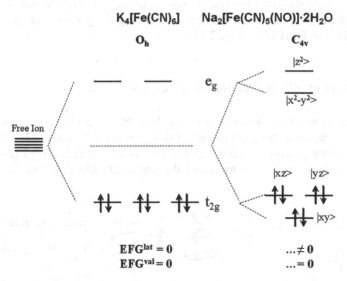

Fig. 2.10 $K_4[Fe(CN)_6]$ is a $3d^6$ LS complex with O_h symmetry, where all six electrons are accommodated in the three t_{2g} orbitals. Both contributions (EFG)^{val} and (EFG)^{lat} vanish; there is no quadrupolar interaction. $Na_2[Fe(CN)_5NO]\cdot2H_2O$ has C_{4v} symmetry with d-orbital splitting as shown on the right. Its LS behavior requires that all six electrons are accommodated in the lowest three orbitals arising from the tetragonal splitting of the former cubic t_{2g} (O_h) subgroup. (EFG)^{val} is still zero, but (EFG)^{lat} ≠ 0 arises from the ligand replacement in he iron coordination sphere

Compression is preferred because in this case the ground state is an orbital singlet with the doubly occupied xy orbital being lowest in energy. Suppose all d-orbitals are singly occupied, as for instance in the case of $[Fe(H_2O)_6]^{3+}$ (HS), $(EFG)^{val}$ would be zero. But the sixth electron placed in the lowest xy orbital in the case of $[Fe(H_2O)_6]^{2+}$ accumulates more charge in the xy plane than along the z-axis and, thus, causes a large $(EFG)^{val} \neq 0$ and the observed quadrupole splitting.

2.3.1.2 Structure of Fe$_3$(CO)$_{12}$

On the basis of single crystal X-ray diffraction measurement three possible molecular structures had been suggested for the metalorganic compound Fe$_3$(CO)$_{12}$ [18]. In all cases the iron atoms form a triangle, but with different surroundings by the CO groups. In the upper two structures of Fig. 2.11 the three iron atoms have identical surroundings, the Mössbauer spectrum is expected to show only one type of resonance signal. The lower structure has two identical iron positions and a different one for the third iron atom. In this case the Mössbauer spectrum is expected to show two different types of resonance signals with an area ratio of 2:1. A Mössbauer effect study performed by Greatrex and Greenwood in 1969 [19] indeed showed two types of resonance signals, a quadrupole doublet A and a singlet B with an area ratio of 2:1 confirming the presence of two types of iron positions in Fe$_3$(CO)$_{12}$ (Fig. 2.11).

2.3.1.3 Effect of π-Backdonation in $[Fe(CN)_5X^{n-}]^{(3+n)-}$

The following example demonstrates that Mössbauer spectroscopy can help to characterize chemical bond properties. Taking from the literature [7] the isomer shift data for the pentacyano complexes of Fe^{II} with a different sixth ligand X and normalizing the isomer shifts to that of the pentacyanonitrosylferrate complex as

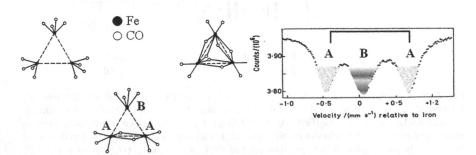

Fig. 2.11 The Mössbauer spectrum of Fe$_3$(CO)$_{12}$ shows a quadrupole doublet A and a singlet B with area ratio 2:1, which confirms that the structure with two inequivalent lattice sites of iron, A and B, is the correct structure [19]

Table 2.2 Effect of π-back donation in $[Fe(CN)_5X^{n-}]^{(3+n)-}$

Ligand X	$\delta/mm\ s^{-1}$	$d_\pi \rightarrow p_\pi$	d-shield	$\rho(0)$
NO^+	0.00			
CO	+0.15			
SO_3^{2-}	+0.22			
$P(C_6H_5)_3$	+0.23			
NO_2^-	+0.26			
$Sb(C_6H_5)_3$	+0.26			
NH_3	+0.26			
$As(C_6H_5)_3$	+0.29			
H_2O	+0.31			

$$\delta \sim \rho(0)\ \frac{\Delta R}{R} \qquad {}^{57}Fe:\ \frac{\Delta R}{R} < 0$$

zero point, one finds the ordering given in Table 2.2 which expresses the varying effects of d_π–p_π *backdonation* for the different sixth ligand X.

The isomer shift values become more positive on going from NO^+ to H_2O. The reason is that in the same ordering the strength of d_π–p_π *back donation* decreases causing an increasing d-electron density residing near the iron center and thus effecting stronger shielding of s-electrons by d-electrons, which finally creates lower s-electron density at the nucleus in the case of H_2O as compared to NO^+. The fact that the nuclear factor $\Delta R/R$ is negative for ${}^{57}Fe$ explains the increasingly positive isomer shift values in the given sequence from NO^+ to H_2O.

2.3.1.4 Effect of Ligand Electronegativity

In Fig. 2.12 isomer shift values of ferrous halides taken from the literature [7] are plotted as a function of Pauling electronegativity values. The electronegativity

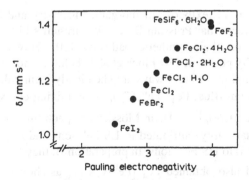

Fig. 2.12 The graph shows the influence of electronegativity on the isomer shift of ferrous halides. The electronegativity increases from iodine to fluorine. In the same ordering the 4s electron population decreases and as a direct consequence the s-electron density at the iron nucleus decreases, and due to the fact that $(R_a^2 - R_g^2) < 0$ for ${}^{57}Fe$ the isomer shift increases from iodide to fluoride

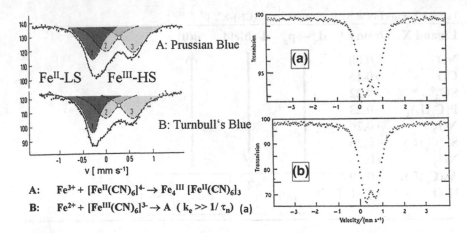

Fig. 2.13 a Mössbauer spectra of Prussian blue and Turnbull's Blue [21]. **b** Mössbauer spectra (77 K) of "Prussian Blue" prepared from (A) $^{57}Fe_2^{III}(SO_4)_3$ (enriched) + $K_4[Fe^{II}(CN)_6]$ (unenriched) and (B) $^{57}Fe^{II}Cl_2$ (enriched) + $K_3[Fe^{III}(CN)_6]$ (unenriched) [22]. The spectra A and B are identical and are indicative of the Fe^{III} cation outside the hexacyano complex. This confirms the occurrence of fast electron transfer in preparation B

increases from iodine to fluorine. In the same ordering the 4s electron population decreases and as a direct consequence the s-electron density at the iron nucleus decreases, and due to the fact that $\left(R_a^2 - R_g^2\right) < 0$ for ^{57}Fe the isomer shift increases from iodide to fluoride.

2.3.1.5 Prussian Blue vs. Turnbull's Blue

It has been stated in old textbooks of inorganic chemistry and hence it was recognized for many years that Prussian Blue and Turnbull's blue are different substances depending on their synthetic pathways [20]. However, a Mössbauer experiment originally performed by Fluck et al. [21] has demonstrated that the two end products prepared in different ways are chemically identical. In preparation A they prepared Prussian Blue, $Fe_4^{III}\left[Fe^{II}(CN)_6\right]_3$, by mixing equivalent amounts of ionic Fe^{III} and $\left[Fe^{II}(CN)_6\right]^{4-}$. Their Mössbauer spectrum shown in Fig. 2.13a reveals a singlet (dark grey) attributed to LS Fe^{II} ions and a quadrupole doublet (light grey) attributed to HS Fe^{III} ions. In preparation B, they mixed ionic Fe^{II} with $\left[Fe^{III}(CN)_6\right]^{3-}$ and also obtained $Fe_4^{III}\left[Fe^{II}(CN)_6\right]_3$ as shown in the Mössbauer spectrum (Fig. 2.13a). Indeed, a fast electron transfer occurs from Fe^{II} to Fe^{III} during mixing the two components, the rate constant for electron transfer being

$k_e \gg 1/\tau_n$, where τ_n is the lifetime of the 14.4 keV state of ^{57}Fe. This experiment was elegantly confirmed a few years later by Maer et al. (Fig. 2.13b) [22].

2.3.2 Switchable Molecules: Spin Crossover

2.3.2.1 Thermally Induced Spin State Switching in Fe^{II} Compounds

Thermally induced spin state transition from a HS state with maximum unpaired electrons to a LS state with minimum unpaired electrons can be encountered in certain transition metal compounds with d^4 up to d^7 electron configurations. Scheme 2.1 sketches the phenomenon in the case of Fe^{II} compounds with six valence electrons in the 3d shell. In Fe^{II} compounds with relatively weak ligands coordinated to the iron ions, e.g. water molecules, the 3d electrons are accommodated spin-free according to Hund's rule of maximum spin of $S = 2$. Such compounds, called *HS complexes*, are paramagnetic and are generally weakly colored. In Fe^{II} compounds with relatively strong ligands like CN$^-$ ions, the six electrons are arranged spin-paired with total spin $S = 0$. Such compounds are called *LS complexes*; they are generally diamagnetic and often colored. If the right kinds of ligands are chosen, e.g. derivatives of tetrazole or triazole, one may observe spin state transition solely by varying the temperature, applying pressure or under irradiation with light [23–25].

Thermally induced spin crossover (SCO) in Fe^{II} compounds is reflected by changes in the electron configuration. In the notation of ligand field theoretical

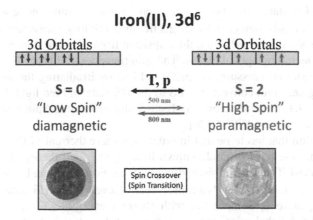

Iron(II), 3d^6

3d Orbitals		3d Orbitals

$S = 0$ $\xleftarrow{\text{T, p}}$ $S = 2$

$\underset{\text{800 nm}}{\overset{\text{500 nm}}{\rightleftarrows}}$

"Low Spin" — "High Spin"

diamagnetic — paramagnetic

Spin Crossover
(Spin Transition)

Scheme 2.1 Classification and physical properties of octahedral Fe^{II} coordination compounds. Most of them (an estimate of >95 %) show either HS or LS behavior depending on the ligand field strength set up at the metal center by the coordinated ligand molecules. Less than 5 % (about 200 as a rough estimate), mainly those with FeN$_6$ core, exhibit thermally induced spin crossover. Spin crossover is also possible by application of pressure and irradiation with light

Fig. 2.14 Jablonski diagram depicting potential wells of the LS and HS sates for an Fe^{II} complex, illustrating the LIESST and reverse-LIESST mechanism. These phenomena refer to the possibility to reversibly address spin states of Fe^{II} complexes by light irradiation

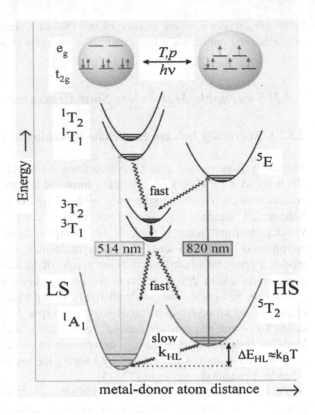

concepts, the electron configuration changes from $(t_{2g})^4(e_g)^2$ in the HS state to $(t_{2g})^6$ in the LS state. This phase transition between paramagnetic and (nearly) diamagnetic is easily detected by magnetic susceptibility measurements. As the color changes simultaneously, too, the transition from one spin state to the other is also easily detected by optical means. The spin transition (ST) can also be afforded by applying external pressure, magnetic field or by irradiating the material with light, where green light converts the LS to the HS state and red light the HS to the LS state. This has become known as Light-Induced Excited Spin State Trapping (LIESST) and reverse-LIESST [26].

The condition that has to be met in order to observe thermal SCO is sketched in Fig. 2.14 using the term symbols known from ligand field theory. Thermal SCO may be observed if the ligand field strength of an Fe^{II} compound is such that the difference between the lowest "vibronic" energy levels of the HS state $^5T_2$ and the LS state $^1A_1$ state is comparable with thermal energy k_BT (k_B = Boltzmann constant). The ST behavior can be influenced by chemical alteration of the material, e.g. ligand replacement, change of non-coordinating anion and solvent molecule, substitution of spin state changing metal by another metal (e.g. iron by zinc). For a comprehensive coverage of chemical and physical influences on the ST behavior see Refs. [23–25].

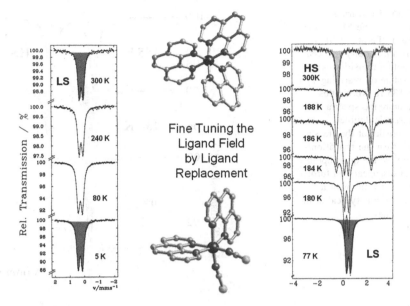

Fig. 2.15 [Fe^{II}(phen)$_3$]X$_2$ (phen = 1,10-phenanthroline) is a typical LS compound with characteristic Mössbauer spectra shown on the *left*. If one of the relatively strong phen ligands is replaced by two monofunctional NCS$^-$ groups, the average ligand field strength becomes weaker such that now the condition for thermally induced SCO, viz. $\Delta E_{HL} \approx k_B T$ is met and the compound [Fe^{II}(phen)$_2$(NCS)$_2$] adopts HS character at room temperature. The temperature dependent Mössbauer spectra shown on the *right* confirm that [Fe^{II}(phen)$_2$(NCS)$_2$] undergoes ST between S = 2 and S = 0 near 180 K

The influence of the ligand molecules on the spin state of the central Fe^{II} ions is demonstrated with the two examples and their temperature dependent Mössbauer spectra shown in Fig. 2.15. [Fe^{II}(phen)$_3$]X$_2$ (phen = 1,10-phenanthroline) is a typical LS compound at all temperatures under study as confirmed by the characteristic Mössbauer spectra on the left with isomer shift of ca. 0.2 and quadrupole splitting ca. 0.5 mm s^{-1} independent of temperature. If one of the relatively strong phen ligands, which occupies two coordination positions of the octahedron, is replaced by two monofunctional isothiocyanato groups, the average ligand field strength becomes weaker than the mean spin pairing energy and the compound [Fe^{II}(phen)$_2$(NCS)$_2$] adopts HS character at room temperature. The Mössbauer spectrum at 300 K shows the typical features of an Fe^{II} HS compound with isomer shift of ca. 1 mm s^{-1} and a large quadrupole splitting of ca. 3 mm s^{-1}. However, the compound [Fe^{II}(phen)$_2$(NCS)$_2$] fulfils the condition for thermal SCO to occur, viz. $\Delta E_{HL} \approx k_B T$. As the temperature is lowered, the compound changes spin state from HS to LS near 180 K as is well documented by the Mössbauer spectra as a function of temperature on the right hand side of Fig. 2.15, which was first reported by Dezsi et al. in 1967 [27]. Since then more than 200 SCO compounds of Fe^{II} and Fe^{III} have been studied by Mössbauer spectroscopy (see e.g. [25]).

Fig. 2.16 The variable temperature ^{57}Fe Mössbauer spectra of $[Fe(ptz)_6](BF_4)_2$, reveals a thermally induced SCO with a transition temperature on warming $T^{\uparrow}_{1/2}$ of ca. 135 K. Above this temperature the spectra show a quadrupole doublet characteristic for the HS state of Fe^{II}. Near the transition temperature the quadrupole doublet disappears on further cooling at the favor of a singlet which is typical of the LS state [28]

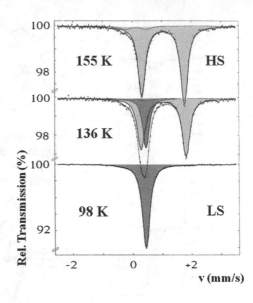

$[Fe(ptz)_6](BF_4)_2$ (ptz = 1-propyl-tetrazole), is another model Fe^{II} coordination compound exhibiting thermal SCO with a transition temperature on warming $T^{\uparrow}_{1/2}$ of ca. 135 K. The ^{57}Fe Mössbauer spectra of Fig. 2.16, recorded on warming from 98 K, clearly confirm the transition at this temperature between the LS phase (singlet shown in dark grey) and the HS phase (quadrupole doublet shown in grey).

2.3.2.2 Light Induced Spin State Switching

With $[Fe(ptz)_6](BF_4)_2$ it was observed for the first time that the ST can also be induced by irradiating the crystals with light; green light converts the LS state to the HS state, which can have very long lifetimes, e.g. on the order of days at temperatures below ca. 20 K [26]. Figure 2.17 shows a single crystal of $[Fe(ptz)_6](BF_4)_2$ (size ca. 3×3 cm^2), which is colorless at room temperature. The absorption spectrum shows only a weak absorption band at ca. 12,000 cm^{-1}. At 80 K the crystal has changed totally to the LS state by thermal SCO and is now red and absorbs relatively strongly at 18,000 and 26,000 cm^{-1}. Irradiating the crystal at ca. 10 K converts the LS state to the metastable HS state. The optical spectrum of the white spot (ca. 1 mm in diameter) is practically identical to the one recorded at 300 K.

Mössbauer spectroscopy is ideally suited to follow the light-induced spin state conversion in $[Fe(ptz)_6]BF_4)_2$ as exemplified in Fig. 2.18. A polycrystalline sample of $[Fe(ptz)_6](BF_4)_2$ was cooled to 15 K. Before irradiation, the sample is in the LS state and shows the typical Mössbauer spectrum of Fe^{II}-LS (upper left). After irradiating with green light (Xe lamp with filters or 514 nm band of an Ar ion laser) at 15 K the sample is quantitatively converted to the metastable HS state (middle left). The asymmetry in the intensity of the two components of the

Fig. 2.17 A single crystal of [Fe(ptz)$_6$](BF$_4$)$_2$ (size ca. 3 × 3 cm^2) is colorless at room temperature. The absorption spectrum shows only a weak absorption band at ca. 12,000 cm^{-1}. At 80 K the crystal has changed totally to the LS state by thermal SCO and is now *red* and absorbs relatively strongly at 18,000 and 26,000 cm^{-1}. Irradiating the crystal at ca. 10 K converts the LS state to the metastable HS state. The optical spectrum of the white spot (ca. 1 mm in diameter) is practically identical to the one recorded at 300 K

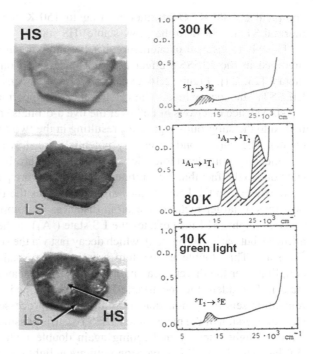

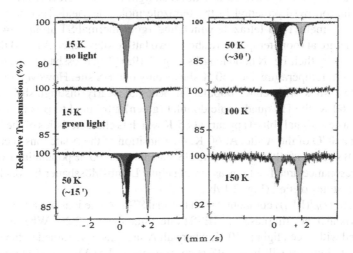

Fig. 2.18 Light induced spin state switching in [Fe(ptz)$_6$](BF$_4$)$_2$ followed by ^{57}Fe Mössbauer spectroscopy [26]

quadrupole doublet is due to the plate-like shape of the crystals (texture effect). Thermal relaxation on a 15 min-timescale sets in at 50 K (lower left and upper right: the sample was heated for 15 min at 50 K and then cooled to the measuring temperature of 15 K in two runs). Thermal relaxation to the stable LS state is complete at 100 K. On further heating to 150 K the sample undergoes again thermal ST at 135 K to the (now stable) HS state.

This photophysical phenomenon has become known as LIESST. The processes involved in the LIESST effect are well understood on the basis of ligand field theory [23, 24]. Figure 2.14 explains the mechanisms of LIESST and reverse-LIESST. The energy level scheme shows in the uppermost part the distribution of the six valence electrons of Fe^{II} over the five d-orbitals split in an octahedral ligand field into the subgroups t_{2g} and $e._g$ resulting in the two spin states LS (left) and HS (right). The corresponding energy potentials are drawn in the lower part of the scheme. The complex molecules in the HS state are bigger than those in the LS state due to the fact that the antibonding e_g orbitals are partially occupied in the HS state, whereas in the LS state they are empty. Thus the HS potentials are placed in positions of larger metal-donor atom distances as compared to the LS potentials.

Green light (514 nm) excites the LS state ($^1A_1$) to the $^1T_1$ and $^1T_2$ states (spin-allowed, but parity-forbidden), which decay fast via the spin triplet states $^3T_{1,2}$ to the $^5T_2$ state. This double intersystem crossing decay path is favored by spin–orbit coupling over the direct decay path back to $^1A_1$. Decay of the $^5T_2$ state to the $^1A_1$ state is forbidden, the metastable HS state is trapped until radiationless thermal relaxation sets in by nonadiabatic multiphoton processes [23, 24]. Light-induced back conversion of the metastable LIESST state is possible by irradiating the sample with red light, thereby undergoing again double intersystem crossing processes similar to the LS to HS conversion with green light (reverse-LIESST) [29].

Replacing propyl by methyl in the tetrazole molecule yields $[Fe(mtz)_6](BF_4)_2$ with mtz = 1-methyl-1H-tetrazole which undergoes thermal ST around 74 K. The crystal structure at room temperature shows two lattice sites called A and B that only differ slightly by their Fe–N bond lengths (Fig. 2.19a) [30]. Mössbauer spectroscopy between room temperature and 160 K shows only one HS site. However, on further cooling, two sites with equal population become clearly distinguishable. Upon cooling to 60 K, the HS quadrupole doublet (green) attributed to B is not affected. However, a new signal (blue) appears at 85 K which is characteristics of the LS state indicating a SCO of the A site. At 60 K, the transition of the A site ions is complete while the B site ions fully remain in the HS state with ca. 50 % population evaluated from the resonance area fractions assuming equal Lamb-Mössbauer factors for both HS and LS states of Fe^{II} (Fig. 2.19b).

The $[Fe(mtz)_6](BF_4)_2$ compound with thermal ST at A site ions but HS behavior at B site ions is an interesting case for LIESST effect studies (Fig. 2.20). When the sample is irradiated with green light at 20 K, where all A site ions have turned to the LS state but all B site ions are still in the HS state, a complete LS(A) → HS(A) photo-conversion is observed. At around 65 K, thermal relaxation back to the LS state is observed for the A site ions. When the sample is irradiated with red light at 20 K, the signal corresponding to B site ions disappears almost totally, and a new resonance signal appears which is characteristic for Fe^{II} ions in the LS state. This is the first observation of a light induced excitation of a stable HS state into a long lived meta-stable LS state. Again, upon warming to ca. 65 K, complete thermal relaxation is observed back to the stable HS(B) state [31].

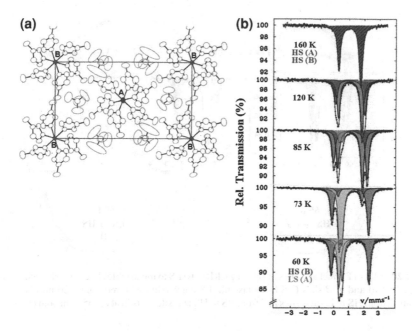

Fig. 2.19 **a** View of the crystal structure of $[Fe(mtz)_6](BF_4)_2$ showing two lattice sites, A and B at 300 K occupied by Fe^{II} ions in the HS state with metal to ligand bond distances Fe(A) − N = 2.181(5), 2.181(5), 2.181(7) Å and Fe(B)-N = 2.161(5), 2.197(5), 2.207(4) Å [30]. **b** Temperature dependent Mössbauer spectra of $[Fe(mtz)_6](BF_4)_2$ over the temperature range 160–60 K [31]. From room temperature down to ca. 160 K, only one quadrupole doublet arising from iron(II) ions in HS(A) and HS(B) state is observed, i.e. A and B site ions are not distinguishable in the Mössbauer spectrum. On further cooling, the A site ions undergo HS to LS transition whereas the B site ions remain in the HS state at all temperatures under study

2.3.2.3 Mechanism of Spin State Switching in Fe^{II} Dinuclear Compounds

The dinuclear Fe^{II} compound [Fe(bpym)(NCS)$_2$]$_2$bpym (bpym = bipyrimidine) belongs to a family of materials that can combine both antiferromagnetic (AF) coupling and SCO phenomena. Its crystal structure, shown in Fig. 2.21a, consists of two Fe^{II} centers, each of them coordinated to a bidentate bpym and two monodentate isothiocyanato anions, and bridged by a bipyrimidine ligand to form a dinuclear unit [32]. This compound does not show thermal ST, but the magnetic susceptibility measurements reveal weak AF coupling ($J \approx -4.1$ cm^{-1}). Replacing the NCS group by NCSe with slightly stronger ligand field strength causes temperature induced ST near 120 K as seen in Fig. 2.21b, which shows the temperature dependent magnetic properties of [Fe(bpym)(NCSe)$_2$]$_2$bpym as a plot of $\chi_M T$ vs. T, where χ_M is the magnetic susceptibility corrected for diamagnetic contributions. It reveals the presence of weakly coupled HS Fe^{II} ions in the region between room temperature and ca. 120 K. Around this temperature, a dramatic decrease of $\chi_M T$ due to thermally induced ST is observed to reach a plateau with a

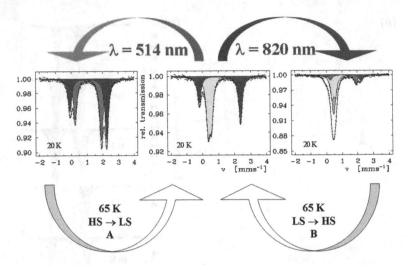

Fig. 2.20 LIESST effect at *A* sites of [Fe(mtz)₆](BF₄)₂ (LS to metastable HS conversion with green light, 514 nm) and at *B* sites (HS to metastable LS conversion with red light, 820 nm). Thermal relaxation from HS to LS at *A* sites and from LS to HS at *B* sites sets in at ca. 65 K in both cases [31]

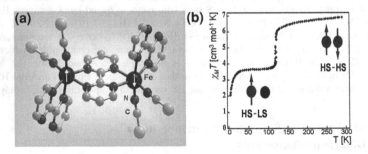

Fig. 2.21 **a** View of the crystal structure of [Fe(bpym)(NCS)₂]₂bpym [32]. **b** Plot of $\mathsf{I}_M T$ vs. *T* for [Fe(bpym)(NCSe)₂]₂bpym, where I_M is the magnetic susceptibility corrected for diamagnetic contributions [33]

$\chi_M T$ value of ca. 3.5 cm³ mol⁻¹ K which indicates that about 50 % of the Fe^{II} ions have undergone ST to the LS state (Fig. 2.21b) [33]. The question, however, arises concerning the nature of the plateau region: Does the plateau originate from only [HS–LS] pairs or from a 1:1 mixture of [HS–HS] and [LS–LS] pairs?

Neither can this question be answered by magnetic susceptibility measurements nor by conventional Mössbauer spectroscopy. The latter would yield identical spectra for the HS state of Fe^{II} atoms in [HS–LS] and [HS–HS] spin pairs as seen in Fig. 2.22a. However, the Mössbauer spectrum recorded in applied magnetic field can give a conclusive answer based on the different effective hyperfine magnetic fields arising from different expectation values of the spin state of the

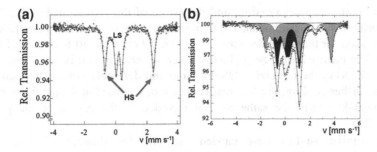

Fig. 2.22 Mössbauer spectra of [Fe(bpym)(NCSe)$_2$]$_2$bpym at: **a** 4.2 K without applied magnetic field and **b** 4.2 K in applied magnetic field of 5 T. LS in [HS–LS] and [LS–LS] pairs (*black*), HS in [HS–LS] pairs (*grey*), HS in [HS–HS] pairs (*white*) [34, 35]

Fe^{II} ions [34, 35]. The effective hyperfine field H_{eff} at the iron nuclei of a paramagnetic non-conducting sample in an external field H_{ext} may be estimated as $H_{eff} \approx H_{ext} - [220 - 600(g - 2)]\langle S \rangle$ where $\langle S \rangle$ is the expectation value of the atomic spin moment and g the Landé splitting factor [36, 37]. The difference between the expectation values of S for the Fe^{II} atom in the LS and in the HS states in [HS–LS] and [HS–HS] pairs enables one to distinguish unambiguously between the dinuclear units consisting of two possible spin states in an external magnetic field. To do so, the strength of the external magnetic field should be sufficiently high and the temperature sufficiently low in order to avoid magnetic relaxation taking place within the characteristic time window of a Mössbauer experiment. The zero-field spectrum of the [Fe(bpym)(NCSe)$_2$]$_2$bpym recorded at 4.2 K reflects, in agreement with the magnetic measurements, the nearly "one-half" ST according to the area fractions of the HS (48.0 %) and LS (52.0 %) components with parameters $\delta^{HS} = 0.86(1)$ mms^{-1}, $\Delta E_Q^{HS} = 3.11(2)$ mms^{-1}, and $\delta^{LS} = 0.22(1)$ mms^{-1}, $\Delta E_Q^{LS} = 0.36(1)$ mms^{-1}, respectively (Fig. 2.22a). The measurement in a magnetic field of 50 kOe (5 T) at 4.2 K yields a spectrum that consists of three components as can be seen in Fig. 2.22b. One of them (shown in black) with relative intensity (area fraction) 52.0 % and with isomer shift and quadrupole splitting indicative of Fe^{II} in LS state shows an expected effective hyperfine field that is practically the same as the applied field, $H_{eff} \approx H_{ext}$ (because S = 0). This subspectrum arises from LS-Fe^{II} ions in [HS–LS] and [LS–LS] pairs. The second component (shown in white) with area fraction of ca. 4.0 % is a doublet with isomer shift and quadrupole splitting values characteristic of Fe^{II} in HS state. The resonance lines are magnetically broadened through magnetic dipole interaction with a local effective hyperfine field of $H_{eff} = 14$ kOe, which is rather small and originates from a small $\langle S \rangle$ value as a result of antiferromagnetic coupling in [HS–HS] pairs. The third component (shown in grey) with relative intensity of ca. 44.0 % and isomer shift and quadrupole splitting values indicative of Fe^{II} in HS state can be unambiguously assigned to [HS–LS] pairs; because the measured effective magnetic field at the iron nuclei of 81 kOe clearly stems from a spin quintet ground state of Fe^{II} (S = 2). As a result, the complete distinction between dinuclear units becomes possible and the nature of the plateau in Fig. 2.22b has been clarified: it is predominantly due to the presence of [HS–LS] pairs [34, 35].

The reason for the sharp decrease of $\chi_M T$ vs. T below ca. 30 K (Fig. 2.21b) has also become clear with the help of Mössbauer spectroscopy. It is due to zero-field splitting (ZFS) of the remaining HS–Fe^{II} ions and definitely not a continuation of ST upon further cooling; the Mössbauer spectrum recorded at temperatures below the plateau is nearly the same as that recorded in the region of the plateau (Fig. 2.22a).

Very similar studies were carried out with the dinuclear Fe^{II} complex $[Fe(phdia)(NCS)_2]_2$phdia   (phdia = 4,7-phenanthroline-5,6-diamine),   which exhibits an almost complete two-step thermally induced ST with a plateau around 100 K in the $\chi_M T$ vs. T magnetization curve [38]. After rapidly cooling the sample from the plateau down to 4.2 K the Mössbauer spectrum was recorded at that temperature in applied magnetic field of 5 T. The analysis of the magnetically perturbed spectrum indicated that the plateau consisted mainly of [HS–LS] pairs.

In another Mössbauer effect study the mechanism of the half-way ST in the dinuclear complex of formula $[Fe_2^{II}(pmatrz)_2](BF_4)_4 \cdot DMF$ with pmatrz = 4-amino-3,5-bis{[(2-pyridyl-methyl)amino]methyl}-4H-1,2,4-triazole could be elucidated by recording the spectra solely as a function of temperature in zero field. The results are in full agreement with the analysis of the structure determination [39]. The structure of this dinuclear compound is shown in Fig. 2.23 [39]. The Fe^{II} ions are in the HS state (shown in black) at room temperature as expected for [HS–HS] pairs. On cooling, only half of the Fe^{II} ions switch to the LS state as confirmed by the magnetic susceptibility measurements (Fig. 2.23). This is possible in two ways: (1) Both Fe^{II} complex ions have undergone ST but in only 50 % of the original [HS–HS] pairs; (2) Only one Fe^{II} center in all [HS–HS] pairs has switched. Mössbauer spectroscopy has confirmed that the second switching manner holds, viz. a ST between [HS–HS] to [LS–HS] pairs. Such a ST from [HS–HS] to [LS–HS] pairs on cooling is in agreement with the temperature dependence of the magnetic properties (Fig. 2.23). The HS to LS transition of half of the Fe^{II} ions occurs around 225 K to reach a plateau of very large length. The drop in $\chi_M T$ at very low temperature is definitely not due to a further HS to LS transition, because the Mössbauer spectrum recorded at 4.2 K shows a 50 % HS population. The sharp decrease of $\chi_M T$ at low temperatures is due to ZFS of HS Fe^{II} ions [40]. The Mössbauer spectra recorded as a function of temperature and shown in Fig. 2.24 not only confirm a 50 % spin state conversion, but also give insight into structural distortion that accompanies the spin transition (Fig. 2.23). The quadrupole doublet measured at 298 K refers to the original [HS–HS] pairs. At 240 K new signals appear corresponding to LS Fe^{II} ions (blue) in [HS–LS] pairs and HS Fe^{II} ions (dark red) in [HS–LS] pairs, both with the same area fraction. Upon further cooling to 180 K, a complete transition of HS ions in [HS–HS] pairs to a 50 % mixture of LS and HS ions in [HS–LS] pairs is identified, for the first time in zero-field Mössbauer spectroscopy [41].

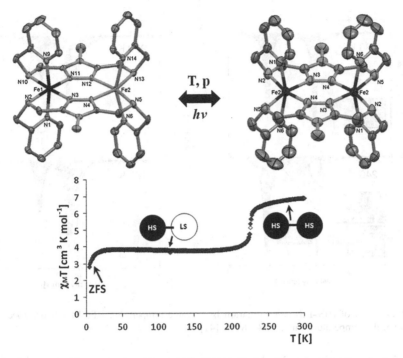

Fig. 2.23 View of the crystal structure of $\left[\text{Fe}_2^{\text{II}}(\text{pmatrz})_2\right](\text{BF}_4)_4\cdot\text{DMF}$ at 298 K in HS state (*right*) and at 123 K in LS state (*left*), and the magnetic susceptibility as a function of temperature for $\left[\text{Fe}_2^{\text{II}}(\text{pmatrz})_2\right](\text{BF}_4)_4\cdot\text{DMF}$ [39]

As seen from the 180 K spectrum the original quadrupole doublet (red) from [HS–HS] pairs has totally disappeared at the favor of the poorly resolved LS quadrupole doublet (blue) from [HS–LS] pairs and the new HS doublet (dark red) from [HS–LS] pairs. Its quadrupole splitting is significantly smaller than that in the original [HS–HS] pairs, and clearly points to the occurrence of a molecular distortion caused by the spin transition. In other words, the ST occurring in one Fe^{II} center is rapidly (in times characteristic of molecular vibrations) transmitted through the bridging ligand to the neighboring Fe^{II} center where it causes some molecular distortion such that the ligand field strength is weakened to such an extent that no more thermal SCO takes place with this Fe^{II} ion on further cooling. This is indicated in the structure (Fig. 2.23) by the black color for the Fe^{II} center remaining permanently in HS state. Clearly, Mössbauer spectroscopy has revealed such extreme subtleties in this study which is hardly possible with any other physical technique.

2.3.2.4 Thermal Spin Transition in a Trinuclear Fe^{II} Compound

The trinuclear Fe^{II} compound of formula $\left[\text{Fe}_3(\text{iptrz})_6(\text{H}_2\text{O})_6\right](\text{CF}_3\text{SO}_3)_6$ (iptrz = 4-isopropyl-1,2,4-triazole) was synthesized and its structure and magnetic

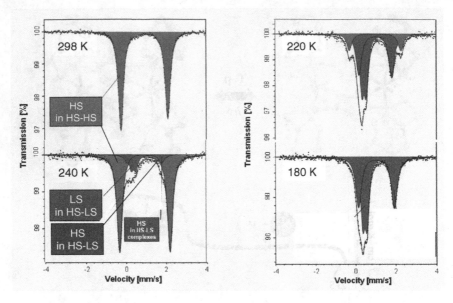

Fig. 2.24 Proof of [HS–LS] pair formation by variable temperature ^{57}Fe Mössbauer spectroscopy over the temperature range 298–180 K [41]

behavior characterized [42]. A perspective view of the structure is depicted in Fig. 2.25a. The central Fe^{II} is coordinated to three 1,2,4-triazole units and thus has FeN_6 core suitable to undergo thermal spin transition. The outer two Fe^{II} sites are each coordinated to three 1,2,4-triazole units and are capped by three water molecules, thus each having an FeN_3O_3 core, the ligand field strength of which is too weak for Fe^{II} to undergo thermal spin transition.

All three Fe^{II} centers are HS at 300 K as confirmed by Mössbauer spectroscopy (Fig. 2.25b). The quadrupole doublet shown in red, marked HSc, with the large splitting refers to the central Fe^{II}. The large splitting arises mainly from the valence electron contribution to the EFG; the lattice contribution is small due to the relatively high symmetry (close to O_h) of the FeN_6 core. The quadrupole doublet shown in yellow, marked HSo, refers to the two outer two Fe^{II} sites. The considerably smaller quadrupole splitting in this case arises from a relatively large lattice contribution to the EFG due to the low symmetry of the FeN_3O_3 core. This is opposite in sign to the valence electron contribution and thus reduces the total EFG giving a smaller total quadrupole splitting. The intensity ratio of the doublets of HSo to HSc is 1:2 as expected.

As the temperature is lowered, thermal ST from HS to LS takes place only at the central Fe^{II}. While the intensity of the yellow doublet of HSo remains unchanged, that of the red doublet of the central Fe^{II} diminishes at the favor of a new signal, shown in blue, which arises from LS state central Fe^{II} ions, marked by LSc. At the same time, a new quadrupole doublet (green) arising from the outer Fe^{II}–HS sites in HS–LS–HS trinuclear species, denoted as HS_o^{SC}, with twice the

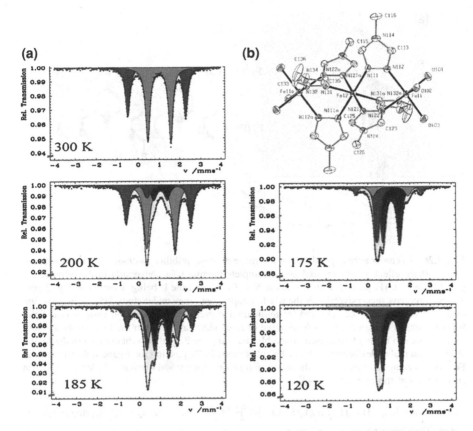

Fig. 2.25 a Projection showing the crystal structure of $[Fe_3(iptrz)_6(H_2O)_6](CF_3SO_3)_6$ [42]. **b** Temperature dependent ^{57}Fe Mössbauer spectra of $[Fe_3(iptrz)_6(H_2O)_6](CF_3SO_3)_6$ [42]. The *yellow*, *red* and *green* doublets correspond to HS_O, HS_c and HS_O^{SC}, respectively. The *blue* signal corresponds to LS_c

intensity of the blue LSc resonance signal. This 2:1 intensity ratio for HS_O^{SC}: LSc (green to blue) remains on further lowering the temperature. The same 2:1 intensity ratio remains for HSo: HSc (yellow to red) at any temperature until they entirely disappear after the ST is complete. The Mössbauer spectrum recorded at 120 K shows only the resonance signals for HS–LS–HS species. Clearly, the outer Fe^{II}–HS sites "feel" the process of thermal ST at the center, which is being communicated through the three rigid 1,2,4-triazole bridges and cause a noticeable structural distortion at the outer Fe^{II}–HS sites without influencing the spin state.

This is another "textbook example" for the high sensitivity of Mössbauer spectroscopy similar to the Mössbauer study of the $[Fe_2^{II}(pmatrz)_2](BF_4)_4 \bullet DMF$ discussed above. The tiny small distortion felt by the outer Fe^{II}–HS sites, which is caused by the ST from HS to LS in the center, is hardly detectable by diffraction techniques. Similarly, HS↔LS relaxation effects that are not seen by magnetic susceptibility measurements were detected at room temperature for the trinuclear

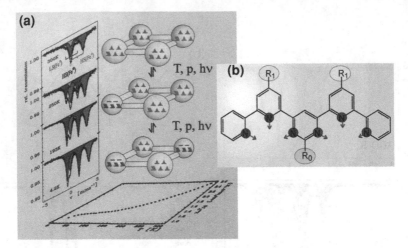

Fig. 2.26 a Temperature dependence of the magnetic susceptibility (*bottom*) and the Mössbauer spectra (*left*) reflect a very gradual and incomplete thermal spin conversion in the tetranuclear grid-like system $[Fe_4L_4]X_8$, with the anion $X = ClO_4^-$ and the L being depicted in Fig. 2.26b. The ST proceeds successively over the whole temperature range under study [45]. Spin transition in this system can also be achieved by application of pressure or irradiation with (*green*) light. **b** Four nitrogen donating ligands (derivative of terpyridine) as shown in the figure hold together four Fe^{II} ions to form a tetranuclear assembly, termed $[2 \times 2]$ grid, as schematized in the figure. With ligand molecules bearing CH_3 or Ph substituents in R_0 position the ligand field strengths at Fe^{II} ions becomes comparable to the mean spin pairing energy and thermally induced ST sets in successively at the iron centers [44]

SCO complex $[Fe_3(hyetrz)_6(H_2O)_6](CF_3SO_3)_6$ (hyetrz = 4-(2'-hydroxyethyl)-1,2,4-triazole) [43].

2.3.2.5 Spin Transition in "Grid-like" Tetranuclear Fe^{II} Complexes

The magnetism of a series of tetranuclear complexes of the $[Fe_4L_4]X_8$ $[2 \times 2]$-grid-type was investigated, revealing the occurrence of ST behavior within this class of compounds. Four L-type ligands embrace four Fe^{II} ions as schematized in Fig. 2.26b. The magnetic behavior depends directly on the nature of the substituent R_0 in the 2-position of the central pyrimidine group of the ligand L. All Fe^{II} ions in compounds with R_0 substituents favoring strong ligand fields ($R_0 = H$; OH) remain completely in the diamagnetic LS state. Only complexes bearing R_0 substituents attenuating the ligand field strength by steric (and to a lesser extent electronic) effects ($R_0 = Me$; Ph) exhibit ST behavior triggered by variation of temperature. In general, gradual and incomplete transitions without hysteresis were observed for these magnetically active complexes. The spin conversion takes place successively at the iron centers which causes the very gradual decrease of the magnetic susceptibility stretched over the whole temperature range under study [44]. Figure 2.26a shows on the left some representative ^{57}Fe Mössbauer spectra recorded between 300 and 4.2 K. The red quadrupole doublet refers to the HS–Fe^{II} ions, the one shown in blue to the LS–Fe^{II} ions. The approximately equal resonance areas of both doublets at 4.2 K indicates that the ST is incomplete and apparently has occurred only at two of the four Fe^{II} ions of a [2 × 2] grid unit. The question arises whether the two Fe^{II} ions exhibiting SCO are direct neighbors or placed diagonal in opposite positions of the grid. In relation to what one has learned from the studies of ST in dinuclear and trinuclear compounds (see Sects. 2.3.2.3 and 2.3.2.4), viz. that ST can cause a molecular distortion of a neighboring complex with accompanying weakening of the ligand field, it is not unreasonable to conclude that a similar process occurs in the present grid system. Thus it is likely that the two complex molecules which have converted to the LS state at low temperature are placed in diagonal positions as sketched in Fig. 2.26 [45].

2.3.2.6 Spin Transition in a 'Pentanuclear' Fe^{II} Coordination Compound

An unique pentanuclear Fe^{II} compound with the formula $[Fe_2L_5(NCS)_4]_2[FeL_2(NCS)_2(H_2O)_2] \cdot H_2O$, where L = 4-(p-tolyl)-1,2,4-triazole, has been synthesized and structurally characterized [46]. The structure was found to consist of two types of iron-containing structural units, a mononuclear unit and a dinuclear one (Fig. 2.27). The mononuclear unit has a crystallographic inversion center, and is coordinated by two NCS anions, two triazole nitrogen atoms, and two water molecules, each hydrogen-bonded to one of the two dinuclear units. The dinuclear units

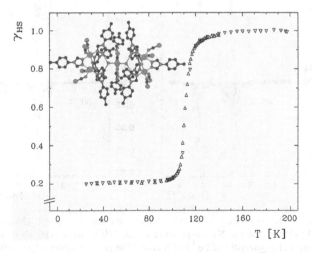

Fig. 2.27 Molar fraction of HS molecules of the pentanuclear Fe^{II} compound of formula $[Fe_2L_5(NCS)_4]_2[FeL_2(NCS)_2(H_2O)_2] \cdot H_2O$, derived from magnetic susceptibility measurements. A sharp ST between HS and LS state of Fe^{II} occurs near 111 K without hysteresis [46]. (Insert) Structure of the pentanuclear Fe^{II} compound $[Fe_2L_5(NCS)_4]_2[FeL_2(NCS)_2(H_2O)_2] \cdot H_2O$. Thermal ST takes place at the four outer Fe^{II} ions with FeN_6 core, but not at the central Fe^{II} ion with FeN_4O_2 core, where the ligand field strength is too weak such that the Fe^{II} ion remains in the HS state

consist of two Fe^{II} ions bridged by three triazole ligands in a 1,2-fashion. The coordination spheres of both iron ions are completed by two NCS anions and one monodentate 1,2,4-triazole ligand. The monodentate triazole ligands are connected through the non coordinating N atom to the mononuclear iron unit by hydrogen bonds from its coordinated water molecule. Magnetic susceptibility measurements indicate a ST only for the iron ions in the dinuclear units, centered at around $T_{1/2} = 111$ K. The transition takes place within a relatively narrow T range. The mononuclear iron ion with relatively weak ligand field strength of the FeN_4O_2 core remains in the HS state even at very low temperature, yielding a ratio of 4:1 of LS to HS iron ions. Temperature-dependent Mössbauer spectroscopy and magnetic susceptibility measurements confirm these results as seen from Figs. 2.27 and 2.28 [46].

2.3.2.7 Spin Transition in Metallomesogens

Metallomesogens belong to a class of metal-containing compounds which exhibit liquid–crystal (LC) properties. The possibility of combining LC properties with SCO behavior and accompanying change of color and magnetism in advanced functional materials has been a major objective of research in this field in recent years. In addition, the possibility of tuning the physical (mesomorphic, optical,

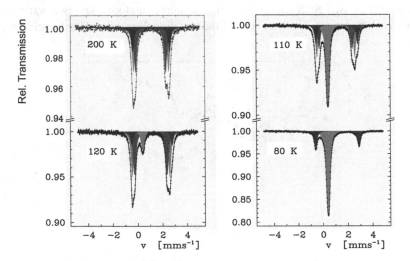

Fig. 2.28 Variable temperature ^{57}Fe Mössbauer spectra of $[Fe_2L_5(NCS)_4]_2[FeL_2(NCS)_2(H_2O)_2]\cdot H_2O$ [44]. Above the ST temperature of ca. 110 K the spectra show three doublets (*red, green, orange*) characteristic of Fe^{II} in HS state. The *green* doublet refers to the central Fe^{II} site, which remains in the HS state over the whole temperature range under study. The *red* and *orange* doublets with nearly equal intensity arise from the HS- Fe^{II} ions in the outer dinuclear units; the slight differences in quadrupole splitting point at slightly different molecular distortions. These two doublets decrease in intensity on lowering the temperature at the favor of a new signal (singlet shown in *blue*) with increasing intensity arising from the four LS-Fe^{II} ions in the dinuclear units after thermal spin transition

magnetic) properties of metallomesogens is significantly extended, since the organic ligand of these systems can be varied. Liquid crystalline materials in which a SCO center is incorporated into the mesogenic organic skeleton establish a separate class of compounds for which an interplay of structural transitions and liquid crystallinity is expected. This may lead to advantages in practical applications, for example, processing SCO materials in the form of thin films, enhancement of ST signals, switching and sensing in different temperature regimes, or achievement of photo- and thermochromism in metal-containing liquid crystals. The change of color in coexistence with liquid crystallinity is certainly a phenomenon of particular interest in the field of materials sciences.

Galyametdinov et al. [47] reported on temperature-dependent Mössbauer and magnetic susceptibility measurements of an Fe^{III} compound which exhibits liquid crystalline properties above and thermal ST below room temperature. This was the first example of SCO in metallomesogens. Later, different families of Fe^{II} and Co^{II} systems were also investigated [48–57]. The question whether the solid–liquid crystal phase transition provokes the spin-state change in SCO metallomesogens has been addressed in several series of Fe^{II} systems employing a variety of physical measurements [54, and references therein]. In all these studies ^{57}Fe Mössbauer spectroscopy has been extremely helpful, e.g. in controlling the completeness of ST in both the high and low temperature regions, where $\chi_M T$ data are often not reliable due to calibration difficulties. Also, one can unambiguously decide whether a significant decrease of the $\chi_M T$ vs. T plot towards lower temperatures is due to SCO or zero-field splitting. An example is the study of the one-dimensional (1D) 1,2,4-triazole-based compound $[Fe(C_{10}-tba)_3](4-MeC_6H_4-SO_3)_2 \cdot nH_2O$, with $C_{10}-tba = 3,5$-bis(decyloxy)-N-(4H-1,2,4-triazol-4-yl)benzamide, n = 1 or 0 [55, 56]. This system exhibits a spin state change on warming as a result of solvent release with a concomitant change of color between white (HS state) and purple (LS state) [58]. The magnetic properties of the pristine compound (n = 1) and the dehydrated sample (n = 0) are depicted in the form of $\chi_M T$ vs. T plots in Fig. 2.29.

At 300 K the value of $\chi_M T = 0.20$ cm^3 K mol^{-1} indicates that the compound is in the LS (purple) state. The Mössbauer spectrum recorded at 4.2 K is in agreement with the magnetic data, i.e. the HS population is 4.8 % and the LS population 95.2 % (Fig. 2.30a). Upon heating $\chi_M T$ increases abruptly within a few degrees, reaching the value of 3.74 cm^3 K mol^{-1} at 342 K. This clearly shows that a spin state change from LS to HS has occurred. The thermo-gravimetric analysis (TGA) of this system showed that dehydration takes place in the same temperature region where the spin state change occurs. The magnetic susceptibility of the dehydrated compound (n = 0) was recorded in a temperature loop, i.e. from 375 K down to 10 K and then up again to 375 K. The dehydrated complex reveals incomplete SCO, accompanied by hysteresis and color change (from purple in the LS state to white in the HS state), in the temperature region of 250–300 K. Around 50 % of Fe^{II} sites have changed the spin state as can be inferred from the value of $\chi_M T$ at 200 K. The Mössbauer spectrum recorded at 200 K (Fig. 2.30b) yields 49.3 % of Fe^{II} in the LS state and 50.7 % in the HS state. The further decrease of the

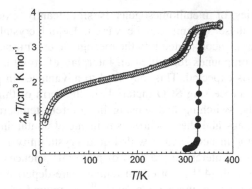

Fig. 2.29 Magnetic properties in the form of $\chi_M T$ vs. T of the one-dimensional 1,2,4-triazole-based mesogen of formula $[Fe(C_{10}-tba)_3](4-MeC_6H_4SO_3)_2 \cdot nH_2O$, with $n = 1$ (*filled circles*) and $n = 0$ (*open circles*) [55, 56]. The pristine compound is in the LS state at 300 K, but loses crystal water on heating accompanied by sharp ST to the HS state. The dehydrated sample ($n = 0$) shows a entirely different SCO behavior

$\chi_M T$ value below 100 K, particularly below 50 K, is due to the ZFS of those iron atoms which remain in the HS state even at very low temperature, as derived from the Mössbauer spectrum at 4.2 K (HS population is 48.1 %, LS population is 51.9 %) (Fig. 2.30c).

Another member of the tba-based mesogen complexes is the room temperature operational 1D coordination polymer $[Fe(C_{12}-tba)_3](CF_3SO_3)_2$ (C_{12}-tba = 3,5-bis(dodecyloxy)-N-(4H-1,2,4-triazol-4-yl)benzamide) which has been studied by Mössbauer spectroscopy [53]. The disk-like cations are self-assembled in columns, where the Fe^{II} ions are stacked on top of each other in the middle of the column as sketched in Fig. 2.31. The distance between the iron ions, determined by powder X-ray diffraction and Extended X-ray Absorption Fine Structure (EXAFS) measurements, varies with temperature and influences the spin state of the Fe^{II} ions as controlled by Mössbauer spectroscopy (Fig. 2.31). The variation of the chain length of the alkoxy substituent on the triazole ligand in $[Fe(C_n-tba)_3](BF_4)_2 \cdot H_2O$ with C_n-tba = 5-bis(alkoxy)-N-(4H-1,2,4-triazol-4-yl)benzamide) can dramatically affect the spin state population as illustrated in Fig. 2.32 where the dark grey and light grey signals refer to the LS and HS states, respectively [53].

2.3.2.8 Nuclear Decay-Induced Excited Spin State Trapping (NIESST): Mössbauer Emission Spectroscopy

In conventional Mössbauer spectroscopy one uses a single-line source, e.g. ^{57}Co embedded in a rhodium matrix in the case of ^{57}Fe spectroscopy, and the iron containing material under study as absorber. This technique is termed *Mössbauer Absorption Spectroscopy* (MAS) in order to distinguish it from the so-called source experiment, also known as *Mössbauer Emission Spectroscopy* (MES). In a MES

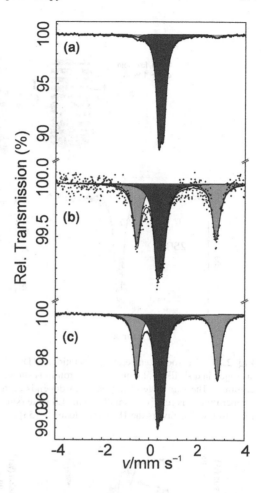

Fig. 2.30 ^{57}Fe Mössbauer spectra of the mesogen [Fe(C$_{10}$–tba)$_3$](4-MeC$_6$H$_4$SO$_3$)$_2$·nH$_2$O, with n = 1 at 4.2 K (**a**) of n = 0 at 200 K (**b**) and of n = 0 at 4.2 K (**c**) [53, 54]

experiment one uses a single-line absorber (free of electric quadrupole and magnetic dipole interaction), e.g. K$_4$[Fe(CN)$_6$], and a ^{57}Co doped sample under study as source. In this case, the recorded Mössbauer spectrum refers to the hyperfine inter-actions in the source material, i.e. it yields information on the chemical and physical properties of the excited ^{57}Fe atoms before they decay to the ground state. The MES technique has been widely used to investigate chemical and physical *after-effects* of nuclear decay in various materials, particularly in coordination compounds [59]. The electron capture decay of radioactive ^{57}Co, whereby an electron from the K-shell is captured by the ^{57}Co nucleus leading to ^{57}Fe, ^{57}Co(EC)^{57}Fe, may lead to a variety of after-effects like bond rupture, ligand radiolysis, change of charge states and excited ligand field states to name the most important consequences of nuclear decay in solid coordination compounds. These after-effects may have lifetimes on the order of 10–500 ns. It is possible to cover this time regime and study the relaxation kinetics of such short-lived after effect species with *time-integral* and *time-differential* Mössbauer emission spectroscopy [60, 61].

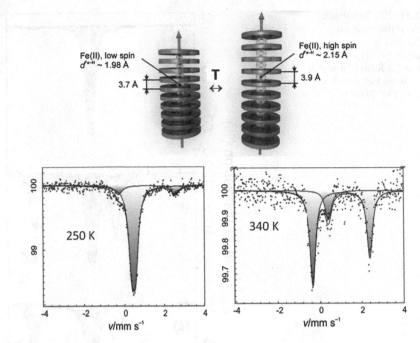

Fig. 2.31 The room temperature operational 1D polymeric mesogen [Fe(C$_{12}$-tba)$_3$](CF$_3$SO$_3$)$_2$ arranges in disk-like columns with the iron ions stacked on *top* of each other in the *middle* of the columns. The spin state of the Fe^{II} ions depends on the distance d between the disks. At room temperature, d is relatively small and the LS state (singlet) is favored. On warming the distance d increases which favors the HS state (doublet) [53]

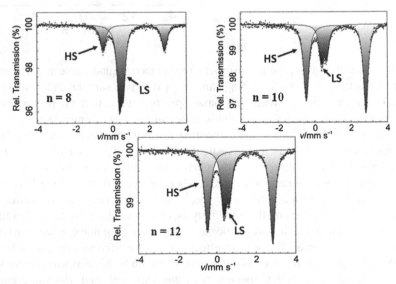

Fig. 2.32 Effect of chain length on the spin state population of [Fe(Cn-tba)$_3$](BF$_4$)$_2$·H$_2$O (n = 8, 10, 12) investigated by Mössbauer spectroscopy at 80 K [53]

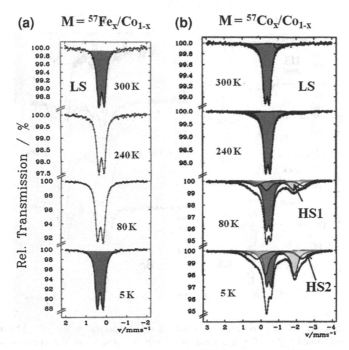

Fig. 2.33 **a** *Left* ^{57}Fe Mössbauer absorption spectra of $[^{57}Fe_x/Co_{1-x}(phen)_3](ClO_4)_2$ as a function of temperature vs. ^{57}Co/Rh (295 K) as source (x = 0.001) **b** *Right* Time-integral ^{57}Fe Mössbauer emission spectra of a $[^{57}Co_x/Co_{1-x}(phen)_3](ClO_4)_2$ source as a function of temperature vs. $K_4[Fe(CN)_6]$ (295 K) (x = 0.001). In (*a*) the source was moved relative to the absorber and in (*b*) the absorber was moved relative to the fixed source mounted in the crystal [63]

As an example, the coordination compound $[Co^{II}(phen)_3](ClO_4)_2$ doped with 0.1 % of ^{57}Fe was studied by Mössbauer absorption spectroscopy using a ^{57}Co/Rh source (Fig. 2.33; spectra on the left). The three phen ligands create a relatively strong ligand field at the Fe^{II} center, the compound shows LS behavior at all temperatures under study. The same system, however doped with ^{57}Co as Mössbauer source, was studied by Mössbauer emission spectroscopy using $K_4[Fe(CN)_6]$ as absorber. The MES spectra (Fig. 2.33, spectra on the right) also show the typical Fe^{II}–LS signal at 300 K down to ca. 200 K. On further cooling, however, the intensity of this signal decreases and at the same time two Fe^{II}–HS doublets, HS1 and HS2, appear with increasing intensity [62]. These unusual spin states are excited ligand field states with temperature-dependent lifetimes on the order of ca. 100 ns.

Similar experiments were carried out with systems whose corresponding Fe^{II} compounds possess intermediate ligand field strengths and show thermal spin crossover. $[Fe(phen)_2(NCS)_2]$ undergoes thermal ST as already discussed above (Sect. 2.3.2.1). The temperature dependent MAS spectra are shown on the left of Fig. 2.34. The analogous Co^{II} compound doped with ^{57}Co and used as Mössbauer source (or the corresponding iron compound doped with ^{57}Co as source which

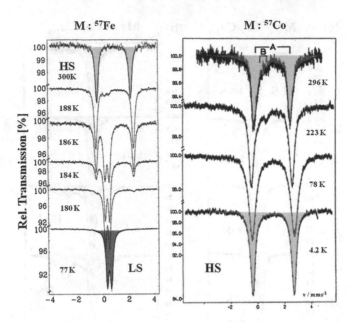

Fig. 2.34 [Fe(phen)$_2$(NCS)$_2$] undergoes thermal ST (MAS spectra on the *left*). The analogous Co^{II} compound doped with ^{57}Co and used as Mössbauer source (or the corresponding iron compound doped with ^{57}Co as source which gives the same results) yield the temperature dependent Mössbauer emission spectra (MES) shown on the *right*. The main result is that in the temperature region, where the MAS spectra reflect the transition to the LS state, the MES spectra still show the typical HS signals arising from short-lived excited ligand field states [64]

gives the same results) yield the temperature dependent MES spectra shown on the right [64]. The main result is that in the temperature region, where the MAS spectra reflect the transition to the LS state, the MES spectra still show the typical HS signals arising from excited ligand field states.

The mechanism of the formation of the excited ligand field states as a consequence of the EC nuclear decay of radioactive ^{57}Co is well understood. It is much related to that of the LIESST phenomenon (Fig. 2.14) and has therefore been termed *Nuclear Decay-Induced Excited Spin State Trapping* (NIESST) [59]. The main difference between the two phenomena lies in the primary step of excitation, which is the application of a light source in the case of LIESST, whereas in NIESST the nuclear decay serves quasi as an intrinsic molecular excitation source. The lifetime or probability of observing the metastable HS ($^5T_2$) state within the Mössbauer time scale (given by the lifetime of the 14.4 keV nuclear state of ^{57}Fe) at a given temperature is governed by the ligand field strength felt by the nucleogenic ^{57}Fe ion: The weaker the ligand field strength, the longer the lifetime of the metastable HS ($^5T_2$) state. It is worth emphasizing that this technique, Mössbauer emission spectroscopy, is most effective for the study of chemical and physical after-effects of nuclear decay processes, referred to as "Hot Atom Chemistry", in solids [63].

The NIESST effect was also studied in Co^{II} SCO compounds, viz.: $[^{57}Co/Co(terpy)_2]X_2 nH_2O$ (X = ClO_4^-, n = $\frac{1}{2}$, X = Cl^-, n = 5), where terpy is the tridentate ligand terpyridine [65]. The perchlorate salt shows thermal SCO with $T_{1/2}$ around 200 K and a HS fraction of nearly 100 % at room temperature, whereas the chloride salt possesses a somewhat stronger ligand field giving rise to thermal ST at much higher temperatures (the HS fraction starts to rise around 200 K, reaches ca. 20 % at 320 K and obviously would increase further) [65]. Conventional Mössbauer absorption measurements were performed on the corresponding systems doped with 5 % Fe^{II}, which was found to be in the LS (S = 0) state at all temperatures under study [65]. The emission spectra of the ^{57}Co-labelled cobalt complexes were measured using a home-made resonance detector, which operates as conversion-electron detector with count rates 10–20 times higher than those of a conventional detector. At room temperature, the nucleogenic ^{57}Fe ions were found to have relaxed to the stable $^{1}A_1$ LS ground and gave the same MAS Mössbauer spectrum like the corresponding Fe^{II} compound. On lowering the temperature a doublet from a metastable Fe^{II}–HS state appears in the MES spectra with increasing intensities. The perchlorate derivative with the weaker ligand field strength shows, at comparable temperatures, a considerably higher amount of Fe^{II}–HS fraction than the chloride derivative with the stronger ligand field. For instance, the emission spectra recorded at 100 K displayed in Fig. 2.35 demonstrate this effect very clearly. Thus, it turns out that the lifetime of the nuclear decay-induced metastable HS state of Fe^{II} is short in strong ligand field surroundings and long in weak ligand fields. In other words in relation to Fig. 2.14, the stronger the ligand field, the larger is the difference between the lowest vibronic energy levels of HS and LS states, and the shorter is the lifetime at a given temperature. This is known as "reduced energy gap law" which holds for all these NIESST studies [59].

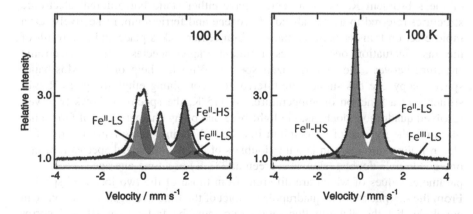

Fig. 2.35 Mössbauer emission spectra of $[^{57}Co/Co(terpy)_2]X_2 nH_2O$ (X = ClO_4^-, n = $\frac{1}{2}$; X = Cl^-, n = 5) as source material vs. $K_4[Fe(CN)_6]$ as absorber (which was kept at 298 K) recorded at 100 K with a conversion-electron detector. *Left* X = ClO_4^-, n = $\frac{1}{2}$. *Right* X = Cl^-, n = 5 (from [65])

2.3.3 Mixed-Valence Compounds, Intramolecular Electron Transfer

Mixed-valence compounds are known to possess two or more transition metal ions with different oxidation states. Intramolecular electron transfer may take place between (generally two) heterovalent metal centers over suitable bridging atoms or ligand molecules. There may be three cases observed in the Mössbauer spectrum. (1) If the electron fluctuation rate τ_f is less than the reciprocal of the Mössbauer time window (given by the lifetime τ_n of the nuclear excited state of the Mössbauer probe), $\tau_f < \tau_n$, the spectrum will show two resolved resonance signals arising from the static ("localized") oxidation states. (2) If $\tau_f > \tau_n$, the spectrum will show a single time-averaged resonance signal which is different from the two signals of case (1) and points at an oxidation state in between those of the two static components. One speaks of a "delocalized" mixed-valence compound. (2) If $\tau_f \approx \tau_n$, the Mössbauer spectrum will be the sum of overlapping heavily broadened resonance lines of cases (1) and (2); the spectrum will be very complicated and difficult to analyze. As the fluctuation rate τ_f is temperature dependent, one often observes a transition between the localized and delocalized cases during variation of temperature.

In the following, we shall briefly discuss three examples of studies of mixed-valency systems with Mössbauer spectroscopy.

2.3.3.1 Mixed-Valence Biferrocenes

In Fig. 2.36 is schematized the molecular structure of biferrocenes which possess two iron atoms with different valence electron structures. Depending on the nature of the substituent R, the iron centers have either static but different electronic structures referred to as "localized" ferrocene and ferricinium, respectively. Or a rapid electron transfer between the two iron centers takes place and as a result of the fast fluctuation one observes a time-averaged species with an electronic structure between the two localized species. With the help of ^{57}Fe Mössbauer spectroscopy one has studied the biferrocenes containing ethyl groups as R substituents as a function of temperature. At 115 K, the spectrum shows two well resolved quadrupole doublets, the light-blue one being characteristic of ferrocene-like iron, the red one of ferrocinium-like iron. With increasing temperature a fast electron fluctuation sets in and the doublets of the two localized species turn into one quadrupole doublet (shown in green at 287 K) of a time-averaged species, the parameter values of which are different from those of the two localized species. From the sharpness of the quadrupole doublet of the time-averaged species one can conclude that the electron fluctuation rate must be faster than 10^7 s^{-1} corresponding to the time-window of ^{57}Fe Mössbauer spectroscopy. This example shows a thermally induced transition between localized and delocalized electronic structures in a mixed-valence organometallic compound.

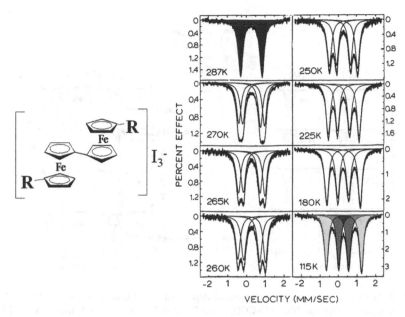

Fig. 2.36 Mixed valence biferrocene with R = Et shows temperature dependent electron fluctuation between the two iron centers. At low temperatures the fluctuation rate is comparatively slow (less than the reciprocal of the Mössbauer time window of ca. 100 ns) and the Mössbauer spectra show two subspectra indicative of "localized" ferrocene (*grey*) and ferrocinium (*dark grey*), respectively. At higher temperatures the fluctuation rate becomes so fast that the Mössbauer spectrum reflects a time-averaged "delocalized" species (*black*) which is neither ferrocene nor ferrocenium [66, 67]

2.3.3.2 Effect of Crystal Solvents Molecules on the Valence Detrapping of Mixed-Valence [Fe₃O(O₂CCH₃)₃)₆(3-Et-py)₃]·S

The molecular structure of the mixed-valence compound $[Fe_3O(O_2CCH_3)_3)_6(3\text{-Et-py})_3]\cdot S$ is visualized in Fig. 2.37. The molecule accommodates two HS–Fe^{III} ions and one HS–Fe^{II} ion which is confirmed by Mössbauer spectroscopy (Fig. 2.37) [68]. In all spectra the more intense quadrupole doublet (red) corresponds to the two HS Fe^{III}–HS ions and the less intense doublet (green) is for the one HS Fe^{II} ion. The ratio of the area fractions of Fe^{III} to Fe^{II} is close to 2 at low temperatures. Towards higher temperatures it tends to become larger than 2, which is due to the larger Lamb-Mössbauer factor of Fe^{III} compared to that of Fe^{II}. It is found [68] that the mixed valency properties of this compound depend on the nature of the crystal solvents molecules. Compounds A (with S = 0.5 benzene) and B (with S = CH_3CN) appear to be valence-trapped over the whole temperature range up to room temperature. The quadrupole doublets arising from HS–Fe^{III} (red) and HS–Fe^{II} (green) are well resolved and sharp. Thus the lifetimes of these trapped (localized) species are longer than the lifetime of the 14.4 keV nuclear excited state. Compound C (with S = CH_3CCl_3) are valence-trapped at low temperatures.

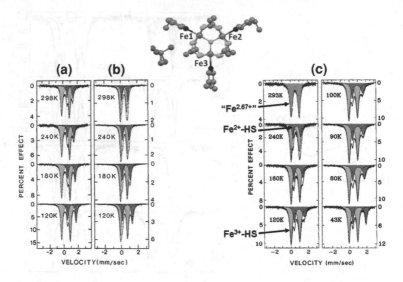

Fig. 2.37 Molecular structure and temperature dependent ^{57}Fe Mössbauer spectra of the mixed valence complex compound [Fe$_3$O(O$_2$CCH$_3$)$_6$(3-Et-py)$_3$]·S with S = 0.5 benzene (**a**), CH$_3$CN (**b**) and CH$_3$CCl$_3$ (**c**). Solvates A and B are "valence-trapped" (localized) with two HS–Fe^{III} ions (*red* doublet) and one HS–Fe^{II} ion (*green* doublet) in the molecular unit; the quadrupole doublets are well resolved without significant line-broadening. **c** Shows a transition from electron localization to delocalization (valence-trapped to detrapped) with increasing temperature. The sharp doublets of the trapped species begin to broaden on heating and finally disappear at the favor of a time-averaged new signal (*blue*) from a species with averaged oxidation state, "Fe$^{2.67+}$", due to fast electron fluctuation within the Fe$_3$O *triangle* [68]

The resonance lines of the two quadrupole doublets begin to broaden around 50 K. Further increase of temperature leads to valence-detrapping (delocalization) near room temperature, where the red and green doublets have disappeared at the favor of a new doublet (blue) which is a time-averaged resonance signal of a species with an "average" oxidation state of "Fe$^{2.67+}$" due to rapid electron circulation within the Fe$_3$O triangle of the complex.

2.3.3.3 Valence Fluctuation in a Trinuclear Cationic Complex

Another interesting example of valence fluctuation and temperature dependent transition between localized and delocalized electronic structures in a trinuclear transition metal compound was reported by Glaser et al. [69]. In this case the electron fluctuation takes place between two iron centers of different oxidation states and separated by a diamagnetic Co^{III} ion. The ^{57}Fe Mössbauer spectra clearly show that at sufficiently low temperatures, i.e. 5 K, the two iron centers are reflected as localized oxidation states with a well resolved doublet for LS–Fe^{III} (red) and a poorly resolved doublet for LS–Fe^{II} (light-blue). At higher temperatures the

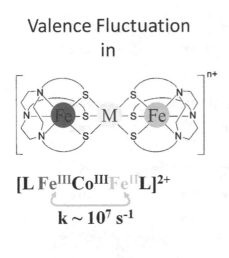

Valence Fluctuation
in

$$[L \; Fe^{III}Co^{III}Fe^{II}L]^{2+}$$

$$k \sim 10^7 \; s^{-1}$$

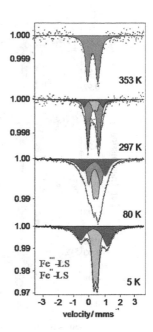

Fig. 2.38 Schematized molecular structure of a trinuclear cationic complex which accommodates LS-Fe^{III} (*red*) and LS-Fe^{II} (*blue*) ions separated by bridging LS-Co^{III} ion. At low temperatures two well resolved spectra are observed assigned to the localized LS-Fe^{III} and the LS-Fe^{II} species. With increasing temperature the intensities of these two doublets decrease at the favor of a new doublet (*green*) from a time-averaged species with intermediate oxidation state of the two iron sites. This species is the only one remaining at room temperature [69]

intensities of these two resonance signals decrease at the favor of a new resonance signal, another quadrupole doublet (blue), arising from a time-averaged species where both iron centers are in an intermediate oxidation state as a result of fast electron fluctuation through the bridging diamagnetic central Co^{III} ion (Fig. 2.38).

2.3.3.4 Valence Fluctuation in EuNiP

Mixed-valency and thermally induced transition between localized and delocalized valence states was observed with ^{151}Eu Mössbauer spectroscopy of the intermetallic compound EuNiP [70] (Fig. 2.39).

High temperature ^{151}Eu Mössbauer measurements between room temperature and ca. 500 K provide proof for mixed-valent behavior in the pnictide EuNiP. Two well resolved signals are observed at room temperature, one being typical for Eu^{2+} (dark grey) and the other one for Eu^{III} (light grey). With increasing temperature the electron fluctuation between Eu^{II} and Eu^{III} becomes faster than the time-window of ^{151}Eu Mössbauer spectroscopy and above ca. 450 K the resonance signals of localized Eu^{II} (dark grey) and Eu^{III} (light grey) gradually disappear at the favor of a new signal with time-averaged oxidation state of $Eu^{2.5+}$ (non shaded) [70].

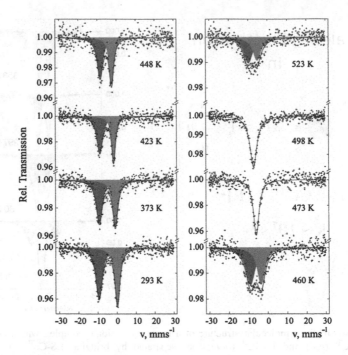

Fig. 2.39 Variable temperature ^{151}Eu Mössbauer spectra of the intermetallic compound EuNiP. Shaded subspectra correspond to Eu^{2+} (*dark grey*) and Eu^{3+} (*light grey*) [70]

EuNiP undergoes a Verwey-type charge delocalization transition when heated above 470 K prior to the structural gamma-beta phase transition at approximately 510 K. This finding confirms the results of photoemission spectroscopy of the isostructural compound EuPdP and of TB-LMTO-ASA (Tight-Binding Linear Muffin-Tin Orbital Atomic-Sphere Approximation) band structure calculations. It has been proposed that a van Hove singularity associated with a high density of 4f states close to the Fermi energy plays a particular role in mixed valency pnictide EuNiP [70].

2.3.4 Molecular Magnetism

Mössbauer spectroscopy, primarily employing ^{57}Fe as a nuclear probe, has developed to an enormously helpful complementary tool supporting standard SQUID and ac/dc magnetic measurements in studies of magnetic behavior of solid materials. Magnetic dipole interaction (see Sect. 2.2.3) will lead to a magnetically split Mössbauer spectrum, a sextet in the case of ^{57}Fe Mössbauer spectroscopy, provided a local effective magnetic field pointing sufficiently long (longer than the lifetime of the 14.4 keV nuclear excited state) in a fixed direction. Magnetic behavior of solid material is usually categorized into two classes: (1) *Long-range cooperative magnetism* due to magnetic interactions (stronger than thermal energy) between metal ions with unpaired electrons, like in metals, alloys and simple inorganic compounds (oxides, halides etc.). (2) *Molecular magnetism* with weak or vanishing long-range spin–spin interactions, also known as "molecule-based magnetism" [71]. The latter kind of magnetism has attracted much interest in recent years [72]. Because of limited space we shall discuss only two examples.

2.3.4.1 Photo-Switchable Prussian Blue Analog $K_{0.1}Co_4[Fe(CN)_6]_{2.7} \cdot 18\ H_2O$

The Prussian-blue analog $K_{0.1}Co_4[Fe(CN)_6]_{2.7} \cdot 18\ H_2O$ with paramagnetic building blocks $[Co^{II}\ (S = 3/2)\,-NC\,-\,Fe^{III}(S = 1/2)]$ was found to undergo photo-induced transition by irradiation with blue light, whereby electron transfer takes place to generate the diamagnetic buildings blocks $[Co^{III}(S = 0)\,-NC\,-\,Fe^{II}(S = 0)]$. This transition is reversible by irradiation with red light (Fig. 2.40). It was also found that the transition from $Co^{II}\ (S = 3/2)\,-NC\,-\,Fe^{III}(S = 1/2)]$ to $[Co^{III}(S = 0)\,-NC\,-\,Fe^{II}(S = 0)]$ is favored by increasing potassium concentration or by replacing

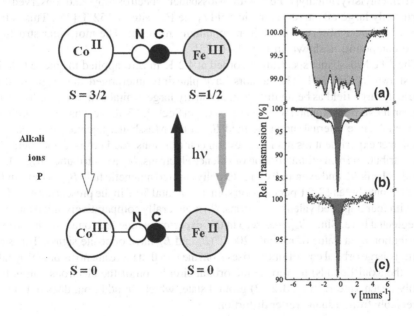

Fig. 2.40 The Prussian-blue analog $K_{0.1}Co_4[Fe(CN)_6]_{2.7} \cdot 18\ H_2O$ contains photosensitive paramagnetic $[Co^{II}\ (S = 3/2)\,-NC\,-\,Fe^{III}(S = 1/2)]$ building blocks which can be converted by irradiation with blue light to diamagnetic $[Co^{III}(S = 0)\,-NC\,-\,Fe^{II}(S = 0)]$ building blocks; back conversion is possible with *red* light. Increase of potassium content or replacement of potassium by bigger caesium ions or application of pressure favors the transition of the $[Co^{III}(S = 0)\,-NC\,-\,Fe^{II}(S = 0)]$ entities to $[Co^{III}(S = 0)\,-NC\,-\,Fe^{II}(S = 0)]$ as confirmed by Mössbauer spectroscopy at 4.2 K under different pressures **a** 1 bar, **b** 3 kbar, and **c** 4 kbar. Shaded subspectra correspond to $Fe^{II}(S = 0)$ in *dark grey* and $Fe^{III}(S = 1/2)$ in ~ *light grey* [73]

potassium by bigger caesium ions. It was speculated that chemical pressure in the crystal lattice plays a decisive role in this process. To prove the validity of this idea, ^{57}Fe Mössbauer spectra of $K_{0.1}Co_4[Fe(CN)_6]_{2.7} \cdot 18\ H_2O$ were recorded at 4.2 K under applied pressure (Fig. 2.40). At ambient pressure the sample contains the paramagnetic $[Co^{II}\ (S = 3/2)\ -NC\ -\ Fe^{III}\ (S = 1/2)]$ building blocks, and the electronic spins of Co^{II} and Fe^{III} sites are oriented sufficiently long in one direction such that the Mössbauer spectrum reflects a well resolved magnetically split sextet with effective field of 165 kOe at the ^{57}Fe nucleus. Increase of pressure to 3 kbar yields a spectrum that shows a dominating singlet (dark grey) characteristic of LS–Fe^{II} with S = 0 at the expense of the collapsing magnetic sextet from remaining LS–Fe^{III} ions. The spectrum recorded under 4 kbar shows only the singlet arising from LS–Fe^{II} ions [73].

2.3.4.2 Orbital Magnetism in a Rigorously Linear Two-Coordinate High-Spin Fe^{II} Compound

Reiff et al. have studied the linear two-coordinate HS Fe^{II} compound Bis (tris(trimethylsilyl)methyl) Fe^{II} with Mössbauer spectroscopy and observed an enormously large effective magnetic field at the Fe^{II} site of 152 T [74]. This is the largest field ever observed in an iron containing material. The molecular structure of the compound is shown in Fig. 2.41 [75].

The ^{57}Fe Mössbauer spectrum recorded at 4.2 K in zero applied magnetic field is also shown in Fig. 2.41. The authors have plausibly interpreted the origin of this extremely large field as being due to an unusually large orbital contribution, B_L, from electron movement around the molecular axis. In Sect. 2.2.3 above it has been outlined that the effective internal magnetic field B_{int} at the Mössbauer nucleus observed in a Mössbauer experiment results from several contributions, the Fermi-contact field B_c, the contribution from orbital motion of valence electrons, B_L, a contribution B_D, called spin-dipolar field, and eventually an externally applied magnetic field B_{ext}. The term B_c roughly contributes 12.5 T per electron spin, i.e. in total 50 T in the present case of HS–Fe^{II} with four unpaired valence electrons. B_D is generally comparatively small and can be neglected here. Since B_{ext} was zero in this experiment, one has found for the orbital contribution B_L a value of roughly 200 T (B_c and B_L have opposite signs). This surprisingly large orbital contribution arises from the fact that there are no in-plane ligands (only the axial ligands) to impede the orbital circulation of the electrons within the doubly degenerate E_g (d_{xy}, $d_{x^2-y^2}$) ground state, which, in addition, does not suffer appreciably from a Jahn–Teller distortion.

2.3.5 Industrial Applications of Mössbauer Spectroscopy

The eminent capability of non-destructive phase analysis with Mössbauer spectroscopy has been used in the multidisciplinary field of materials science, particularly

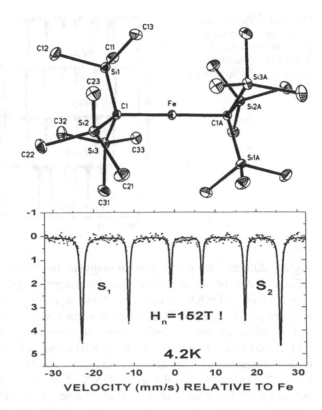

Fig. 2.41 Molecular structure of Bis(tris(trimethylsilyl) methyl)Fe^{II}, a rigorously linear two-coordinate compound of Fe^{II} in HS state with ideal staggered D_{3d} symmetry [75]. ^{57}Fe Mössbauer spectrum of Bis(tris(trimethylsilyl) methyl)Fe^{II}, recorded at 4.2 K in zero applied magnetic field [74]. The internal magnetic field derived from the distance between the two outermost resonance lines, S_1–S_2, is 152 T, the largest field ever observed in an iron compound

for the analysis of iron containing materials of technical relevance such as steel, alloys, pigments, oxides, corrosion products, to name a few. But other "Mössbauer-active" nuclides have also been used. Within the limited scope of this article we must confine the discussion in the following to only one representative example, viz. corrosion studies.

Roughly six different iron oxides and oxyhydroxides of iron are known as corrosion products, which may be formed by corrosion reactions in steel, metallic iron, and iron containing alloys under different conditions. These corrosion products can be distinguished by ^{57}Fe Mössbauer spectroscopy (Fig. 2.42). Magnetite, Fe_3O_4, is an inverse spinel compound of formula $Fe^{III}[Fe^{II}Fe^{III}]O_4$, where half of the Fe^{III} ions (those outside the square brackets) are in tetrahedral sites and the other half (inside the square brackets) in octahedral sites. All Fe^{II} ions are in octahedral sites. According to these three kinds of iron ions one should expect three different resonance signals in the Mössbauer spectrum. This, however, is not the case. Instead one observes at room temperature two overlapping sextets, one arising from Fe^{III} ions in tetrahedral sites, and the other one is a time-averaged sextet arising from Fe^{II} and Fe^{III} ions in octahedral sites with fast electron fluctuations between them (faster than the inverse of the lifetime of the 14.4 keV nuclear level). The oxides α- and γ-Fe_2O_3 show a magnetically split sextet with a

Fig. 2.42 ^{57}Fe Mössbauer
spectra of various corrosion
products which may be
formed under different
conditions. The spectra are
discussed in the text

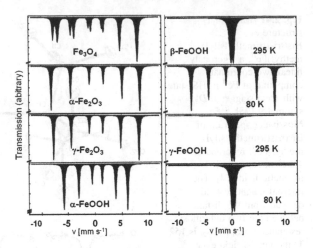

slightly different size of the internal magnetic field. The oxyhydroxides α-, β- and
γ-FeOOH can be distinguished by temperature dependent Mössbauer spectroscopy. While α-FeOOH (Goethite) shows a magnetically split sextet at room
temperature, β- and γ-FeOOH have the same poorly resolved quadrupole doublet
and therefore cannot be distinguished at room temperature. At liquid nitrogen
temperature (ca. 80 K), however, β-FeOOH is magnetically ordered and shows the
typical magnetically split sextet, whereas γ-FeOOH still shows the same poorly
resolved quadrupole doublet as at 295 K. This phase orders magnetically only
below ca. 30 K yielding then a similar sextet as the other modifications. Thus,
Mössbauer spectroscopy enables one to distinguish between these corrosion
products, even in the form of highly dispersed particles (>ca. 10 nm), where
powder X-ray diffraction measurements are no longer applicable.

Figure 2.43 shows an example of routine Mössbauer analysis of finely dispersed
particles formed in the cooling system of a power plant. The particles were collected from the coolant with a special filter and analyzed with standard

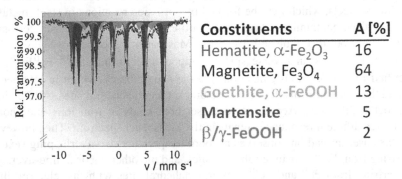

Constituents	A [%]
Hematite, α-Fe$_2$O$_3$	16
Magnetite, Fe$_3$O$_4$	64
Goethite, α-FeOOH	13
Martensite	5
β/γ-FeOOH	2

Fig. 2.43 Routine Mössbauer analysis of finely dispersed corrosion particles formed in the
cooling system of a power plant [76]

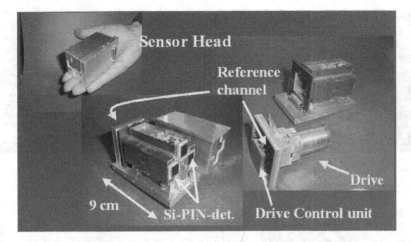

Fig. 2.44 The sensor head of the miniaturized Mössbauer spectrometer MIMOS contains the drive (vibrating) unit, the Mössbauer source, the detectors and amplification electronics. MIMOS is set up in backscattering geometry [77, 78]

transmission Mössbauer spectroscopy. The spectrum is a superposition of the spectra of five different iron-containing species with percentages given in the table on the right in Fig. 2.43. Such measurements are carried out routinely in the context of regular maintenance works in nuclear and conventional power plants.

2.3.6 Miniaturized Portable Mössbauer Spectroscopy

The miniaturization of Mössbauer instrumentation started under the guidance of E. Kankeleit at Technical University of Darmstadt and continued to completion ready for in-field-experiments by G. Klingelhöfer et al. at University of Mainz. The result is a portable miniaturized Mössbauer spectrometer (abbreviated as MIMOS) [77] as displayed in Fig. 2.44. This instrument is set up in backscattering geometry, which renders sample preparation such as the production of powders or thin slices for many applications unnecessary and thus also enables one to perform non-destructive measurements. The instrument can be taken to the field and simply placed, for example, on a rock surface of interest to be analyzed as realized by the MIMOS II instruments on the NASA Mars Exploration Rovers, which act as robotic field geologists. Another example is the analysis of rare and precious samples such as archaeological artefacts. The instrument is still developed further: Currently, for the advanced MIMOS IIa [78], the Si-PIN diode detectors are replaced by Si Drift Detectors, which have a much higher energy resolution. This result in (a) significantly improved signal-to-noise ratios and therefore shorter measurement times, and (b) allows for the simultaneous acquisition of X-Ray Fluorescence spectra for elemental analysis, whereby the ^{57}Co Mössbauer source also acts as the X-ray excitation source.

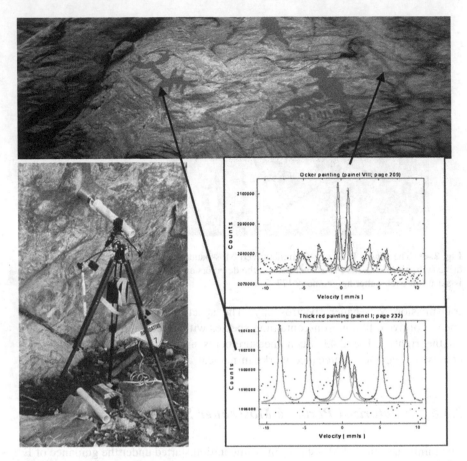

Fig. 2.45 Non-destructive analysis with MIMOS of ancient rock painting in Brazil (near Belo Horizonte). Two Mössbauer backscattering spectra were recorded, one of a darker paint, the other for a light-colored paint. Different iron oxide pigments are responsible for the different colors [80]

2.3.6.1 Archaeometric Applications

Studies of ancient ceramics became the first instance of the use of Mössbauer spectroscopy in what one commonly calls archaeometry—the application of scientific methods in studies of archaeological sites and artefacts. Changes which pottery clays undergo during firing are reflected in the Mössbauer spectra of the fired ceramics and hence can be used, even after millennia of burial, to gain information on the original firing conditions and thus on the techniques. Other areas of archaeological interest are, for example, corrosion or steel properties in iron artefacts, slags from prehistoric copper, tin and iron production, and pigments and paintings (e.g. [79], and references therein).

The portability and ability to perform non-destructive analyses with MIMOS have been exploited in the investigation of ancient rock paintings in Brazil (near Belo Horizonte, Minas Gerais, Fig. 2.45). MIMOS was mounted on a tripod and

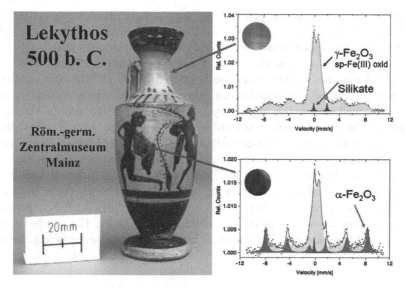

Fig. 2.46 Mössbauer backscattering spectra of a Greek lekythos vase. The different spectra taken from light and dark color positions result from different iron oxide pigments as discussed in the text [81]

Fig. 2.47 The scientific payload of the Mars Exploration Rovers consists of the remote sensing Panoramic Camera (Pancam) and the Miniature Thermal Emission Spectrometer (*Mini-TES*); the in situ or contact instruments Microscopic Imager (*MI*), Alpha Particle X-ray Spectrometer (*APXS*), Mössbauer spectrometer (*MB*), and the Rock Abrasion Tool (*RAT*); the Magnetic Properties Experiment (NASA/JPL/Cornell) [77]

brought in contact with the painting on the wall for measurement. The two Mössbauer spectra in Fig. 2.45, one of a darker paint and the other for a light-colored paint, distinguish between different iron oxide pigments [80].

A similar application is shown in Fig. 2.46. A Greek lekythos vase was investigated non-destructively with the help of MIMOS II [81]. The unpainted surface shows a broad spectrum that can be associated with poorly crystallized iron oxides produced during the firing clay process. The painted surface shows, in addition to the characteristic spectrum of the non-painted area, a well-defined sextet corresponding to a well-crystallized hematite. The Mössbauer spectrum taken over the red painted details shows no significant difference from the non-painted surface. Therefore, the red details are presumably iron-free. However, the room-temperature analysis shows that the vase itself has a poorly crystallized iron oxide (hematite, α-Fe$_2$O$_3$, or maghemite, γ-Fe$_2$O$_3$) and some small Fe^{III} particles (intense lines in the middle of the measured spectral range). The poorly crystallized iron oxide would have been formed from heated clay minerals (e.g. Fe-rich smectites).

2.3.6.2 In Situ Mössbauer Spectroscopy on Mars

The NASA twin Mars Exploration Rovers (MER), Spirit and Opportunity, carry a MIMOS II instrument each [77] (Fig. 2.47). They were launched in June 2003 and landed successfully in January 2004 in Gusev Crater and at Meridiani Planum, respectively. Their nominal mission was only planned to be three months long, but both rovers exceeded expectations considerably and have been actively exploring the martian surface in their eight Earth year at the time of writing this paragraph (March 2011). Both Mössbauer instruments continue to work as well, although their decaying ^{57}Co sources (half-life 270 days) have resulted in significantly longer integration times necessary to obtain statistically good quality spectra than at the beginning of the mission.

The primary MER objective is to explore two sites on the Martian surface where water may once have been present, and to assess past environmental conditions at those sites and their suitability for life [82]. The Red Planet owes its color to Fe-oxides, and surface materials are enriched in Fe relative to Earth. The distribution of Fe between Fe-bearing minerals and its oxidation states constrains the primary rock type (e.g. olivine-bearing vs. non-olivine-bearing basalt), the redox conditions under which primary minerals crystallized (e.g. presence or absence of magnetite), the extent of alteration and weathering (e.g. value of Fe^{III}/Fe$_{Total}$), the type of alteration and weathering products (e.g. oxides vs. sulphates vs. phyllosilicates), and the processes and environmental conditions for alteration and weathering (e.g. neutral vs. acid-chloride vs. acid-sulphate aqueous process under ambient or hydrothermal conditions) [83], making ^{57}Fe Mössbauer spectroscopy an extremely useful tool for Mars exploration.

Fig. 2.48 First Mössbauer spectrum recorded on the Martian surface at Gusev crater (17th January 2004, measuring time 3 h 25 min). The inset shows a view of the Rover and of the MIMOS spectrometer operating on Mars [84, 85]

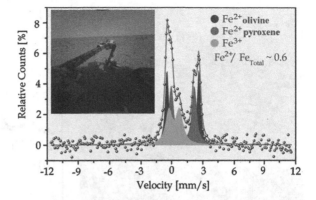

Fig. 2.49 "Blueberry"-like spherules found in *Meridiani Planum* on Mars surface were analyzed by Mössbauer spectroscopy and found to be enriched in haematite, α-Fe_2O_3 [86, 87]

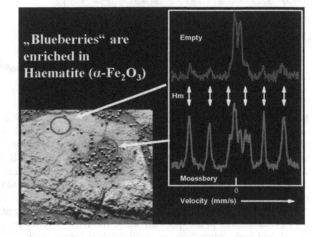

The first Mössbauer spectrum recorded on the surface of Mars was obtained in the Columbia Hills of Gusev Crater [83] close to the Rover landing site (Fig. 2.48). It clearly revealed Fe^{II} containing species (olivine and pyroxene) as well as some Fe^{III} species, not yet identified at that stage. The discovery of olivine, commonly found in lava on earth, was quite informative since this mineral is known to weather to clays and iron oxides in the presence of water [84, 85].

Nests with bluish mineral spherules similar to blueberries [86, 87] (Fig. 2.49) were discovered in a region near the landing site of the Rover *Opportunity* in Meridiani Planum [88, 89]. Mössbauer spectra were recorded in places with and without such 'Blueberry' spherules. The 'blueberries' minerals were found to be enriched in haematite (α-Fe_2O_3).

One of the major discoveries of the MER mission was the identification of the mineral jarosite by Mössbauer spectroscopy in S-rich, layered outcrop rocks at Meridiani Planum [90] (Fig. 2.50). Jarosite is a ferric sulphate hydroxide whose generalized formula can be written $(K,Na,H_3O)(Fe_{3-x}Al_x)(SO_4)_2(OH)_6$, where $x < 1$. The end members $KFe_3(SO_4)_2(OH)_6$, $NaFe_3(SO_4)_2(OH)_6$, and $(H_3O)Fe_3(SO_4)_2(OH)_6$ are jarosite, natrojarosite, and hydronium jarosite, respectively. For

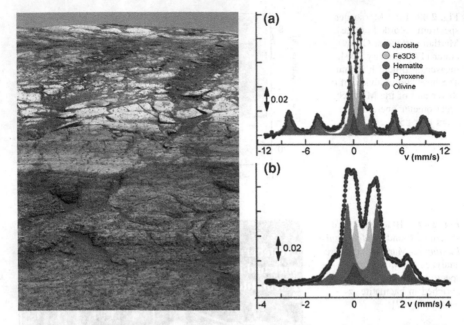

Fig. 2.50 On the left is a false color Pancam image of the 'Karatepe' section, a piece of the layered S-rich outcrop at Meridiani Planum exposed in the wall of Endurance crater. The image was obtained with Mars Exploration Rover Opportunity's Panoramic Camera (Pancam; http://marswatch.astro.cornell.edu/pancam_instrument/images/Sol173B_P2401_L257_false.jpg). One can see the tracks of the rover wheels curving down and where fresh outcrop has been exposed by the Rock Abrasion Tool (*RAT*) for analyses with the APXS and Mössbauer spectrometers. On the right are typical Mössbauer spectra of the outcrop material **a** a spectrum obtained at a velocity range of ± 12 mm/s and **b** at a reduced velocity range of ± 4 mm/s to increase the resolution of the central features in the spectrum where the Jarosite peaks occur. Fe_3D_3 stands for the as yet unassigned ferric mineral phase, possibly super paramagnetic hematite or $Fe(SO_4)(OH)$ [90]

jarosites, the quadrupole splitting ΔE_Q increases in the order $K > Na > H_3O$ and with Al^{III} substitution for Fe^{III}. The ΔE_Q value for Martian jarosite is most consistent with $(K,Na)(Fe,Al)_3(SO_4)_2(OH)_6$ that is Na^+- or K^+-rich jarosite with possible Al substitution.

Jarosite is a mineralogical marker for aqueous processes because it contains the equivalent of ~ 10 wt. % H_2O in its structure as the OH anion. The average S-rich outcrop rock at Meridiani Planum has the equivalent of ~ 2 % H_2O associated with jarosite alone. An important aspect of the jarosite detection is that acidic conditions (pH < 4 at room temperature) are required for its formation. The alteration of basaltic precursor material under oxidizing, acid-sulphate conditions to form jarosite and other phases in the S-rich outcrop rocks at Meridiani Planum could have occurred under conditions provided, for example, by interactions with acid-sulphate, possibly hydrothermal waters and/or condensation of SO_2-rich volcanic emanations [91].

The past environmental conditions characterized by low pH inferred from the detection of jarosite at Meridiani Planum have implications for the suitability for life at Meridiani Planum. While microbial populations on Earth have adapted to low pH levels, they would have challenged prebiotic chemical reactions thought to have played a role in the origin of life on Earth [92].

2.4 Conclusion and Outlook

The examples of chemical applications of Mössbauer spectroscopy discussed in this Tutorial Lecture have mostly been selected from the authors' own research work and can, of course, only provide the reader with ideas about the kind of problems that can be solved with this nuclear resonance technique. Since the discovery of *recoilless nuclear resonance absorption* ("Mössbauer effect") by the german physicist Rudolf Mössbauer more than fifty years ago, Mössbauer spectroscopy has developed to a powerful tool in solid state research, making use of more than twenty "Mössbauer-active" nuclides from the list of more than fourty isotopes for which the Mössbauer effect has been observed [17]. Mössbauer spectroscopy has mostly been employed in conjunction with other physical techniques in order to gain more conclusive information in certain studies, but also in cases where certain problems could not be solved by other techniques.

Two important technical developments have recently opened new pathways in Mössbauer spectroscopy and will undoubtedly play a significant role in future: (1) The realization of a miniaturized portable Mössbauer spectrometer (MIMOS) for material characterization outside the laboratory as briefly described above, and (2) the use of synchrotron radiation for observing nuclear resonance fluorescence. The latter was initiated by E. Gerdau et al. in 1985 who proposed an unconventional Mössbauer technique using synchrotron radiation to observe nuclear resonance in two ways: Nuclear forward scattering (NFS) to study hyperfine interactions, as obtained with conventional Mössbauer spectroscopy, and nuclear inelastic scattering (NIS) to investigate local phonon spectra (partial density of states, PDOS) at the Mössbauer probe nucleus [93]. NFS and NIS, are certainly on their way to a great future [94].

Acknowledgments We thank the Deutsche Forschungsgemeinschaft, the Fonds der Chemischen Industrie and the Fonds National de la Recherche Scientifique (FNRS) for financial support.

Dedicated to Professor Wolfgang Kaim on the occasion of his 60th birthday.

Message to the Next Generation

Looking back over nearly five decades of working with Mössbauer spectroscopy in combination with other physical methods for characterizing inorganic compounds, mainly those exhibiting electronic structure phenomena, I can now state with great satisfaction that it was an excellent decision to learn Mössbauer spectroscopy and apply it as a "forefront tool" in our research projects. I had first learned about it as a young postdoctoral fellow at Brookhaven National Laboratory in USA in the early sixties, shortly after the discovery of the "recoilless nuclear resonance absorption" by the young german physicist Rudolf L. Mössbauer, who made this magnificent discovery while he was working on his doctoral thesis at the Max Planck Institute in Heidelberg. Rudolf Mössbauer was only 32 years old when he received the Nobel in Physics in 1961. It was very fortunate for me to be accepted in a team of excellent physicists at Brookhaven National Laboratory, who had concentrated on applying the Mössbauer effect to characterize inorganic compounds and alloys by measuring the hyperfine interactions. It soon became clear that Mössbauer's unique discovery would develop rapidly to a powerful spectroscopic tool in materials science. Now, nearly five decades later, close to 100,000 published reports dealing with Mössbauer spectroscopy, many textbooks, and more than fifty international conferences, symposia and workshops held so far bear testimony to the firm establishment of this nuclear spectroscopic technique in various branches of solid state research, spreading over physics, chemistry, biology, earth- and geoscience, archaeology and industrial applications. Professor Mössbauer was an excellent speaker, and everybody was fascinated when he spoke about the experiments for his doctoral thesis that led him to the discovery of recoilless nuclear resonance absorption. Also, in many unforgettable personal conversations with him I had the pleasure to learn about details concerning his work. Such occasions never ended without discussions about piano music.

During the many years of my teaching spectroscopy in chemistry and physics, Mössbauer spectroscopy has always been my favourite for several reasons: The students become familiar with fundamental aspects on solid state and experimental physics, cry physics, quantum mechanics and theoretical chemistry to name a few. I consider it therefore highly recommendable, even necessary, that Mössbauer spectroscopy and relevant neighbouring fields are always part of the education in physics and chemistry.

Mössbauer spectroscopy has undoubtedly established as an elegant and versatile tool in materials science, mostly in conjunction with other physical techniques in order to reach deeper and more conclusive information in certain studies, but also in cases where certain problems could not be solved with other techniques. Quo vadis, Mössbauer effect research? Two outstanding developments have opened new pathways in Mössbauer spectroscopy and will definitely play a remarkable role in future: Without quality ranking, (1) the instrumental progress regarding the miniaturization of a Mössbauer spectrometer (MIMOS), and (2) the use of synchrotron radiation for observing nuclear resonance fluorescence. MIMOS has most spectacularly demonstrated its usefulness for extraterrestrial studies, viz. the NASA missions to the planet Mars. There are, of course, also hundreds of possibilities to use it on earth in mobile analytical studies outside the laboratory. A real breakthrough in Mössbauer spectroscopy research was initiated by E. Gerdau et al. in 1985 who proposed an unconventional Mössbauer technique based on the possibility to use synchrotron radiation to observe nuclear resonance.

Nuclear forward scattering (NFS) allows to study hyperfine interactions, as obtained with conventional Mössbauer spectroscopy; nuclear inelastic scattering (NIS) allows to investigate local phonon spectra (partial density of states, PDOS) at the Mössbauer probe nucleus. Compared, for instance, to Raman spectroscopy, NIS can achieve a higher resolution without perturbation of surrounding vibrations. Both synchrotron radiation techniques, NFS and NIS, are certainly on their way to a great future.

References

1. R.L. Mössbauer, Fluorescent nuclear resonance of γ-radiation in Iriridium-191. Z. Physik **151**, 124–143 (1958)
2. R.L. Mössbauer, Kernresonanzabsorption von gammastrahlung in Ir^{191}. Naturwissenschaften **45**, 538–539 (1958)
3. R.L. Mössbauer, Recoiless nuclear resonance absorption of γ-radiation. Science **137**, 731–738 (1962)
4. R.L. Mössbauer, The discovery of the Mössbauer effect. Hyperfine Interact. **126**, 1–12 (2000)
5. O.C. Kistner, A.W. Sunyar, Evidence for quadrupole interaction of Fe^{57m}, and influence of chemical binding on nuclear γ-ray energy. Phys. Rev. Lett. **4**, 412–415 (1960)
6. V.I. Goldanskii, R. Herber (eds.), *Chemical Applications of Mössbauer Spectroscopy* (Academic, New York, 1968)
7. N.N. Greenwood, T.C. Gibb, *Mössbauer Spectroscopy* (Chapman and Hall, London, 1971)
8. U. Gonser (ed.), *Mössbauer Spectroscopy, in Topics in Applied Physics*, vol. 5 (Springer, Berlin, 1975)
9. T.C. Gibb, *Principles of Mössbauer Spectroscopy* (Wiley, New York, 1976)
10. P. Gütlich, R. Link, A.X. Trautwein, *Mössbauer Spectroscopy and Transition Metal Chemistry. Inorganic Chemistry Concepts Series*, vol. 3, 1st edn. (Springer, Berlin, 1978)
11. G.J. Long, *Mössbauer Spectroscopy Applied to Inorganic Chemistry, Modern Inorganic Chemistry Series*, vol. 1 (Plenum, New York, 1984)
12. R.H. Herber, *Chemical Mössbauer Spectroscopy* (Plenum, New York, 1984)
13. G.J. Long, *Mössbauer Spectroscopy Applied to Inorganic Chemistry, Modern Inorganic Chemistry Series*, vol. 2 (Plenum, New York, 1989)
14. G.J. Long, F. Grandjean, *Mössbauer Spectroscopy Applied to Inorganic Chemistry, Modern Inorganic Chemistry Series*, vol. 3 (Plenum, New York, 1989)
15. G.J. Long, F. Grandjean, *Mössbauer Spectroscopy Applied to Magnetism and Materials Science*, vol. 1 (Plenum, New York, 1993)
16. G.J. Long, F. Grandjean, *Mössbauer Spectroscopy Applied to Magnetism and Materials Science*, vol. 2 (Plenum, New York, 1996)
17. P. Gütlich, E. Bill, A.X. Trautwein, (Mössbauer Spectroscopy and Transition Metal Chemistry, Springer, 2011)
18. N.E. Erickson, A.W. Fairhall, Mössbauer spectra of iron in $Na_2[Fe(CO)_4]$ and $Na[Fe_3(CO)_{11}H]$ and comments regarding the structure of $Fe_3(CO)_{12}$. Inorg. Chem. **4**, 1320–1322 (1965)
19. R. Greatrex, N.N. Greenwood, Mössbauer spectra, structure, and bonding in iron carbonyl derivatives. Discuss. Faraday Soc. **47**, 126–135 (1969)
20. E. Müller, Berlin blue and Turnbull's blue. Chemik. Zeit. **38**, 281–282 (1914)
21. E. Fluck, W. Kerler, W. Neuwirth, Der Mößbauer-Effekt und seine bedeutung für die chemie. Angew. Chem. **75**, 461–472 (1963)
22. K. Maer Jr, M.L. Beasley, R.L. Collins, W.O. Milligan, Structure of the titanium-iron cyanide complexes. J. Am. Chem. Soc. **90**, 3201–3208 (1968)

23. P. Gütlich, A. Hauser, H. Spiering, Thermal and optical switching of iron(II) complexes. Angew. Chem. Int. Ed. Engl. **33**, 2024–2054 (1994)

24. P. Gütlich, Y. Garcia, H.A. Goodwin, Spin crossover phenomena in Fe(II) complexes. Chem. Soc. Rev. **29**, 419–427 (2000)

25. P. Gütlich, H.A. Goodwin (eds.), Spin crossover in transition metal compounds. Top. Curr. Chem. 233–235 (2004)

26. S. Decurtins, P. Gütlich, C.P. Köhler, H. Spiering, A. Hauser, Light-induced excited spin state trapping in a transition-metal complex: the hexa-1-propyltetrazole-iron(II) tetrafluoroborate spin crossover system. Chem. Phys. Lett. **105**, 1–4 (1984)

27. I. Dezsi, B. Molnar, T. Tarnoczi, K. Tompa, On the magnetic behaviour of iron(II)-bis-(1,10 phenantroline)-thiocyanate between −190° and 30° C. J. Inorg. Nucl. Chem **29**, 2486–2490 (1967)

28. E.W. Müller, J. Ensling, H. Spiering, P. Gütlich, High-spin ↔ low-ST in hexacoordinate complexes of iron(II) with monodentate 1-alkyltetrazole ligands: a variable-temperature Mössbauer, magnetic susceptibility, and far-infrared study. Inorg. Chem. **22**, 2074–2078 (1983)

29. A. Hauser, Reversibility of light-induced excited spin state trapping in the [Fe(ptz)$_6$](BF$_4$)$_2$, and the [Zn$_{1-x}$Fe$_x$(ptz)$_6$](BF$_4$)$_2$ spin crossover systems. Chem. Phys. Lett. **124**, 543–548 (1986)

30. L. Wiehl, Structures of hexakis(1-propyltetrazole)iron(II) bis(tetrafluoroborate), [Fe(CHN$_4$C$_3$H$_7$)$_6$](BF$_4$)$_2$, hexakis(1-methyltetrazole)iron(II) bis(tetrafluoroborate), [Fe(CHN$_4$CH$_3$)$_6$](BF$_4$)$_2$, and the analogous perchlorates. Their relation to spin crossover behaviour and comparison of Debye-Waller factors from structure determination and Mössbauer spectroscopy. Acta Cryst. B **49**, 289–303 (1993)

31. P. Poganiuch, S. Decurtins, P. Gütlich, Thermal- and light-induced spin transition in [Fe(mtz)$_6$](BF$_4$)$_2$: first successful formation of a metastable low-spin state by irradiation with light at low temperatures. J. Am. Chem. Soc. **112**, 3270–3278 (1990)

32. G. De Munno, M. Julve, J.A. Real, F. Lloret, R. Scopelliti, Synthesis, crystal structure and magnetic properties of the chiral iron(II) chain [Fe(bpym)(NCS)$_2$]$_n$ (bpym = 2,2'-bipyrimidine). Inorg. Chim. Acta **250**, 81–85 (1996)

33. V. Ksenofontov, A.B. Gaspar, J.A. Real, P. Gütlich, Pressure-induced spin state conversion in antiferromagnetically coupled Fe(II) dinuclear complexes. J. Phys. Chem. B **105**, 12266–12271 (2001)

34. V. Ksenofontov, H. Spiering, S. Reiman, Y. Garcia, A.B. Gaspar, N. Moliner, J.A. Real, P. Gütlich, Direct monitoring of spin state in dinuclear iron(II) coordination compounds. Chem. Phys. Lett. **348**, 381–386 (2001)

35. V. Ksenofontov, H. Spiering, S. Reiman, Y. Garcia, A.B. Gaspar, J.A. Real, P. Gütlich, Determination of spin state in dinuclear iron(II) coordination compounds using applied field Mössbauer spectroscopy. Hyperfine Interact. **141/142**, 47–52 (2002)

36. R. Zimmermann, G. Ritter, H. Spiering, Mössbauer spectra of the tetrakis-(1,8-naphthyridine) iron(II) perchlorate in external magnetic fields. Evidence of slow relaxation In paramagnetic iron(II). Chem. Phys. **4**, 133–141 (1974)

37. R. Zimmermann, G. Ritter, H. Spiering, D.L. Nagy, A further example of slow relaxation in high-spin iron(II) compounds: Fe(papt)$_2$•C$_6$H$_6$. J. Phys. **35-C6**, 439 (1974)

38. V. Ksenofontov, A.B. Gaspar, V. Niel, S. Reiman, J.A. Real, P. Gütlich, On the nature of the plateau in two-step dinuclear spin crossover complexes. Chem. Eur. J. **10**, 1291–1298 (2004)

39. Klingele, M. H., Moubaraki, B., Cashion, J. D., Murray, K. S., Brooker, S.: The first X-ray crystal structure determination of a dinuclear complex trapped in the [LS-HS] state [Fe$_2^{II}$(PMAT)$_2$](BF$_4$)$_4$·DMF. Chem. Comm. 8, 987-989 (2005)

40. Y. Garcia, C.M. Grunert, S. Reiman, O. van Campenhoudt, Gütlich P (2006) The two-step spin conversion in a supramolecular triple helicate dinuclear iron(II) complex studied by Mössbauer spectroscopy. Eur. J. Inorg. Chem. **17**, 3333–3339 (2006)

41. C.M. Grunert, S. Reiman, H. Spiering, J.A. Kitchen, S. Brooker, P. Gütlich, Mixed spin-state [HS–LS] pairs in a dinuclear spin transition complex: confirmation by variable-temperature ^{57}Fe Mössbauer Spectroscopy. Angew. Chem. Inter. Ed. **47**, 2997–2999 (2008)

42. J.J.A. Kolnaar, G. vanDijk, H. Kooijman, A.L. Spek, V.G. Ksenofontov, P. Gütlich, J.G. Haasnoot, J. Reedijk, Synthesis, structure, magnetic behavior, and Mössbauer spectroscopy of two new iron(II) spin transition compounds with the ligand 4-Isopropyl-1,2,4-triazole. X-ray structure of [Fe$_3$(4-isopropyl-1,2,4-triazole)$_6$(H$_2$O)$_6$](tosylate)$_6$·2H$_2$O. Inorg. Chem. **36**, 2433–2440 (1997)

43. Y. Garcia, P. Guionneau, G. Bravic, D. Chasseau, J.A.K. Howard, O. Kahn, V. Ksenofontov, S. Reiman, P. Gütlich, Synthesis, crystal structure, magnetic properties and ^{57}Fe Mossbauer spectroscopy of the new trinuclear [Fe$_3$(4-(2′-hydroxyethyl)-1,2,4-triazole)$_6$(H$_2$O)$_6$](CF$_3$SO$_3$)$_6$ spin crossover compound. Eur. J. Inorg. Chem. **7**, 1531–1538 (2000)

44. M. Ruben, E. Breuning, J.M. Lehn, V. Ksenofontov, F. Renz, P. Gütlich, G.B.M. Vaughan, Supramolecular Spintronic Devices: Spin transitions and magnetostructural correlations in [Fe$_4^{II}$L$_4$]$^{8+}$ [2 × 2]-grid-type complexes. Chem. Eur. J. **9**, 4422–4429 (2003)

45. E. Breuning, M. Ruben, J.-M. Lehn, F. Renz, Y. Garcia, V. Ksenofontov, P. Gütlich, E. Wegelius, K. Rissanen, Spin crossover in a supramolecular Fe$_4^{II}$ [2x2] grid triggered by temperature, pressure, and light. Angew. Chem. Inter. Ed. **39**, 2504–2507 (2000)

46. J.J.A. Kolnaar, M.I. de Heer, H. Kooijman, A.L. Spek, G. Schmitt, V. Ksenofontov, P. Gütlich, J.G. Haasnoot, J. Reedijk, Synthesis, structure and properties of a mixed mononuclear/dinuclear iron(II) spin crossover compound with the ligand 4-(p-tolyl)-1,2,4-triazole. Eur. J. Inorg. Chem. **5**, 881–886 (1999)

47. Y. Galyametdinov, V. Ksenofontov, A. Prosvirin, I. Ovchinnikov, G. Ivanova, P. Gütlich, W. Haase, First example of coexistence of thermal spin transition and liquid-crystal properties. Angew. Chem. Int. Ed. **40**, 4269–4271 (2001)

48. T. Fujigaya, D.L. Jiang, T. Aida, Switching of spin states triggered by a phase transition: spin-crossover properties of self-assembled iron(II) complexes with alkyl-tethered triazole ligands. J. Am. Chem. Soc. **125**, 14690–14691 (2003)

49. S. Hayami, K. Danjobara, K. Inoue, Y. Ogawa, N. Matsumoto, Y. Maeda, A photoinduced spin transition iron(II) complex with liquid-crystal properties. Adv. Mater. **16**, 869–872 (2004)

50. S. Hayami, R. Moriyama, A. Shuto, Y. Maeda, K. Ohta, K. Inoue, Spin transition at the mesophase transition temperature in a cobalt(II) compound with branched alkyl chains. Inorg. Chem. **46**, 7692–7694 (2007)

51. S. Hayami, N. Motokawa, A. Shuto, N. Masuhara, T. Someya, Y. Ogawa, K. Inoue, Y. Maeda, Photoinduced spin transition for iron(II) compounds with liquid-crystal properties. Inorg. Chem. **46**, 1789–1794 (2007)

52. M. Seredyuk, A.B. Gaspar, V. Ksenofontov, Y. Galyametdinov, J. Kusz, P. Gütlich, Iron(II) metallomesogens exhibiting coupled spin state and liquid crystal phase transitions near room temperature. Adv. Funct. Mater. **18**, 2089–2101 (2008)

53. M. Seredyuk, A.B. Gaspar, V. Ksenofontov, Y. Galyametdinov, M. Verdaguer, F. Villain, P. Gütlich, One-dimensional iron(II) compounds exhibiting spin crossover and liquid crystalline properties in the room temperature region. Inorg. Chem. **47**, 10232–10245 (2008)

54. M. Seredyuk, A.B. Gaspar, V. Ksenofontov, Y. Galyametdinov, J. Kusz, P. Gütlich, Does the solid–liquid crystal phase transition provoke the spin-state change in spin-crossover metallomesogens? J. Am. Chem. Soc. **130**, 1431–1439 (2008)

55. M. Seredyuk, A.B. Gaspar, V. Ksenofontov, S. Reiman, Y. Galyametdinov, W. Haase, E. Rentschler, P. Gütlich, Room temperature operational thermochromic liquid crystals. Chem. Mater. **18**, 2513–2519 (2006)

56. M. Seredyuk, A.B. Gaspar, V. Ksenofontov, S. Reiman, Y. Galyametdinov, W. Haase, E. Rentschler, P. Gütlich, Multifunctional materials exhibiting spin crossover and liquid-crystalline properties. Interplay between spin crossover and liquid-crystal properties in iron(II) coordination complexes. Hyperfine Interact. **166**, 385–390 (2006)

57. A.B. Gaspar, M. Seredyuk, P. Gütlich, Spin crossover in metallomesogens. Coord. Chem. Rev. **253**, 2399–2413 (2009)

58. Y. Garcia, P.J. Van Koningsbruggen, R. Lapouyade, L. Fournes, L. Rabardel, O. Kahn, V. Ksenofontov, G. Levchenko, P. Gütlich, Influences of temperature, pressure, and lattice solvents on the spin transition regime of the polymeric compound [Fe(hyetrz)$_3$]A$_2$·3H$_2$O

(hyetrz = 4-(2'-hydroxyethyl)-1,2,4-triazole and A⁻ = 3-nitrophenylsulfonate). Chem. Mater. **10**, 2426–2433 (1998)

59. P. Gütlich, Nuclear decay induced excited spin state trapping (NIESST). Top. Curr. Chem. **234**, 231–260 (2004)

60. R. Grimm, P. Gütlich, E. Kankeleit, R. Link, Time and temperature dependence of aftereffects in $[^{57}Co(phen)_3](ClO_4)_2 \cdot 2H_2O$ from time-differential Mössbauer emission spectroscopy. J. Chem. Phys. **67**, 5491 (1977)

61. R. Albrecht, M. Alfen, P. Gütlich, Z. Kajcsos, R. Schulze, H. Spiering, F. Tuczek, A new spectrometer for time-differential Mössbauer emission spectroscopy (TDMES). Nucl. Instr. Methods **257**, 209–214 (1987)

62. J. Ensling, B.W. Fitzsimmons, P. Gütlich, K.M. Hasselbach, Anomalous spin states of iron(II) in $[Fe(phen)_3](ClO_4)_2 \cdot 2H_2O$. Angew. Chem. Int. Ed. **9**, 637 (1970)

63. H. Sano, P. Gütlich, Hot atom chemistry in relation to Mössbauer emission spectroscopy, in *Hot Atom Chemistry*, ed. by T. Matsuura (Kodansha Ltd., Tokyo, 1984), p. 26

64. J. Ensling, P. Gütlich, K.M. Hasselbach, B.W. Fitzsimmons, Anomalous spin states of iron(II) in ^{57}Fe Mössbauer emission spectra of $[^{57}Co(phen)_2(NCS)_2]$ and $[^{57}Co(bipy)_2(NCS)_2](cis)$. Chem. Phys. Lett. **42**, 232–236 (1976)

65. A. Oshio, H. Spiering, V. Ksenofontov, F. Renz, P. Gütlich, Electronic relaxation phenomena following $^{57}Co(EC)^{57}Fe$ nuclear decay in $[Mn^{II}(terpy)_2](ClO_4)_2 \cdot 1/2H_2O$ and in the spin crossover complexes $[Co^{II}(terpy)_2]X_2 \cdot nH_2O$ (X = Cl and ClO_4): a Mössbauer emission spectroscopic study. Inorg. Chem. **40**, 1143–1150 (2001)

66. S. Iijima, R. Saida, I. Motoyama, H. Sano, The temperature dependence of the trapped and averaged-valence state in mono-oxidized dialkylbiferrocenes. Bull. Chem. Soc. Jpn. **54**, 1375–1379 (1981)

67. M.J. Cohn, M.D. Timken, D.N. Hendrickson, Mössbauer spectroscopy of mixed-valence biferrocenes in high magnetic fields. J. Am. Chem. Soc. **106**, 6683–6689 (1984)

68. C.C. Wu, H.G. Jang, A.L. Rheingold, P. Gütlich, D.N. Hendrickson, Solvate molecule effects and unusual ^{57}Fe Mössbauer line broadening in the valence detrapping of mixed-valence $[Fe_3O(O_2CCH_3)_6(3-Et-py)_3] \cdot S$. Inorg. Chem. **35**, 4137–4147 (1996)

69. T. Glaser, T. Beissel, E. Bill, T. Weyhermüller, V. Schünemann, W. Meyer-Klaucke, A.X. Trautwein, K. Wieghardt, Electronic structure of linear thiophenolate-bridged heterotrinuclear complexes [LFeMFeL]n + (M = Cr, Co., Fe; n = 1–3): localized vs delocalized models. J. Am. Chem. Soc. **121**, 2193–2208 (1999)

70. V. Ksenofontov, H.C. Kandpal, J. Ensling, M. Waldeck, D. Johrendt, A. Mewis, P. Gütlich, C. Felser, Verwey type transition in EuNiP. Europhys. Lett. **74**, 672–678 (2006)

71. O. Kahn, *Molecular Magnetism* (VCH Publishers Inc, Weinheim, 1993)

72. J.S. Miller, M. Drillon (eds.), *Magnetism: Molecules to Materials*, vol. I–III (Wiley, Weinheim, 2001)

73. V. Ksenofontov, G. Levchenko, S. Reiman, P. Gutlich, A. Bleuzen, V. Escax, M. Verdaguer, Pressure-induced electron transfer in ferrimagnetic Prussian blue analogs. Phys. Rev. B **68**, 024415 (2003)

74. W.M. Reiff, A.M. LaPointe, E. Witten, Virtual free ion magnetism and the absence of Jahn–Teller distortion in a linear two-coordinate complex of high-spin iron(II). J. Am. Chem. Soc. **126**, 10206–10207 (2004)

75. A.M. LaPointe, $Fe[C(SiMe_3)_3]_2$: synthesis and reactivity of a monomeric homoleptic iron(II) alkyl complex Inorg. Chim. Acta **345**, 359–362 (2003)

76. J. Ensling, P. Gütlich, Laboratory Report, University of Mainz, 1985; Gütlich, P., Schröder, C.: Mössbauer spectroscopy. Bunsen-Magazin **1**, 4–22 (2010)

77. G. Klingelhöfer, R.V. Morris, B. Bernhardt, D. Rodionov, P.A. de Souza Jr, S.W. Squyres, J. Foh, E. Kankeleit, U. Bonnes, R. Gellert, C. Schröder, S. Linkin, E. Evlanov, B. Zubkov, O. Prilutski, The Athena MIMOS II Mössbauer spectrometer investigation. J. Geophys. Res. **108**, 8067 (2003)

78. G. Klingelhöfer, D. Rodionov, M. Blumers, L. Strüder, P. Lechner, B. Bernhardt, H. Henkel, I. Fleischer, C. Schröder, J. Girones Lopez, G. Studlek, J. Maul, J. Fernandez-Sanchez, C.

d'Uston, The advanced miniaturised Mössbauer spectrometer MIMOS IIa: increased sensitivity and new capability for elemental analysis. Lunar Planet. Sci. **39**, 2379 (2008)

79. E. Wagner, A. Kyek, Mössbauer spectroscopy in archaeology: introduction and experimental considerations. Hyperfine Interact. **154**, 5–33 (2004)

80. G. Klingelhöfer, G.M. da Costa, A. Prous, B. Bernhardt, Rock paintings from Minas Gerais, Brasil, investigated by in situ Mössbauer spectroscopy. Hyperfine. Interact. C **5**, 423–426 (2002)

81. P.A. de Souza Jr, B. Bernhardt, G. Klingelhöfer, P. Gütlich, Surface analysis in archaeology using the miniaturized Mössbauer spectrometer MIMOS II. Hyperfine Interact. **151**, 125–130 (2003)

82. S.W. Squyres, R.E. Arvidson, E.T. Baumgartner, J.F. Bell III, P.R. Christensen, S. Gorevan, K.E. Herkenhoff, G. Klingelhöfer, M.B. Madsen, R.V. Morris, R. Rieder, R.A. Romero, Athena Mars rover science investigation. J. Geophys. Res. **108**, 8062 (2003)

83. S.W. Squyres, R.E. Arvidson, J.F. Bell III, J. Brückner, N.A. Cabrol, W. Calvin, M.H. Carr, P.R. Christensen, B.C. Clark, L. Crumpler, D.J. Des Marais, C. d'Uston, T. Economou, J. Farmer, W. Farrand, W. Folkner, M. Golombek, S. Gorevan, J.A. Grant, R. Greeley, J. Grotzinger, L. Haskin, K.E. Herkenhoff, S. Hviid, J. Johnson, G. Klingelhöfer, A. Knoll, G. Landis, M. Lemmon, R. Li, M.B. Madsen, M.C. Malin, S.M. McLennan, H.Y. McSween, D.W. Ming, J. Moersch, R.V. Morris, T. Parker, J.W. Rice Jr, L. Richter, R. Rieder, M. Sims, M. Smith, P. Smith, L.A. Soderblom, R. Sullivan, H. Wänke, T. Wdowiak, M. Wolff, A. Yen, The spirit Rover's Athena science investigation at Guser crater, Mars. Science **305**, 794–799 (2004)

84. G. Klingelhöfer, E. De Grave, R.V. Morris, A. Alboom, V.G. Resende, P.A. Souza Jr, D. Rodionov, C. Schroeder, D.W. Ming, A. Yen, Mössbauer spectroscopy on Mars: goethite in the Columbia Hills at Gusev crater. Hyperfine Interact. **166**, 549 (2006)

85. R.V. Morris, G. Klingelhöfer, B. Bernhardt, C. Schröder, D.S. Rodionov, P.A. Jr de Souza, A. Yen, R. Gellert, E.N. Evlanov, J. Foh, E. Kankeleit, P. Gütlich, D.W. Ming, F. Renz, T. Wdowiak, S.W. Squyres, R.E. Arvidson, Mineralogy at Gusev crater from the Mössbauer spectrometer on the spirit rover. Science **305**, 833–836 (2004)

86. S.W. Squyres, R.E. Arvidson, J.F. Bell III, J. Brückner, N.A. Cabrol, W. Calvin, M.H. Carr, P.R. Christensen, B.C. Clark, L. Crumpler, D.J. Des Marais, C. d'Uston, T. Economou, J. Farmer, W. Farrand, W. Folkner, M. Golombek, S. Gorevan, J.A. Grant, R. Greeley, J. Grotzinger, L. Haskin, K.E. Herkenhoff, S. Hviid, J. Johnson, G. Klingelhöfer, A.H. Knoll, G. Landis, M. Lemmon, R. Li, M.B. Madsen, M.C. Malin, S.M. McLennan, H.Y. McSween, D.W. Ming, J. Moersch, R.V. Morris, T. Parker, J.W. Rice Jr, L. Richter, R. Rieder, M. Sims, P. Smith, L.A. Soderblom, R. Sullivan, H. Wänke, T. Wdowiak, M. Wolff, A. Yen, The opportunity rover's Athena science investigation at Meridiani planum, Mars. Science **306**, 1698–1703 (2004)

87. L.A. Soderblom, R.C. Anderson, R.E. Arvidson, J.F. Bell III, N.A. Cabrol, W. Calvin, P.R. Christensen, B.C. Clark, T. Economou, B.L. Ehlmann, W.H. Farrand, D. Fike, R. Gellert, T.D. Glotch, M.P. Golomber, R. Greeley, J. Grotzinger, K.E. Herkenhoff, D.J. Jerolmack, J.R. Johnson, B. Joliff, G. Klingelhöfer, A.H. Knoll, Z.A. Learner, R. Li, M.C. Malin, S.M. McLennan, H.Y. McSween, D.W. Ming, R.V. Morris, J.W. Rice Jr, L. Richter, R. Rieder, D. Rodionov, C. Schröder, F.P. Seelos IV, J.M. Soderblom, S.W. Squyres, R. Sullivan, W.A. Watters, C.M. Weitz, M.B. Wyatt, A. Yen, J. Zipfel, Soils of Eagle crater and Meridiani planum at the opportunity rover landing site. Science **306**, 1723–1726 (2004)

88. S.W. Squyres, J.P. Grotzinger, R.E. Arvidson, J.F. Bell III, W. Calvin, P.R. Christensen, B.C. Clark, J.A. Crisp, W.H. Farrand, K.E. Herkenhoff, J.R. Johnson, G. Klingelhöfer, A.H. Knoll, S.M. McLennan, H.Y. McSween Jr, R.V. Morris, J.W. Rice Jr, R. Rieder, L.A. Soderblom, In situ evidence for an ancient aqueous environment at Meridiani planum, Mars. Science **306**, 1709–1714 (2004)

89. S.W. Squyres, A.H. Knoll, R.E. Arvidson, B.C. Clark, J.P. Grotzinger, B.L. Jolliff, S.M. McLennan, N. Tosca, J.F. Bell III, W. Calvin, W.H. Farrand, T.D. Glotch, M.P. Golombek, K.E. Herkenhoff, J.R. Johnson, G. Klingelhöfer, H.Y. McSween, A.S. Yen, Two years at Meridiani planum: results from the opportunity rover. Science **313**, 1403–1407 (2006)

90. G. Klingelhöfer, R.V. Morris, B. Bernhardt, C. Schroeder, D.S. Rodionov, P.A. Jr de Souza, A. Yen, R. Gellert, E.N. Evlanov, B. Zubkov, J. Foh, U. Bonnes, E. Kankeleit, P. Gütlich, D.W. Ming, F. Renz, T. Wdowiak, S.W. Squyres, R.E. Arvidson, Jarosite and hematite at Meridiani Planum from Opportunity's Mossbauer spectrometer. Science **306**, 1740–1745 (2004)
91. R.V. Morris, G. Klingelhöfer, C. Schröder, D.S. Rodionov, A. Yen, D.W. Ming, P.A. De Souza Jr, T. Wdowiak, I. Fleischer, R. Gellert, B. Bernhardt, U. Bonnes, B.A. Cohen, E.N. Evlanov, J. Foh, P. Gütlich, E. Kankeleit, T. McCoy, D.W. Mittlefehldt, F. Renz, M.E. Schmidt, B. Zubkov, S.W. Squyres, R.E. Arvidson, Mössbauer mineralogy of rock, soil, and dust at Meridiani Planum. Mars: Opportunity's journey across sulfate-rich outcrop, basaltic sand and dust, and hematite lag deposits. J. Geophys. Res. **111**, 1215 (2006)
92. A.H. Knoll, M. Carr, B. Clark, D.J. Des Marais, J.D. Farmer, W.W. Fischer, J.P. Grotzinger, S.M. McLennan, M. Malin, C. Schröder, S. Squyres, N.J. Tosca, T. Wdowiak, An astrobiological perspective on Meridiani Planum. Earth Planet. Sci. Lett. **240**, 179–189 (2005)
93. E. Gerdau, R. Rüffer, H. Winkler, W. Tolksdorf, C.P. Klages, J.P. Hannon, Nuclear Bragg diffraction of synchrotron radiation in yttrium iron garnet. Phys. Rev. Lett. **54**, 835–838 (1985)
94. P. Gütlich, Y. Garcia, Mössbauer spectroscopy: elegance and versatility in chemical diagnostics. J. Phys. Conf. Ser. **217**, 012001 (2010)

Curriculum Vitae

Philipp Gütlich, Professor Dr.-Ing.
Born August 5th, 1934 in Rüsselsheim/Germany Married, 2 children

1941–1946	Elementary School in Rüsselsheim and Karben (Germany)
1946–1955	High school (Gymnasium) in Rüsselsheim
1955–1961	Technische Hochschule Darmstadt (Inorganic, Organic, Analytical, Technical and Physical Chemistry, Physics, Mathematics) Diplom-Ingenieur "Mit Auszeichnung" Master's thesis in Physical Inorganic Chemistry Academic Prize of Technical University of Darmstadt for best thesis

1961–1963 Ph.D. thesis "Surface Investigations Using the BET-Method and Heterogeneous Isotope Exchange on Barium Sulfate", Ph.D."Mit Auszeichnung"

1964 Postdoctoral research stay at "Centre d'Etudes Nucléaires à Saclay in France (Paris), 6 months

1964–1965	Postdoctoral research stay in Brookhaven National Laboratory (USA), work on Mössbauer spectroscopy of transition metal compounds
1966	Return to Technische Hochschule Darmstadt, research work towards "Habilitation" ("Applications of Mössbauer Spectroscopy in Inorganic Chemistry")
1976–1970	Participation in Theoretical Chemistry Schools C.A. Coulson (Oxford), P.O. Löwdin (Uppsala), H. Hartmann (Frankfurt)
1969	Habilitation in Inorganic Chemistry and Nuclear Chemistry
1972	Professor of Theoretical Inorganic Chemistry at Technische Hochschule Darmstadt
1974	2 offers for Full-Professorships a) Chair of Inorganic and Analytical Chemistry, University of Mainz b) Chair of Theoretical Inorganic Chemistry, University of Hamburg
1975	Professorship at Mainz University accepted
1989-1991	Dean of the Department of Chemistry and Pharmacy
1996-2001	Director of the Institute of Inorganic and Analytical Chemistry
April 2001	Professor Emeritus

Memberships

Gesellschaft Deutscher Chemiker
Bunsengesellschaft für Physikalische Chemie
Deutsche Gesellschaft für Metallkunde (until 2001)
Deutsche Physikalische Gesellschaft
American Chemical Society (for ca. 30 years, until 2001)
Chemical Society of London (for ca. 30 years, until 2001)

Guestprofessorships

University of Geneva and Bern in Switzerland (1985)
University of Lima/Peru (for UNESCO, 1985)
Tamkang University Taipei (1986)
University of Louvain-la-Neuve/Belgium (1991)
University of Fribourg/Switzerland (1994)
Ochanomizu University Tokyo (1994)
Université Pierre et Marie Curie, Paris (1997)
Vienna University of Technology (1998)
Toho University Tokyo (2007)

Awards and Honors

1961   Academic Prize of Technische Hochschule Darmstadt
1964   Fellowship of German Gouvernment for research stay in France
1965   Fellowship of Volkswagen-Foundation for research stay in USA
1989   Research Award of Japanese Society for the Promotion of Science
1993   Max Planck Research Award
2002   Honorary Member of the Internat. Board on the Applications of the
       Müssbauer Effect
2002   Foreign Member of the Russian Academy of Natural Science
2003   Honorary Doctor and Professor of the University of Budapest
2007   Member of Academia Europaea
2007   Honorary Doctor of Toho-University Tokyo

Teaching Experience

- Physical Methods in (Inorganic) Chemistry
- Theoretical Inorganic Chemistry

– Introduction to Quantum Mechanics
– Electrons in Atoms and Molecules
– Ligand Field Theory
– Molecular Orbital Theory
– Group Theory

- Coordination Chemistry
- Magnetochemistry

Research Fields

- Electronic structure (static and dynamic) of transition metal compounds

– Bond and valence state properties
– Molecular structure
– Spin crossover (thermally, optically and pressure-induced)
– Magnetic properties
– Photochemical and photophysical phenomena
– Phase transformations
– Kinetic and thermodynamic properties
– Chemical and physical aftereffects of nuclear transformation

- Surface physics and chemistry

– Thin metallic layers by Langmuir–Blodgett technique
– Corrosion
– Structure and reactivity on glass surfaces

- Industrial applications of Mössbauer spectroscopy
- Magnetic and optical recording material
- Battery material
- Glasses
- Corrosion

Research equipment

- ca. 8 Mössbauer spectrometers for transmission, scattering, CEMS, DEC-EMS, time-integral and time-differential measurements, also under pressure or in applied magnetic field and variable temperature (≥ 2 K)
- 2 Magnetometers (Foner, SQUID) for $350 \geq T \geq 2$ K
- ESCA-spectrometer
- AUGER-spectrometer
- UV/Vis spectrometer, also for single crystals in the range $300 \geq T \geq 5$ K

Raman spectrometer, T-dependent in the range $1500 \geq T \geq 4$ K

- FT-FIR spectrometer, T-dependent in the range $300 \geq T \geq 4$ K
- Single crystal X-ray diffractometer, T-dependent $300 \geq T \geq 10$ K, CCD

Publications

- ca. 475 original papers
- 1 book ("Mössbauer Spectroscopy and Transition Metal Chemistry")
- Chapters in ca. 15 books
- Editor (together with N. Sutin, USA) of "Comments on Inorganic Chemistry", which was launched in 1981)
- Editor of proceedings of 5 Seeheim Workshops on Mössbauer Spectroscopy
- Editor (with H.A. Goodwin) of the series of Spin Crossover Transition Metal Compounds, 3 volumes (Nr. 233, 234, 235) in Topics in Current Chemistry (Springer), 2004
- Book "Mössbauer Spectroscopy and Transition Metal Chemistry", Second Edition, Springer-Verlag, 2010

Seminars and Presentations

- More than 300 Invited Talks/Plenary Lectures
- Ca. 500 contributions to conferences (oral and posters)

Yann Garcia

Communication: French, English, Spanish, German (basic knowledge)

Research domains: Supramolecular functional (photo)magnetic materials, Crystal growth and design of MOFs and coordination polymers, Applications of Mössbauer and muon spin relaxation spectroscopies

Teaching area:
- Analytical chemistry and instrumental methods in solid state physics/chemistry
- 122 international publications, 9 book chapters, 2 patents, h index: 22
- Guest editor of three special issues (Eur. J. Inorg. Chem., Möss. Eff. Ref. Data J.)
- Author of 4 cover page articles
- 93 oral communications and 169 posters
- 104 oral communications and 191 posters
- Invited lectures at conferences (23) and universities (14)

Positions

2009	Professor at UCL
2004–2009	Associate professor at UCL
2001–2004	Tenure track assistant professor at UCL
1999–2001	EU-TMR Post-doctoral research fellow with Prof. Dr. P. Gütlich Institut für Anorganische und Analytische Chemie, Mainz (Germany).
1995–1999	Doctoral fellow of the University of Bordeaux I –Laboratory of Molecular Sciences, Bordeaux Institute of Condensed Matter Chemistry, CNRS (France) - supervisor: Prof. Dr. O. Kahn, academician.

Education

At the University of Bordeaux I (France):

1995–1999	Doctoral thesis in Materials Science and Solid State Chemistry (highest distinction), Bordeaux Institute of Condensed Matter Chemistry—CNRS
1994–1995	DEA of Materials Science
1993–1994	MSc in Physical Sciences
1992–1993	BSc in Physical Sciences
1990–1992	DEUG A in Chemistry and Physics
1987–1990	High school, Bachelor C (Mathematics, Physics, Chemistry), Bergerac

Services

Since 2007	President, Groupe Francophone de spectrométrie Mössbauer (GFSM)
Since 2009	Member, International Board on the Applications of the Mössbauer Effect (IBAME)
Since 2012	President of the thematic doctoral school of the French speaking community of Belgium 'Molecular, supramolecular and functional chemistry'

Referees for international grant applications, doctoral thesis outside UCL and one academic position

Editorial boards

Since 2005	Member, International advisory board of Eur. J. Inorg. Chem. (Wiley-VCH)
Since 2007	Member, International advisory board of Open Inorg. Chem. J. (Bentham Sc. Pub).
Since 2011	Editor-in-Chief of Curr. Inorg. Chem. (Bentham Sc. Pub).
Since 2011	Associate Editor of MERDJ Regular referee for high impact factor international chemistry journals (Angew. Chem. Int. Ed., J. Am. Chem. Soc., Chem. Comm., Chem. Eur. J., Dalton Trans., CrystEngComm, Eur. J. Inorg. Chem. ...)

Award

2009   Nominated as emerging leader of the future Mössbauer community by the Mössbauer Effect Data Center (Vienna, July 2009).

Chapter 3
Application of Mössbauer Spectroscopy in Earth Sciences

Robert E. Vandenberghe and Eddy De Grave

Abstract Iron being the fourth most abundant element in the earth crust, ^{57}Fe Mössbauer spectroscopy has become a suitable additional technique for the characterization of all kind of soil materials and minerals. However, for that purpose a good knowledge of the spectral behavior of the various minerals is indispensable. In this chapter a review of the most important soil materials and rock-forming minerals is presented. It starts with a description of the Mössbauer spectroscopic features of the iron oxides and hydroxides, which are essentially present in soils and sediments. Further, the Mössbauer spectra from sulfides, sulfates and carbonates are briefly considered. Finally, the Mössbauer features of the typical and most common silicate and phosphate minerals are reported. The chapter ends with some typical examples, illustrating the use and power of Mössbauer spectroscopy in the characterization of minerals.

3.1 Introduction

The discovery of the resonant absorption of gamma-rays in ^{191}Ir by Rudolf Mössbauer in 1958 was a milestone in nuclear physics because it was formerly assumed that such a phenomenon could never occur due to the large recoil energies involved. The interest for this new finding was nevertheless still limited in the first years, but, a real breakthrough emerged from the fact that many isotopes showed a larger and much more sensitive effect. So, a new technique, called Mössbauer spectroscopy, was born. The most important feature of this spectroscopic method is the extreme sharpness of the emission line which can easily be

R. E. Vandenberghe (✉) · E. De Grave
Department of Physics and Astronomy, Ghent University,
Proeftuinstraat 86, 9000 Ghent, Belgium

Y. Yoshida and G. Langouche (eds.), *Mössbauer Spectroscopy*,
DOI: 10.1007/978-3-642-32220-4_3, © Springer-Verlag Berlin Heidelberg 2013

varied in energy by giving the source velocities of the order of cm/s. This enables
to study the different hyperfine interactions related to the electronic shells of the
Mössbauer active atoms in solids which, in a further stage, yield valuable information about structural and magnetic properties of materials. Up to now, the
Mössbauer effect has been observed for nearly 100 nuclear transitions in about 80
isotopes, distributed over 43 elements. Of course, as with many other spectroscopic methods, not all of these transitions are suitable for practical studies and
about twenty elements remain for applications. However, it is a gift of nature that
the resonant absorption effect is easily achieved in the iron-57 isotope which has
an abundance of 2.14 % in natural iron. Considering the importance of the element
iron in many branches of science and technology, it is obvious that since 1960, this
new spectroscopic technique has proven to be very useful in the study of all kinds
of iron-bearing materials. In particular, the abundance of iron in the earth's crust
(4th element in wt %) renders this kind of spectroscopy extremely suitable for the
characterization of soil materials and minerals. Moreover, a major advantage of
Mössbauer spectroscopy is the fact that it probes the influences on the iron nucleus
locally. This means that, through the determination of the hyperfine parameters,
not only the different iron-bearing components in a sample can be distinguished,
but also the different "types" of iron present in a mineral can be detected. This
leads to a variety of applications in geology and soil sciences such as the qualitative and quantitative analysis with respect to the various mineralogical compounds and the determination of the oxidation state and coordination of iron in
minerals.

From experimental point of view, the equipment for Mössbauer spectroscopy
(MS) is nowadays relatively simple and not expensive. Especially, since the data
collection can be achieved by a compact electronic unit connected to a PC, the
latter not necessarily being of high performance, a complete Mössbauer set-up
with low-temperature facilities is cheaper than say, an X-ray diffraction apparatus.
However, most of solid-state laboratories or particularly geological institutes in the
present case, do not possess Mössbauer spectrometers in their standard equipment.
This is often due to the severe rules which are enforced with radioactive source
acquisition and handling. Moreover, the relatively serious recurrent cost of sources, having a half-life time of 270 days in the case of ^{57}Fe demands a permanent
operation of the spectrometer, which is not evident in case of limited need. Even
though the application for the characterization of soils and minerals does not
require a deep fundamental physical knowledge, MS is for all the aforementioned
reasons not so popular as a standard technique in the community of geologists and
soil scientists as it should be. Therefore, a sound and permanent cooperation
between Mössbauer laboratories and institutes related to earth sciences is necessary and should still further be promoted.

Another feature that lowers to some extent the general popularity of MS in
comparison with other techniques is related to the spectral analysis. Mössbauer
spectroscopy is not a so-called push-button technique, yielding directly consistent
results after each measurement. As in many techniques, the spectra need to be
refined using home-made or commercial computer programs. But, the analyzing

procedure is not always straightforward leading often after trial and error to uncertain results. This is particularly true for complex spectra of minerals and soil materials exhibiting in many cases strongly overlapping absorption lines or showing distributed hyperfine interactions. It is therefore imperative that the Mössbauer spectroscopist uses the appropriate fitting procedures, has a good insight in the various spectra that can occur, and relies if necessary on the results of complementary techniques. This tutorial aims to introduce the reader into those various aspects of the application of MS in earth sciences. Together with a description of the spectral behavior expected for the most relevant minerals, a number of examples shall be given, which illustrates the analytical power of MS.

3.2 Mössbauer Spectroscopy Applied to Earth Sciences

As already mentioned in the introduction, MS leads to applications in earth sciences and particularly in mineralogy, which are very important with respect to qualitative and quantitative analysis of the samples and the determination of the oxidation state and coordination of iron in the involved minerals. Moreover, this technique additionally provides in some cases a crude insight in the morphological and chemical features of the minerals.

3.2.1 Qualitative and Quantitative Analyzing Power of MS

For qualitative analyses of rocks, soils, sediments, ores, etc., MS consists of the recognition of typical "fingerprint" spectra of the various iron-bearing species present in the sample. The standard fingerprint spectra are usually obtained from pure natural or synthetic samples. The appearance of the typical spectra, defined by their specific hyperfine parameters, enables in many cases to assign immediately the unknown components in the sample. The doublet spectra of non-magnetic (paramagnetic or superparamagnetic) materials are defined by two hyperfine parameters, i.e. the isomer shift δ_{Fe} and the quadrupole splitting Δ. The sextet spectra of magnetically ordered materials are defined by three hyperfine parameters, i.e. the isomer shift δ_{Fe}, the quadrupole shift 2ε and the magnetic hyperfine field B. Such spectra are particularly obtained in the case of oxides at RT and of (oxy)hydroxides at lower temperatures. The hyperfine field is then a welcome extra parameter for qualitative phase characterization. The magnetic transition temperature at which the sextet is expected to change into doublet is often also of prominent value.

However, one may not overestimate the direct qualitative analyzing power of Mössbauer spectroscopy. Indeed, many situations occur in which the spectra do not give such a decisive qualitative information. This is particularly true in the cases where merely paramagnetic doublet spectra are obtained. For instance, many minerals have Fe^{3+} in octahedral environment which all yield similar doublets with a quadrupole splitting in the range 0.4–0.8 mm/s and therefore they cannot be unambiguously identified. Moreover, the hyperfine parameters can slightly be altered by morphological and chemical influences or may show a distributive behavior, which prevents a clear-cut assignment of the spectral components. As will be shown further, many minerals with poor crystallinity or small-particle morphology exhibit a range of hyperfine field values and the sextet spectra often consist of asymmetrically broadened absorption lines. For all those cases it is obvious that a more elaborated Mössbauer analysis combined with results of other techniques are necessary.

Quantitative information is obtained from the spectra through the relative area of each subspectrum. This area is essentially proportional to the concentration of each kind of iron with its valence in a specific environment. So, the distribution of various types of iron among different sites in a mineral or the concentration of different iron-bearing compounds in a multi-phase assemblage can in principle be determined.

The area ratio A_A/A_B of the spectra of two Fe species A and B is given by

$$\frac{A_A}{A_B} = \frac{\Gamma_A R(T_A) T_A}{\Gamma_B R(T_B) T_B} \tag{3.1}$$

where Γ is the width at half maximum of the absorption peaks and $R(T)$ is a thickness reduction function due to saturation effects which depends on the absorber thickness T. The absorber thickness is not a real physical thickness, but a scalar defined by $T = nf\sigma_0$, where n is the number of Mössbauer active atoms per cm^2 in the absorber, f the Mössbauer fraction (recoilless fraction) and σ_0 the cross section at resonance equal to 2.35×10^{-18} cm^2 for ^{57}Fe. If we set $\Gamma_A = \Gamma_B$ and in a first approximation $R(T_A) = R(T_B)$ the ratio N_A/N_B of the amount of iron atoms of both types A an B can then be calculated from

$$\frac{N_A}{N_B} = \frac{n_A}{n_B} = \frac{f_B A_A}{f_A A_B} \tag{3.2}$$

The Mössbauer fraction f of each kind of iron species is mainly governed by the lattice dynamics in the crystal. Therefore, f is dependent on the coordination and can differ slightly from mineral to mineral. Moreover, the Mössbauer fraction is particularly very sensitive to the valence state of iron and is for Fe^{2+} considerably lower than for Fe^{3+}. Because of the relationship with lattice vibrations, f is also strongly temperature dependent. This means that at RT a large difference in f values is observed and only to a lesser extent at 80 K. The f values for some iron-containing minerals, determined from the temperature dependence of the isomer shift (second order Doppler shift), are listed in Table 3.1.

So far, we did not take into account any thickness effects. The aforementioned formulas are only valid in the case of small thicknesses, because only in that case the transmission integral function, which mathematically describes the spectrum, can be replaced by a sum of Lorentzian lines. The latter profile can be considered

Table 3.1 Mössbauer fractions f at RT and 80 K for some minerals (after Ref. [3])

Mineral	Formula	Val.	Coord.	f_{RT}	f_{80}
Oxides and hydroxides					
Hematite	α-Fe$_2$O$_3$	3+	O$_6$	Natural[a] 0.837–0.851	0.923–0.928
				Synthetic[b] 0.811–0.860	0.914–0.960
Magnetite	Fe$_3$O$_4$	3+	O$_4$	0.889	0.941
		2.5+	O$_6$	~0.80	—
Maghemite	γ-Fe$_2$O$_3$	3+	(O$_4$,O$_6$)$_{av}$	0.814	0.915
Goethite[c]	α-FeOOH	3+	(O,OH)$_6$	0.804	0.912
Akaganeite	β-FeOOH	3+	O$_2$(OH)$_4$	0.877	0.937
Lepidocrocite	O$_3$(OH)$_3$			0.835	0.922
	γ-FeOOH	3+	(O,OH)$_6$	0.793	0.908
Ilmenite	FeTiO$_3$	2+	O$_6$	0.650	0.856
Spinel	Mg$_{0.9}$Fe$_{0.1}$AlO$_4$	2+	O$_4$	0.697	0.875
Silicates					
Ferridiopside	Ca$_{1.03}$Mg$_{0.77}$Fe$_{0.20}$Si$_{1.96}$O$_6$	3+	O$_6$(M1)	0.899	0.945
		3+	O$_4$(Te)	0.862	0.932
Diopside	Ca$_{0.89}$Mg$_{0.81}$Fe$_{0.04}$Si$_{2.12}$O$_6$	2+	O$_6$(M1)	0.747	0.893
		3+	O$_6$(M1)	0.862	0.932
Hedenbergite	Ca$_{1.00}$Mg$_{0.15}$Mn$_{0.3}$Fe$_{0.76}$Al$_{0.03}$Si$_{2.00}$O$_6$	2+	O$_6$(M1)	0.708	0.879
		2+	O$_6$(M1)	0.700	0.876

(continued)

Table 3.1 (continue)

Mineral Val.	Coord.			f_{RT}	f_{80}
Aegerine	$Na_{1.02}Ca_{0.06}Mg_{0.04}Fe_{0.97}-Al_{0.06}Ti_{0.06}Si_{1.84}O_6$	3+	O_6 (M1)	0.874	0.936
Enstatite	$Mg_{1.65}Fe_{0.27}Al_{0.05}Si_{2.02}O_6$	2+	O_6(M2)	0.675	0.882
		2+	O_6 (M1)	0.773	0.902
Olivine[d]	$(Mg,Fe)_2SiO_4$	2+	O_6	0.744	0.893
Some others					
Ankerite	$CaFe_{0.5}Mg_{0.5}(CO_3)_2$	2+	O_6	0.711	0.822
Siderite	$FeCO_3$	2+	O_6	0.743	0.891

[a] Range for 3 samples, [b] Range for 10 samples, [c] Well crystallized [d] Unknown composition

to be reasonably valid up to $T = 2$. Taking $f = 0.8$ and knowing that natural iron contains 2.14 % ^{57}Fe, this thickness corresponds to a n value of 4.6 mg/cm^2 Fe. This value can be a guide for absorber preparation for a single-line spectrum. For a doublet spectrum the area is divided over two lines and this value may be doubled (≈ 10 mg/cm^2) and for a sextet spectrum with (3:2:1:1:2:3)/12 it may be increased fourfold (≈ 20 mg/cm^2) if we consider the outer lines. It is worth to remark that the ideal n value for absorber preparation, based on γ-ray absorption [1] or determined from thickness considerations by Rancourt et al. [2] is similar and leads to 5–10 mg/cm^2 of Fe as a rule of thumb.

However, in the case of a thin absorber without appreciable thickness effects, there is still a deviation from the true ratio N_A/N_B which comes from the thickness reduction $R(T)$. This function was called "saturation" function by Bancroft [3], although the original saturation function $L(T)$ was defined by Lang [4]. The latter was the reduced thickness itself, and hence $L(T) = R(T)T$. Therefore, in our concept, the reduction function $R(T)$ tends to unity when the thickness is zero, but decreases steeply with increasing thickness. If the amount of the different iron species is similar, thus leading to similar thicknesses, the values of the $R(T)$ function are nearly equal and will therefore not lead to a significant error in the value of N_A/N_B. However, if one of the subspectra has a small area relative to the other, the $R(T)$ value for the large area and thus with large thickness will deviate more from unity than the one for the small area. Consequently this leads to a substantial error in the determination of the relative amount of iron. From a thorough numerical analysis based on simulated spectra, Rancourt [5] has calculated the overestimation of the area of a minority line as a function of the area fraction which he represents in a graph for different thicknesses. To give an idea of his results, the true ratio $N_A/N_B = 0.2$ with a nominal thickness $T = 1$ will result in an observed area ratio of about $A_A/A_B = 0.22$ (10 % overestimation) and $N_A/N_B = 0.1$ will result in $A_A/A_B = 0.116$ (16 % overestimation). Using these data it is possible to correct for these effects.

In conclusion, MS has a great potential for determining the relative amount of each kind of iron with its specific valence and with its specific environment if one takes the aforementioned considerations into account. The relative amount of each Fe-bearing mineral in multi-phase assemblages can be similarly calculated from the relative areas. However, the main drawback resides in the analysis of the spectra having strongly overlapping lines in most cases. It is therefore indispensable that an appropriate fitting procedure is used providing the most adequate description of the spectrum. Such a procedure demands often information from other techniques and from Mössbauer measurements at different temperatures or even in applied magnetic fields.

3.2.2 Determination of the Oxidation State and Coordination of Iron

The determination of the valence states of iron in minerals is of extreme importance in geology. In contrast to most of the other abundant elements, iron has in its high-spin state predominantly two valence states which can easily transform from one into another through oxidation or reduction. In this way the valence state may be indicative for the geological history of the minerals (weathering, pressure and temperature changes, ...) and even the color can be associated to the valence states or to transitions between them. For the determination of the oxidation states of iron, MS is commonly used because most techniques are not able to distinguish between Fe^{2+} and Fe^{3+}, and chemical analyses often result in unreliable results due to oxidizing or reducing side effects. Among the hyperfine parameters, the isomer shift is very sensitive to the valence and enables to discern readily the various valence states of iron in minerals. Fe^{3+} usually shows a relatively small isomer shift δ_{Fe} in the range 0.3–0.6 mm/s whereas Fe^{2+} covers the range 0.7–1.2 mm/s (Fig. 3.1).

The quadrupole splitting, on the other hand, is generally large for divalent iron, but depends also strongly on the coordination (Fig. 3.2). So, the combination of both the isomer shift and quadrupole splitting values can give some idea about the coordination of iron. Moreover, Δ in the case of Fe^{3+} is merely determined by the lattice contribution, and therefore the quadrupole splitting is also a good measure for the local distortions in the lattice.

An important application in that respect is distinguishing cis and trans configurations of an octahedral $O_4(OH)_2$ (or $(OH)_4O_2$) Fe^{3+} coordination, which quite often occurs in mineralogical systems (Fig. 3.3). Simple point-charge calculations show that the quadrupole splitting should follow the relation $\Delta_{trans} = 2\Delta_{cis}$.

Although the ratio is in practice never exactly 2, due to other effects such as the influences of more distant charges, the measurement of the quadrupole splitting enables the direct determination of those two types of isomers. In the case of Fe^{2+}, the lattice contribution to the electric field gradient is usually opposite to the large valence contribution of the iron cation itself yielding $\Delta_{trans} < \Delta_{cis}$.

In the case of magnetic spectra, the magnetic hyperfine field B is generally also a direct indication of the oxidation state in addition to the isomer shift. Far below the magnetic transition temperature the hyperfine field of Fe^{3+} amounts to 45–55 T

Fig. 3.1 Isomer shift (δ_{Fe}) values at RT versus coordination number for *low-spin* (II, III) and *high-spin* (2+, 3+) Fe in compounds and minerals

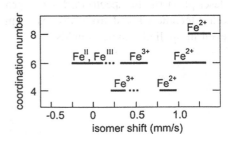

Fig. 3.2 Quadrupole splitting (Δ) values at RT versus coordination number for *low-spin* (II, III) and *high-spin* (2+, 3+) Fe in compounds and minerals

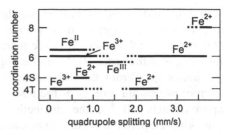

Fig. 3.3 *Trans* and *cis* configuration of a $O_4(OH)_2$ octahedron

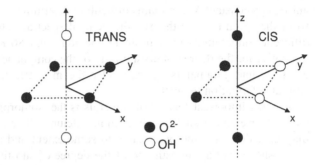

whereas it is usually much lower than 40 T for Fe^{2+} due to the reducing effect of the orbital contribution in B which is generally more pronounced for Fe^{2+} than for Fe^{3+}. However, the Fe^{2+} magnetic spectra are often very complex due to the perturbation of a relatively strong quadrupole interaction and/or to structural and magnetic disorder effects.

3.2.3 Morphological Effects and Isomorphous Substitution

In soils, iron oxides and oxyhydroxides are commonly poorly crystallized and as such occurring as agglomerates of ultra fine particles. In various techniques, such a behavior becomes a perturbing factor in the characterization of iron minerals. In X-ray diffraction, for instance, the line width of the reflections may increase considerably, often inhibiting a clear-cut assignment and also hindering an accurate determination of the other well-crystallized components. In MS the poor crystallinity is mainly reflected by a superparamagnetic behavior, which results in a magnetic transition (sextet-doublet transition), that occurs at much lower temperatures than in the well-crystallized material. However, at very low temperatures the small-particle system still reproduces to a large extent the hyperfine parameter values of the bulk compound so that from the point of view of qualitative analysis the compound can still be relatively well determined. The superparamagnetic fluctuations of the particles' magnetic moments are described by a relaxation time given by

$$\tau = \tau_0 \exp(KV/kT) \qquad (3.3)$$

where KV is the energy barrier that has to be bridged in order to change the direction of the magnetization, with K the effective anisotropy energy density, V the particle volume and kT is the thermal energy (k Boltzmann's constant). The factor τ_0 is of the order 10^{-9} s and is slightly dependent on the magnetization and the anisotropy. From this relaxation formula, and taking τ to be equal to the Larmor precession time τ_L of the nuclear spin in a local magnetic field, which is approximately 10^{-8} s, the magnetic transition temperature, which is called 'blocking temperature' in this case, can in principle be related to the particle size. However, due to all kinds of uncertainties, in particular regarding the value of the anisotropy constant K, it remains difficult to determine this size quantity accurately from the spectra. Nevertheless, the general spectral behavior observed at a few different temperatures (e.g. room temperature and 80 K) may already be fairly indicative for the degree of crystallinity of the compound. Moreover, the values of the hyperfine parameters, in particular the magnetic hyperfine field, may give similar information as well.

Another important feature of minerals is the isomorphous substitution for iron by another element such as aluminum, calcium, silicon, etc. Such a substitution may alter the hyperfine parameters to some extent and relationships can then be established enabling an estimate of the degree of substitution from the hyperfine parameters. Unfortunately, it often happens that both morphological effects and substitution concurrently give rise to similar changes in the spectrum, thus leading to ambiguous conclusions. In the next sections the possibilities of the morphological characterization by the Mössbauer effect will be discussed for some particular minerals.

3.3 Characterization of Iron Oxides and Hydroxides

Iron oxides and hydroxides are the most important iron-bearing constituents of soils, sediments and clays. To characterize the samples, i.e. the identification of the different minerals present and the determination of their morphology and chemical composition, a variety of standard techniques are commonly used such as X-ray and electron diffraction, chemical analyses, optical and electron microscopy, infrared spectroscopy and thermal analysis (DTA, DTC,…). Most of these techniques are further applied in conjunction with selective dissolution or other separation methods in order to obtain more specific information about particular components in the complex soil system. In addition to all those characterization methods, MS has proven to be a valuable complementary technique for the study of these kinds of materials and in particular for the characterization of iron oxides and hydroxides which are usually poorly crystallized.

Most oxides and hydroxides in soils are indeed known to be less well or rather poorly crystallized. This feature results in the first place in a lowering of the magnetic transition temperature yielding a doublet at temperatures were normally a sextet is expected. Therefore, for characterization purposes, it is mostly recommended to measure at different temperatures. For practical reasons, Mössbauer spectra are usually collected at two specific temperatures, i.e. room temperature (RT) and liquid nitrogen temperature (~ 80 K) from which already valuable results can be obtained. The more expensive low-temperature measurements down to 4 K or the time-consuming detailed temperature scanning of the spectra are only necessary in those cases where important additional information can be expected.

Secondly, the magnetic sextet spectra of these materials at lower temperature exhibit usually asymmetrically broadened lines. Several mechanisms have been suggested to explain this behavior, but that discussion would lead us far beyond the scope of this paper. Fortunately, for characterization purposes, such spectra can be adequately fitted by a few sextets or even more accurately by considering a distribution of hyperfine fields [6–8]. In this case the spectra are no longer fitted with a single sextet of Lorentzian lines but by a set of such sextets. A versatile distribution fitting procedure which uses several tens of elemental sextets with the necessary smoothing constraint, has been developed by Hesse and Rübartch [9] and was further improved by Le Caër and Dubois [10] or extended by Wivel and Mørup [11] and has been combined in one analyzing method [12]. In this way a much more accurate description can be obtained of partly overlapping sextets, such as those occurring for goethite-hematite associations as for instance demonstrated by the analysis of a Tunisian soil profile [8].

For characterization purposes, MS studies on natural soil samples usually rely on the results of a variety of systematic studies on pure natural or synthetic compounds. Such systematic studies provide a description of the spectral behavior of the different Fe-bearing oxides and hydroxides and a determination of the hyperfine parameters in relation to morphological and chemical features. Many reviews on this subject are available in literature [13–19]. In what follows a brief survey will be given of the spectral features of the various iron oxides and (oxy)hydroxides in relation to their identification and characterization in natural soil samples.

3.3.1 Goethite (α-FOOH)

Goethite is by far the most encountered Fe-bearing compound in soils, sediments and clays, and has therefore been intensively studied in the past forty years. In its most ideal mineral form it is antiferromagnetic with a Néel temperature $T_N = 400$ K [20, 21]. The magnetic hyperfine field amounts to 38.1 T at RT, 50.0 T at 80 K and saturates to 50.7 T at 4 K. However, the well-crystallized form is of rare occurrence and has only been found in particular sites such as the Harz Mountains (Germany) and Lostwithiel (Cornwall, UK) due to the presence of the required extreme hydrothermal formation conditions.

In soils goethite is usually obtained as a weathering product of Fe^{2+} silicates and to a lesser extent of sulfides, carbonates, oxides, etc. This results in a poorly crystalline

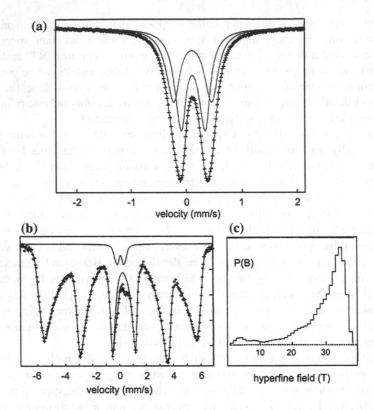

Fig. 3.4 RT Mössbauer spectra of poorly crystallized goethite fitted with two doublets (**a**) and a relatively well crystallized goethite (**b**) with corresponding hyperfine field distribution (**c**)

form, consisting of conglomerations of small, usually needle-shaped crystallites. Consequently, the spectral behavior is mainly governed by superparamagnetic relaxation (Eq. 3.3) with an anisotropy constant K of the order 10^3 Jm^{-3}. The spectra of goethite exhibit simultaneously a sextet and doublet over a wide temperature range, which can be attributed to the non-uniformity of particle morphology within the goethite sample. A doublet spectrum is obtained at room temperature for mean crystallite sizes smaller than about 15 nm. Such a doublet consists of somewhat broadened lines (Fig. 3.4a) indicating the presence of a distribution in the quadrupole splitting with an average value of about 0.55 mm/s (Fig. 3.8). This doublet resembles that of many other paramagnetic or superparamagnetic Fe^{3+}-bearing species and is therefore practically not useful for identification of goethite. However, the magnetically split spectrum of poorly-crystalline goethite, commonly obtained at lower temperatures, is usually recognized by its asymmetrically-broadened lines (Fig. 3.4b). Moreover, its average hyperfine field B_{av} is usually lower than that of well-crystallized goethite and decreases with increasing temperature in different ways depending on the particle size. Therefore its value at a certain temperature

Table 3.2 Summary of hyperfine parameters for goethite

Crystallinity	T (K)	Spectrum	B_{av} (T)	B_p (T)	2ε or Δ (mm/s)	δ_{Fe} (mm/s)
Very high	RT	S	38.1	38.1	−0.28	0.37
(e.g. Harz)	80	S	50.0	50.0	−0.26	0.48
	4	S	50.7	50.7	−0.26	0.49
Related high	RT	S	30–35	31–38	−0.26	0.37
	80	S	47–49	49–50	−0.26	0.47
Moderate	RT	coll S	20–25	25–33	−0.25	0.37
		+D	–	–	0.55	0.36
	80	S	43–47	48–49	−0.25	0.47
Poor	RT	D	–	–	0.55–0.6	0.36
		+ coll S	<20	–	−0.25(f)	0.36
	80	S	40–43	47–49	−0.25	0.47
Very poor	RT	D	–	–	0.6	0.36
	80	S	35–40	47–48	−0.24	0.46
		+D (Fh?)			0.6	0.46

S sextet, *Coll S* collapsing sextet, *D* doublet, *f* value fixed in the fit

could roughly serve as a measure for the average particle size or crystallinity of the goethite phase in a sample. A summary of the hyperfine parameters for goethite obtained from own research is given in Table 3.2. The remaining doublet at 80 K is in many cases most probably ferrihydrite. Examples of different spectra will be shown in Sect. 3.7.

Unfortunately, similar field-reducing and distributive effects are also caused by isomorphous substitutions which are frequently encountered in natural samples. Because of the high abundance of Al in nature the substitution of Al for Fe in goethite has attracted most attention and has been intensively studied [22–25]. Substitution by diamagnetic Al first of all lowers T_N so that a doublet is obtained at RT for an Al substitution of 12 at % and larger [26]. Secondly, this substitution reduces the supertransferred contribution to B, leading to a reduction and a distribution of B. From systematic studies on synthetic aluminous goethites an average field reduction of about 0.05 T at 4 K and of about 0.14 T at 80 K per at % Al has been derived. Because the degree of crystallinity plays a similar role in the dependence of the hyperfine field, both Al concentration and a crystallinity parameter have been introduced in various linear equations of the type,

$$B = B_0 - \alpha\, C(\%Al) - \beta\, S\,(\mathrm{m^2/g}) \text{ or } \beta/MCD_{111} \tag{3.4}$$

where α and β are given coefficients. Either the specific surface area S from the BET method or the mean crystallite diameter MCD_{111} obtained from XRD broadening of the [27] reflection has been taken as a measure for the crystallinity [22, 24, 25].

These relationships were initially considered as promising complementary means to characterize goethite in natural samples. From the most probable hyperfine field B_p or better, the average hyperfine field B_{av}, either the Al concentration or a measure for the crystallinity can in principle be determined if the other parameter is known from complementary experiments. Although it has been shown that these equations could be successfully applied to certain series of well-characterized natural goethite samples [22, 24], they completely failed for many other natural and synthetic sample series [28]. Friedl and Schwertmann [29] investigated 33 natural goethite samples from different origins, which according to their formation conditions could be divided into 24 samples from tropical and subtropical soils and 9 samples from lake ores. Comparison of the observed hyperfine fields at 4 K with the respective values calculated from the correlation equations resulted in significant deviations, with a somewhat different general behavior for the goethites occurring in the tropical soils than for those occurring in the lake ores. This was corroborated by the linear regressions separately obtained for the two kinds of samples, resulting in different coefficients for both the Al content C and the crystallinity as reflected by the inverse MCD. From a thorough study on well-defined samples of Al goethite, it has been shown that other structural parameters, such as water content, excess hydroxyl ΔOH and structural defects, all having some influence on the lattice parameters [30, 31], play also a substantial role in the magnitude of the magnetic and electrical hyperfine parameters [32–36]. These additional structural parameters are mainly determined by goethite formation factors such as crystallization rate, temperature, OH concentration, etc., and are to some extent related to each other [30].

The situation becomes even more complicated if other elements are involved in the goethite formation. In view of the similar structure of α-MnOOH (groutite), Mn has also been found to substitute for Fe in goethite to a large extent [37, 38]. Because Mn^{3+} is a paramagnetic ion the hyperfine field is less reduced than in the case of Al [38]. Other elements such as Si and P show the tendency to adsorb to the goethite crystallites rather than to substitute for iron in the structure [39–41]. This surface "poisoning" not only opposes the crystal growth during the formation but may also reduce B due to surface effects [28].

It can be concluded that MS is a very suitable tool with respect to the qualitative and to some extent quantitative analysis of goethite in soils and sediments. However, it is clear that MS is still far from being a "magic" analytical technique that provides an in-depth knowledge of the morphological properties of goethite. For the moment, if one compares goethites from a same soil profile, MS can yield some crude indications with respect to crystallinity if equal isomorphous substitutions are expected. Also, other techniques are not so powerful in that respect because natural soil samples mostly contain goethite in more or less close association with other mineral species, likewise hampering to obtain accurate results.

3.3.2 Akaganéite (β-FeOOH)

Akaganéite as the second polymorph of iron oxyhydroxide is by far less abundant in nature in comparison with goethite. In fact, akaganéite requires a small amount of

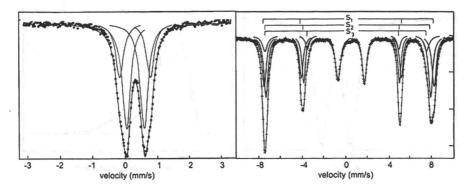

Fig. 3.5 Mössbauer spectrum of a synthetic akaganéite sample at RT with two doublets (*left*) and at low temperature (30 K) with three sextets (*right*) (after Ref. [306])

Table 3.3 Representative hyperfine parameters of oxyhydroxides, other than goethite

Mineral	T (K)	Spectrum	B (T)	2ε or Δ (mm/s)	δ_{Fe} (mm/s)
Akaganeite	RT	D1	–	0.51–0.56	0.37
β-FeOOH		D2	–	0.91–0.96	0.38
	80	S1	∼47	∼−0.1	∼0.46
		S2	∼44	∼−0.5	∼0.47
	4	S1	49.2	−0.10	0.49
		S2	47.7	−0.18	0.48
		S3	47.2	−0.72	0.50
Lepidocrocite	RT	D1	–	0.55–0.70	0.37
γ-FeOOH		D2	–	1.0–1.2	0.36
	4	S	44–46	∼0.02	0.48
Feroxyhite	RT	D	–	0.69	0.37
δ'-FeOOH	4	S1	52.5	0.17	0.48
		S2	50.8	0.07	0.48

chloride or fluoride ions to stabilize the structure. The mineral is an antiferromagnet with a Néel temperature of about 290 K, the latter being often lower, depending on the crystal water content [42]. The RT spectrum consists of a broad doublet with typical asymmetric line shapes (Fig. 3.5). It can be considered as being composed of two discrete doublets, one with $\delta_{Fe} = 0.37$ mm/s and Δ varying between 0.51 and 0.56 mm/s and the second one with $\delta_{Fe} = 0.38$ mm/s and Δ varying between 0.91 and 0.96 mm/s [43, 44]. At 80 K the spectrum exhibits a somewhat broad-lined asymmetric sextet, which for analytical purposes should at least be adjusted with two sextets whereas at lower temperatures three sextets are needed in order to adequately describe the spectrum of akaganéite [45]. The hyperfine parameters are summarized in Table 3.3.

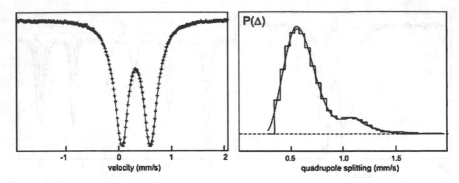

Fig. 3.6 RT Mössbauer spectrum of a lepidocrocite sample (*left*) and corresponding quadrupole distribution (*right*)

3.3.3 Lepidocrocite (γ-FeOOH)

The third oxyhydroxide, lepidocrocite, is an antiferromagnet with a magnetic transition temperature T_N of about 73 K [46]. The magnetic transition is usually not sharp and exhibits a temperature range, often larger than 10 K, in which a sextet and a doublet spectrum coexist. Although this would imply a superparamagnetic behavior [47], similar to other iron oxides and oxyhydroxides, a study of synthetic lepidocrocites of various particle sizes revealed that surface effects play in this case a more significant role, leading to a variety of Néel temperatures which causes the broad transition range [48]. This behavior is probably associated with the typical morphological features of lepidocrocite, which consists of small very thin, raggedly structured platelets [47, 48].

The RT spectrum exhibits a somewhat broadened doublet (Fig. 3.6) with an average quadrupole splitting between 0.55 and 0.7 mm/s, depending on the particle morphology. A more detailed analysis of the doublet yields quadrupole splitting distributions which have more or less two maxima (Fig. 3.6) [47]. The first maximum with $\Delta \sim 0.52$ mm/s is attributed to the bulk part whereas the second with $\Delta \sim 1.1$ mm/s is believed to result from the surface species. At 80 K both distributions are more broadened, so that the second maximum is less pronounced. From diagnostic point of view, these features are not specific enough to discern lepidocrocite from other oxides or oxyhydroxides and one has to rely on magnetically split spectra far below 80 K. At 4 K the magnetic spectrum yields a magnetic hyperfine field of about 44–46 T and a very small quadrupole shift 2ε of about 0.02 mm/s. This hyperfine field is somewhat lower than that of the other iron oxides and oxyhydroxides, which makes lepidocrocite relatively well discernible in complex iron oxide samples. However, a distinction between lepidocrocite and ferrihydrite is much more difficult in view of the broad lines of the latter at low temperatures. The range of the hyperfine parameter values are given in Table 3.3.

Similarly to goethite, lepidocrocite can also contain Al [49]. From a Mössbauer study on a series of synthetic Al-substituted samples [50] it could be confirmed that Al replaces Fe in the structure to a small extent. A reduction of the maximum-probability hyperfine field B_m with about 0.06 T per at %Al was observed.

3.3.4 Feroxyhite (δ-FeOOH or δ'-FeOOH)

δ-FeOOH, which is isostructural with $Fe(OH)_2$, is the fourth polymorph of iron oxyhydroxide and had initially no mineralogical name because it was not found in nature. On the other hand, δ'-FeOOH, which can be considered as a poorly ordered variant of δ-FeOOH with a somewhat different Fe^{3+} arrangement, has been identified as a rare mineral known as feroxyhite (originally written 'feroxyhyte' by Chukhrov et al. [51]). Showing a typical Fe^{3+} doublet at RT with $\delta_{Fe} = 0.37$ mm/s and $\Delta = 0.69$ mm/s this mineral cannot be distinguished from other oxyhydroxides. This is particularly true in soils where biogenic δ'-FeOOH is often associated with ferrihydrite [52], the latter possessing a similar doublet. Feroxyhite orders magnetically only at very low temperatures and shows at 4 K a broad-lined sextet [53, 54], which in fact is composed of two overlapping sextets (Table 3.3).

Nowadays, both the delta-oxyhydroxides are being called feroxyhite, because δ'-FeOOH is considered to be nothing else than the poorly crystallized (super-paramagnetic at RT) variant of δ-FeOOH, the latter being magnetic at RT.

3.3.5 Ferrihydrite

Ferrihydrite is an iron oxyhydroxide with a high surface water content having an approximate composition $Fe_5O_{12}H_9$ written as $Fe_5O_3(OH)_9$ or $Fe_5HO_8.4H_2O$. The most important feature of this mineral is its poor crystallinity, with particle sizes in the range 2–7 nm. XRD patterns consist of 2–6 broad peaks which are not always discernible due to the presence of broad lines of poorly crystalline goethite that often accompanies ferrihydrite in soils. Different models have been proposed for the structure [55–62] and the subject is still under discussion.

Both poor crystallinity and structural disorder lead to a very broad temperature range in which the magnetic order develops and the nature of magnetic order is still in doubt. For a so-called 6-XRD-line ferrihydrite sample the magnetic transition spans a temperature region from 35 to 110 K [63]. Ferrihydrites with less than 6 XRD lines are still paramagnetic at 80 K. It is now believed that these low magnetic transition temperatures result from a kind of superparamagnetic behavior from interacting particles [64], whereas the real antiferromagnetic-paramagnetic transition temperature of ferrihydrite is estimated to be of the order of 500 K [65].

It is obvious that the high degree of structural disorder in such a poorly crystalline system causes broad distributions of hyperfine parameters resulting in broad-lined doublets and sextets. At RT a broad doublet is obtained (Fig. 3.7) with average Δ in the range 0.7–0.9 mm/s. Though the spectra reflect in fact a broad

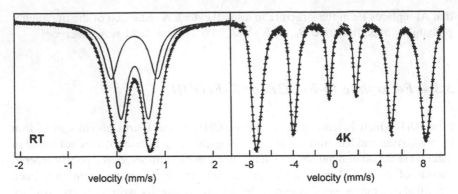

Fig. 3.7 Mössbauer spectrum of ferrihydrite at RT fitted with two doublets (*left*) and at 4 K fitted with a δ-correlated hyperfine field distribution (*right*)

Fig. 3.8 Quadruple distribution in a poorly crystalline sample of goethite and lepidocrocite, and in ferrihydrite

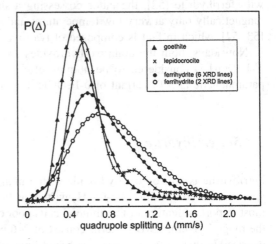

distribution of quadrupole splittings (Fig. 3.8), they are usually satisfactorily described by two doublets (Fig. 3.7) with parameters of about $\delta_{Fe} = 0.37$ mm/s, $\Delta = 0.6$ mm/s (FWHM $\Gamma = 0.3$ mm/s), and $\delta_{Fe} = 0.38$ mm/s, $\Delta = 0.9$ mm/s (FWHM $\Gamma = 0.4$ mm/s). The intensity ratio of these doublets is strongly dependent on the crystallinity and varies from 70/30 to about 30/70 for "better" to poorly crystallized ferrihydrite, respectively.

Such a broad doublet is not only characteristic for ferrihydrite, but also for other Fe^{3+} bearing minerals. Paramagnetic akaganéite, for instance, shows also a broad doublet spectrum at RT, although it is in fact composed of two discrete spectral components. Moreover, akaganéite is readily recognized by the appearance of a magnetically split spectrum at temperatures far above the transition temperature of ferrihydrite. Another compound showing the typically broad doublet at RT with similar distribution is fully oxidized vivianite, often called oxykerchenite.

Similarly, as for most of the ferrihydrites, the spectrum of oxykerchenite remains a doublet down to 80 K [66].

At very low temperatures the sextet of ferrihydrite exhibits broad lines with a rather symmetrical lineshape (Fig. 3.7). At 4 K the average hyperfine field amounts to between 46 and 50 T [67], with a small quadrupole shift 2ε between -0.02 and -0.1 mm/s. The spectra are usually slightly asymmetric in which in particular the 6th line is less deep than the first one. This has been attributed to the presence of additional tetrahedral iron sites [68], although this is still in doubt and correlation effects between B in the distribution on the one hand, and Δ and/or δ on the other hand, which are normally expected in largely disordered structures, may similarly produce such an asymmetry.

As in other iron oxyhydroxides, isomorphous substitution for Fe by Al is expected in natural samples. A study of synthetic samples with Al substitutions by [69] revealed an increasing asymmetry of the doublet lineshape at RT, pointing to an increased average quadrupole splitting. At low temperatures, a decrease of the average hyperfine field and a lowering of the magnetic transition temperature region with increasing Al content is reported. All these features are quite similar to crystallinity effects and are therefore not of practical use for characterization purposes. Also silicon seems to play an important role in ferrihydrite. As in the case of goethite, Si species can easily adsorb on ferrihydrite and thus prevents its further growth [70]. Childs [57] claimed that ferrihydrite can contain up to 9 %at Si, but the question remained if it is adsorbed or incorporated in the structure. Campbell et al. [71] suggested that Si is structurally incorporated and demonstrated the ability of Si to inhibit a transformation to more stable Fe^{3+} oxides or oxyhydroxides.

However, several more recent works have resulted in a complete change in the earlier, more-or-less contradictory ideas about ferrihydrite and its XRD and Mössbauer behavior. Berquó et al. [72] report the possibility of synthesizing Si-ferrihydrites with much better crystallinity. One of the authors' ferrihydrites shows even relatively sharp lines in the XRD pattern and its Mössbauer spectrum even consists of a somewhat collapsed sextet at RT, but with no doublet contribution. The spectrum of a natural sample, showing similarly seven, but somewhat broadened lines in the XRD pattern, exhibits a collapsed sextet at 130 K, but remains exclusively a doublet at RT.

On the other hand, it has recently been shown that some ferrihydrite species still exhibit a magnetic-superparamagnetic transition at very low temperatures. In particular, this seems to happen when ferrihydrite is closely associated to organic carbon [73]. These so-called DOM (dissolved organic matter) ferrihydrites have a lower hyperfine field and are even not completely magnetically ordered at 4 K (see also Sect. 3.5.1).

In conclusion, the very poor crystallinity and the low superparamagnetic blocking temperature in natural ferrihydrites hamper to some extent the characterization of ferrihydrite with MS at standard measuring temperatures (RT, 80 K). From the point of view of identification, ferrihydrite can be recognized by MS as long as it represents the main constituent in soil samples. The broad quadrupole distribution (see Fig. 3.8) with its high average value is already quite indicative for the presence of ferrihydrite, although, clear-cut evidence can only be obtained from the typical sextet spectrum obtained at very low temperatures (<80 K). On the other hand, XRD may provide more direct evidence through the characteristic broad-lines pattern which will be discernible if ferrihydrite is present in sufficient amount in the sample. Unfortunately, in many soils ferrihydrite occurs only as a minor fraction and no technique is able to recognize it directly among the other interfering components such as poorly crystalline goethite. In goethite-ferrihydrite associations, the doublet fraction which persists down to 80 K may account for ferrihydrite as well as for very poorly crystalline and/or highly substituted goethite. However, if the 80 K spectrum exhibit a goethite sextet with rather narrow lines, the remaining doublet can most likely be attributed to ferrihydrite because goethite with two discrete and drastically different crystallinities or substitutions are not commonly expected in the same sample. At very low temperatures (4 K) the hyperfine fields of goethite and ferrihydrite are comparable. Hence, these components cannot be separated merely by their difference in quadrupole shift. Only the broader lines of the ferrihydrite sextet might be indicative.

Another method which is helpful in the characterization of ferrihydrite is selective dissolution in acid ammonium oxalate [74] followed by a so-called differential X-ray diffraction (DXRD) [75], which consists of subtracting the pattern of a treated sample from that of an untreated one, thus isolating and enhancing the typical ferrihydrite diffraction pattern. In MS, this dissolution applied to natural soil samples usually results in a decrease of the doublet at RT and 80 K, and a narrowing of the sextet lines at lower temperatures. However, it has been demonstrated that ammonium oxalate also dissolves organically bound Fe, Fe from magnetite and from poorly crystalline lepidocrocite [76–79], and also attacks vivianite, siderite [66] and even poorly crystalline goethite [80]. So, MS applied in combination with such a treatment is not always decisive unless information about the presence of the above interfering compounds is obtained from other techniques.

3.3.6 Hematite (α-Fe$_2$O$_3$)

Hematite is the most abundant iron oxide in soils and sediments. In comparison to other iron oxides and hydroxides, hematite (α-Fe$_2$O$_3$) exhibits a non-common magnetic behavior. In addition to the normal magnetic-paramagnetic transition at $T_N = 955$ K, pure and well-crystallized hematite transforms at about 265 K from a low-temperature antiferromagnetic (AF) to a high-temperature weakly ferromagnetic (WF) state, known as the Morin transition. This transition consists in fact of a 90° spin reorientation from an antiferromagnetic spin configuration in the c direction into the basal plane (perpendicular to the c-axis) in which the spins are slightly canted resulting in a weak ferromagnetism (Fig. 3.9). This magnetic

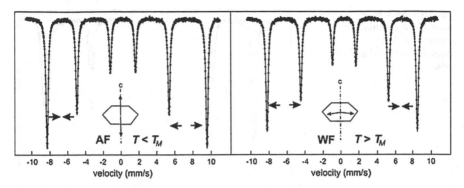

Fig. 3.9 Mössbauer spectra of hematite in the antiferromagnetic state at 80 K (*left*) and in the weakly ferromagnetic state at RT (*right*)

Table 3.4 Representative hyperfine parameters of hematite of various crystallinity

Crystallinity	T (K)	Spectrum	B_{av} (T)	B_p (T)	2ε (mm/s)	δ_{Fe} (mm/s)
High	RT	S(WF)	51.7	$=B_{av}$	−0.19	0.36
	80	S(AF)	54.1	$=B_{av}$	0.38	0.47
	4	S(AF)	54.2	$=B_{av}$	0.39	0.48
Medium	RT	S(WF)	50.0–51.0	$\approx B_{av}$	−0.20	0.37
	80	S(WF)	52.5–53.0	$\approx B_{av}$	−0.19	0.47
		S(AF)	53.5–54.0	$\approx B_{av}$	>0.10	0.47
	4	S(AF)	54.2	$=B_{av}$	0.38	0.48
		S(WF)	53.3	$=B_{av}$	−0.20	0.48
Poor	RT	S(WF)	37.0–48.5	49.5–50.0	−0.21	0.37
	80	S(WF)	52.0–53.0	$\approx B_{av}$	−0.20	0.47
	4	S(WF)	53.2	$=B_{av}$	−0.20	0.48

behavior is well reflected in the Mössbauer spectra. The hyperfine parameters of hematite at different temperatures are given in Table 3.4.

The hyperfine field B in the AF state at 4 and 80 K are nearly equal which is conceivable in view of the high T_N. In the WF state at RT a value of 51.7 T is observed. At the Morin transition temperature T_M, B changes abruptly. The drop of about 0.8 T is explained by the influence of the spin reorientation on the orbital and dipolar contributions to B [81]. The quadrupole shift 2ε, which is only slightly temperature dependent, changes more drastically at T_M. At 80 K a large positive value of 0.38 mm/s is observed whereas the WF state has a negative value of −0.19 mm/s at RT (indicated by arrows in Fig. 3.9). The relation between those two values is consistent with the EFG principal axis lying in the direction of the c-axis. In view of this large difference in 2ε both the WF and AF phase can be separately identified from the fitting, and therefore MS is an extremely powerful tool to study the Morin transition. Moreover, because the latter is very sensitive to microcrystalline effects, lattice imperfections and impurities, it is clear that this technique could offer some possibilities for the characterization of hematite in natural samples.

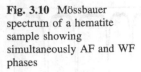

Fig. 3.10 Mössbauer spectrum of a hematite sample showing simultaneously AF and WF phases

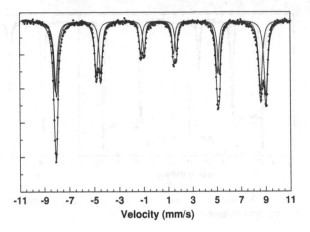

Pedogenic hematite consists mainly of small crystallites and, similar to goethite, the spectral features are governed by superparamagnetic relaxation effects. However, the effective anisotropy constant K of hematite is of the order of 10^4 J/m^3 [82], yielding blocking temperatures which are higher than in the case of goethite. Consequently, in most cases the spectra are still magnetically split at RT. Nevertheless, due to microcrystalline effects, the lines may still be asymmetrically broadened. The origin of the field distributive and reducing effects is similar to that of goethite, although, the reduction of B caused by surface effects has been well established in hematite by MS on ^{57}Fe surface enriched hematite [83, 84] and by surface studies with conversion electrons MS [85, 86]. For very small crystallite sizes with particle dimension $D \ll 8$ nm a doublet is observed at RT with a quadrupole splitting Δ in the range 0.5–1.1 mm/s [87], apparently strongly dependent on the particle size.

Morphological effects such as particle size, lattice imperfections and the presence of micro- and macropores have also a pronounced influence on the Morin transition. First of all, the transition temperature lowers with decreasing particle size [88–90] and for particle sizes smaller than about 20 nm the Morin transition is even completely suppressed, resulting in a single WF phase down to 4 K [82]. Further, the normally sharp Morin transition becomes a temperature region with the coexistence of the two phases in which the AF phase diminishes in favor of the WF phase with increasing temperature. A typical spectrum in this transition region is shown in Fig. 3.10.

This region broadens considerably with decreasing particle dimensions and with increasing structural defects [91], and the coexistence of both the AF and WF phase even can extend down to 0 K. Such a case is represented in Fig. 3.11 where the relative areas (RA) and the quadrupole splitting do not change below 150 K. The Morin transition temperature can then be defined as that temperature at which the amount (RA) of AF phase is reduced to half of its initial value at low temperatures.

The influence of the average particle size on the Morin transition temperature has been mainly investigated for synthetic samples and is represented in Fig. 3.12. From

(a) **(b)**

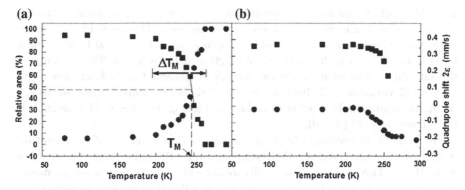

Fig. 3.11 Temperature behavior of the transition region: relative area (**a**) and the quadrupole shift 2ε (**b**) for a small-particle hematite

Fig. 3.12 Morin transition temperature vs. inverse average particle size for differently prepared hematite samples (*Black square* prepared from decomposition of lepidocrocite; for the other symbols, see Ref. [99])

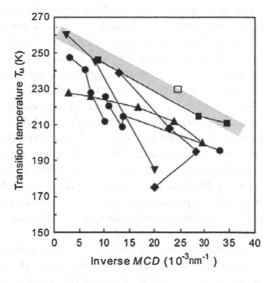

that picture it follows that this transition temperature is not solely dependent on the particle size, but also differs according to the preparation method. Large defects and the presence of hydroxyl groups (OH⁻) probably cause the large fluctuations in T_M [92–94]. However, hematite samples prepared from lepidocrocite generally show the highest transition temperatures [95–97] and it is believed that natural samples, which are mostly formed from ferrihydrite will possess the same features. The shaded band shown in Fig. 3.12 might be a reasonable analytical guideline for the relation between T_M and the average particle dimensions.

Isomorphous Al for Fe substitution in hematite is also a common phenomenon and has been intensively studied [92, 96, 98–101]. Hematite can contain more than 15 at % Al and, similarly to goethite, this diamagnetic substitution has primarily a twofold effect on the magnetic properties, i.e. it lowers the magnetic transition temperature and reduces the saturation hyperfine field. However, in view of the high Néel temperature, the sextet-doublet transition still occurs at high temperatures above RT and is therefore not of practical diagnostic use. The reduction of the saturation value of B, which can already be observed at 80 K in view of the very small variation of B between 80 and 4 K, is also rather small. At 80 K a reduction of about 0.04 T per at % Al has been found, whereas at RT it amounts to 0.08 T per at % Al [92, 102].

Similarly as for goethite, a linear relationship of B as a function of Al content and particle size has been proposed [102] which is valid at RT and for concentrations less than 10 %Al. However, all these results are derived from synthetic samples, mostly obtained from goethite. Therefore, particularly at RT, the magnetic hyperfine field will still be largely influenced by morphological effects. Moreover, most preparation methods, based on the decomposition of oxyhydroxides, result in inhomogeneous Al substitution [96]. A more clear-cut picture for the dependence of the hyperfine field on Al substitution is obtained for hematites prepared from oxinates [103] where a reduction of 0.061 at RT and 0.032 at 80 K per at % Al is observed.

Somewhat more pronounced effect of Al substitution is reflected in the behavior of the Morin transition. With increasing Al content the transition temperature T_M decreases and the transition region becomes significantly broader [97]. Moreover the Morin transition is completely suppressed at about 10at % Al in bulk hematite [99] and even at somewhat lower concentrations (8 at %) for less crystalline hematite [100]. On the other hand, the effect of Al on the Morin transition temperature is smaller in the case of more homogeneous Al substitution in samples prepared from oxinates [103]. Using the aforementioned definition, the Morin transition temperature for as-such obtained hematite species decreases by 8 K per at % Al. Because the spectral implications of Al substitution are quite similar to those of morphological effects, the separation of both effects remains a major problem and additional techniques are necessary for the characterization of natural samples.

Another element which is a possibly abundant candidate for substitution of iron in natural hematite samples is manganese [8]. Mn was substituted for iron up to about 18 % in synthetic samples [104], but, it is believed that in natural samples the substitution is much lower (<5 at %) [105]. In a study of some Mn-hematites prepared from Mn-substituted goethites [106] a somewhat smaller decrease for the hyperfine field of the WF phase at RT is observed in comparison with that of similarly prepared Al-hematites, which is expected in view of manganese being a magnetic ion. On the other hand, manganese reduces the Morin transition temperature more rapidly, and already at about 4 at % Mn the transition is completely suppressed [106]. A still more drastic effect on T_M is caused by Ti substitution which is also abundant in nature. Less than 1 at % Ti completely inhibits the Morin transition [27]. However, Ti occurs rather in high concentration tending more to the isostructural ilmenite ($FeTiO_3$). Another important substitution element could be silicon, although little information in that respect is found in the literature. There are indications that silicon increases T_M slightly. For example, an asymmetric spectrum at RT has been found for a Si-rich hematite sample from Elba, pointing to the presence of the AF phase above 265 K [90].

In conclusion, the characteristic hyperfine parameters and the high magnetic transition temperature T_N and blocking temperature T_B for hematite render MS very suitable for the identification of this material in natural samples. Moreover, when used in conjunction with other techniques, MS yields some indicative information about morphological or substitutional properties through the determination of the magnetic hematite phases present in the sample. Although small-particle effects are not so pronounced as in the spectra of goethite, the spectra of hematite may still show asymmetric lines, particularly at room temperature. Hence, appropriate fitting procedures such as the ones based on a hyperfine field distribution are recommended, especially when the hematite sextet overlaps with other sextets.

3.3.7 Fe–Ti Oxides

Related to hematite, ilmenite, $FeTiO_3$, has a similar rhombohedral structure. Alternating planes are occupied by Ti^{4+} and Fe^{2+}, breaking down the strong magnetic interactions that occur in hematite. It is therefore a weak antiferromagnet with $T_N = 55$ K [107]. Although the whole range between hematite and ilmenite, $Fe_{2-x}Ti_xO_3$ can be synthesized, natural samples of ilmenite possess a relatively high amount of Ti ($0.75 < x < 0.95$). Ilmenite with a higher degree of stoichiometry was found in lunar samples [108]. At room temperature a doublet is observed with a high isomer shift, $\delta_{Fe} = 1.1$ mm/s and a low quadrupole splitting $\Delta = 0.71$ mm/s. These typical hyperfine parameters for ilmenite, make it relatively easy to discern this oxide in the Mössbauer spectrum. At 80 K the quadrupole splitting is larger and amounts to about 1.0 mm/s. At 5 K only a very low magnetic hyperfine field (~ 4 T) is observed due a strong opposite orbital contribution of Fe^{2+} [107, 109].

Deviation from stoichiometry implies the presence of Fe^{3+} which is clearly observed in the spectra by an asymmetry, namely the left absorption line is broader and deeper. The spectra can then be analyzed with two doublets: one from Fe^{2+} having hyperfine parameters very close to those of pure ilmenite, and one from Fe^{3+} with typical hyperfine parameters for trivalent iron [110] (see Table 3.5).

Another well-known mineral is pseudobrookite, which consists of a complete solid solution series between ferrous $FeTi_2O_5$ and ferric Fe_2TiO_5. Also in this case, the Mössbauer spectra reveal iron species that vary from Fe^{2+} over mixed valences to Fe^{3+}. However, the structure contains two different sites (8f and 4c) where iron can be present, resulting in two doublets for each iron valence. The hyperfine parameters according to Guo et al. [111] are summarized in Table 3.5.

Table 3.5 Representative hyperfine parameters for Fe–Ti oxides

Mineral	T (K)	Iron	B (T)	2ε or Δ (mm/s)	δ_{Fe} (mm/s)
Ilmenite $FeTiO_3$	RT	Fe^{2+}	–	0.71	1.1
Ilmenite (non-stoichiom)	RT	Fe^{2+}		0.65–0.70	1.0–1.1
$Fe_{1+x}Ti_{1-x}O_3$		Fe^{3+}		0.3–0.5	0.3
Pseudobrookite (ferrous)	RT	Fe^{2+} 8f		1.10	2.16
$FeTi_2O_5$		Fe^{2+} 4c		1.06	3.15
Pseudobrookite (Intermed.)	RT	Fe^{2+} 8f		1.1–1.2	1.6–2.1
$Fe_{1+x}Ti_{2-x}O_5$		Fe^{2+} 4c		1.04–1.06	2.8–3.1
		Fe^{3+} 8f		0.38–0.41	0.54–0.58
		Fe^{3+} 4c		0.34–0.39	0.85–1.00
Pseudobrookite (ferric)	RT	Fe^{3+} 8f		0.38	0.57
Fe_2TiO_5		Fe^{3+} 4c		0.38	0.92

3.3.8 Magnetite

Magnetite (Fe_3O_4) is a ferrimagnetic spinel oxide with a Néel temperature of 858 K. It naturally occurs often in a fairly crystallized form and its presence can readily be recognized in a Mössbauer spectrum. Magnetite has the following structural formula $\left(Fe^{3+}\right)_A\left[Fe_2^{2.5+}\right]_B O_4$ where the octahedral (B-site) ferrous and ferric ions merge into $Fe^{2.5+}$ due to a fast electron hopping in pairs above the so-called Verwey transition (>125 K). Consequently, the RT spectrum of magnetite exhibits two partly resolved sextet patterns (Fig. 3.13) resulting from tetrahedral (A site) Fe^{3+} with $B = 49.1$ T, $2\varepsilon = 0$ mm/s, $\delta_{Fe} = 0.28$, and octahedral (B site) $Fe^{2.5+}$ with $B = 46.0$ T, $2\varepsilon = 0$ mm/s, $\delta_{Fe} = 0.66$ mm/s. The latter sextet possesses a somewhat broader linewidth ($\Gamma \sim 0.5$ mm/s) because it is in fact composed of two B-site sextets with $B = 45.6$ T, $2\varepsilon = 0.18$ mm/s, $\delta_{Fe} = 0.66$ mm/s, and $B = 46.0$ T, $2\varepsilon = -0.05$ mm/s, $\delta_{Fe} = 0.66$ mm/s respectively, as a result of two different possible directions of the magnetic hyperfine field with respect to the local B-site EFG principal axes [112]. However, from the ferric A-site component, magnetite is readily recognized by MS at RT and for characterization purposes a two-sextet fitting is adequate. For ideal magnetite the sextet area ratio $S(B)/S(A)$ has to be 2:1. In practice, this ratio is somewhat lower and amounts to about 1.8:1 at RT. This is related to the Mössbauer fraction f of the $Fe^{2.5+}$ on the B sites being somewhat lower than that of Fe^{3+} on the A sites.

However, deviations from this ideal ratio are often observed. The main reason is that magnetite may partly be oxidized by replacing Fe^{2+} by Fe^{3+} and introducing vacancies. Because the electron hopping occurs in pairs, this oxidation does not result in another intermediate Fe valence on the octahedral sites, but in a decrease of the $Fe^{2.5+}$ component and the appearing of a B-site Fe^{3+} sextet for which the hyperfine parameters do not appreciably differ from those of the A-site sextet. Together with the introduced vacancies, a decrease of the $Fe^{2.5+}$ B-site sextet area and an apparent increase of that of the A-site sextet are observed. Therefore, one cannot speak anymore about one A-site and one B-site sextet—a mistake that is

Fig. 3.13 Typical spectrum
of magnetite at RT with outer
Fe^{3+} and inner $Fe^{2.5+}$ sextet

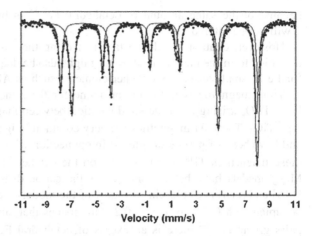

Velocity (mm/s)

Table 3.6 Representative hyperfine parameters of magnetite

Magnetite	T (K)	Spectrum	B_{av} (T)	δ_{Fe} (mm/s)
Pure Fe_3O_4	RT	Fe^{3+} A	49.0	0.28
		$Fe^{2.5+}$ B	45.9	0.66
	130	Fe^{3+} A	50.4	0.36
		$Fe^{2.5+}$ B	48.0	0.76
Oxidized $Fe_{3-x}O_4$	RT	Fe^{3+} A	49.0	0.28
		Fe^{3+} B	~50.0	0.36
		$Fe^{2.5+}$ B	45.9	0.66

often made—but instead should speak about one Fe^{3+} sextet and one $Fe^{2.5+}$ sextet (Table 3.6).

The observed value for the $S(Fe^{2.5+})/S(Fe^{3+})$ ratio of a magnetite phase can be used to determine its degree of oxidation. Oxidized magnetite has the general formula $Fe_{3-x}O_4$ with $0 < x < 0.33$. In that case one can expect the following structural formula, $\left(Fe^{3+}\right)_A[Fe^{2.5+}_{2(1-3x)}Fe^{3+}_{5x}\square_x]_BO_4$, where $\square$ stands for the vacancies and where an equal amount of octahedral Fe^{2+} and Fe^{3+} results in $Fe^{2.5+}$. Considering the ratio $R = S(Fe^{2.5+})/S(Fe^{3+})$ to be about 1.8 for pure magnetite at RT, one can write

$$R = \frac{1.8(1-3x)}{1+5x} \tag{3.5}$$

for oxidized magnetite, leading to

$$x = \frac{1.8-R}{5.4+5R} \tag{3.6}$$

On the other hand, it has been claimed that in the case of oxidation of magnetite, the vacancies might be present on both lattice sites [113, 114]. Anyway, because of the same amount of octahedral $Fe^{2.5+}$ in both cases, the formula for x will not be altered.

However, caution should be taken in using the equation because part of the deviation from the ideal ratio $R = 1.8$ might also be due to isomorphic substitution for Fe by small quantities of other elements such as Al and especially Ti.

Titanomagnetite is the common name for the minerals with general formula $Fe_{3-x}Ti_xO_4$ arising from the solid solution between magnetite and the ulvöspinel, $Fe_2^{2+}TiO_4$. These Ti-magnetites are very common in igneous rocks such as basalts and have been of particular interest in connection with magnetism of the earth. In these magnetites Ti^{4+} substitutes Fe on the octahedral sites creating more Fe^{2+}. Many models have been proposed for the cation distribution (e.g. Pearce et al. [115] and references therein). In most of the models the tetrahedral sites are fully occupied with Fe^{3+} up to $x = 0.2$. This means that, apart from the $Fe^{2+} - Fe3^+$ pairs giving $Fe^{2.5+}$ there is an excess of octahedral Fe^{2+}. This results in a two-sextet magnetite-like spectrum with an additional inner shoulder on the $Fe^{2.5+}$ sextet belonging to a Fe^{2+} sextet [116, 117]. However, natural samples which are chemically inhomogeneous and frequently non-stoichiometric show often more complex Mössbauer spectra.

Natural magnetite may also occur with a small particle morphology yielding a Mössbauer spectrum with asymmetrically shaped lines. Due to the increased overlap of the lines of both sextets in that case, it becomes difficult to determine the ratio $R = S(Fe^{2.5+})/S(Fe^{3+})$ accurately.

At low temperatures, the spectrum of pure magnetite is very complex (Fig. 3.14a) and may be described by at least five subspectra [118]. This is due to the 3d electron localization below the so-called Verwey transition at about 125 K leading to discrete Fe^{2+} and Fe^{3+} spectral contributions of the B sites. However, this transition temperature is lowered in the case of substitution or partial oxidation [119–121]. This is illustrated in Fig. 3.14b where oxidized magnetite ($Fe_{2.944}O_4$)

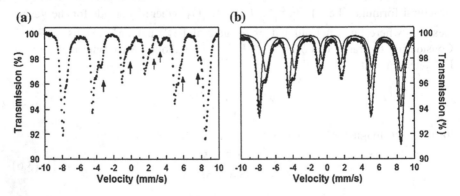

(a)   **(b)**

Fig. 3.14 Spectra below the Verwey transition: **a** spectrum of stoichiometric magnetite at 100 K with visible Fe^{2+} lines (indicated by arrows), and **b** spectrum of non-stoichiometric magnetite $Fe_{2.944}O_4$ at 100 K with the two typical Fe^{3+} and $Fe^{2.5+}$ sextets

still shows the typical $Fe^{2.5+}$ and Fe^{3+} sextets at 100 K which is below the normal Verwey transition for stoichiometric magnetite [35].

3.3.9 Maghemite

Maghemite, which is a fully oxidized form of magnetite, has the structural formula $\left(Fe^{3+}\right)_A[Fe^{3+}_{5/3}\square_{1/3}]_B O_4$ where $\square$ represents again the vacancies on the octahedral sites. The corresponding Mössbauer spectrum consists of a somewhat broad-lined Fe^{3+} sextet, which is in fact composed of two non-resolved sextets from Fe^{3+} in tetrahedral and octahedral sites, respectively (Fig. 3.15a). Only by using an external field, the hyperfine field and the isomer shift of both sextets could be accurately determined [125] (see Sect. 3.6.4). Hyperfine parameter values are given in Table 3.7.

Maghemite is commonly formed from oxidation of fine-course lithogenic magnetite, although its abundance in tropical and subtropical regions can also be explained by the conversion of for instance goethite through fires under reducing conditions. Also fine-particle magnetite produced by bacteria may lead to maghemite after spontaneous oxidation. In most of those cases, maghemite possesses a rather small-particle morphology leading to a superparamagnetic behavior. Maghemite can then appear in the Mössbauer spectrum as a doublet at RT. From the effective anisotropy constant of about 10 Jm^{-3} [123] a superparamagnetic doublet is expected at RT for particle sizes smaller than about 5 nm. This

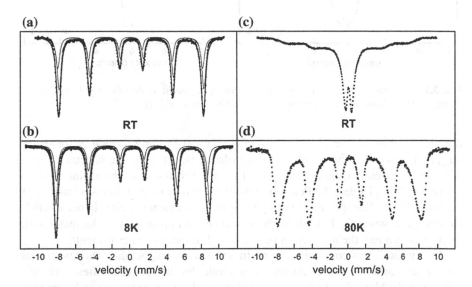

Fig. 3.15 Mössbauer spectra of maghemite: well-crystallized maghemite at RT (**a**) and at 8 K (**b**) fitted with A-site and B-site sextets according to parameters derived from external-field spectra; poorly crystallized maghemite at RT (**c**) and at 80 K (**d**) (after da Costa et al. [122])

Table 3.7 Representative hyperfine parameters of maghemite

Crystallinity	T (K)	Spectrum	B_{av} (T) or Δ (mm/s)	δ_{Fe} (mm/s)
Good	RT	S Fe^{3+} A	49.9	0.24
		S Fe^{3+} B	49.9	0.36
	80	S Fe^{3+} A	49.9	0.36
		S Fe^{3+} B	52.9	0.48
	4	S Fe^{3+} A	52.0	0.36
		S Fe^{3+} B	53.1	0.48
Medium	RT	S Fe^{3+} av	30–40	0.28
	80	S Fe^{3+} av	45–50	0.45
Poor	RT	D Fe^{3+} av	0.65–0.75	0.35
	80	S Fe^{3+} av	30–45	0.45
		D Fe^{3+} av	0.65–0.75	0.43

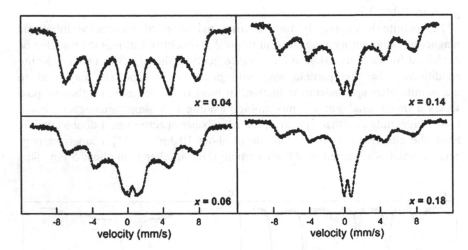

Fig. 3.16 Evolution of the 80 K spectra of poorly crystallized Al-substituted maghemite, γ-$Fe_{2-x}Al_xO_3$ with increasing Al content (after da Costa et al. [126])

doublet exhibits an average Δ in the range 0.65–0.75 mm/s, which is somewhat larger than that of other oxides or oxyhydroxides. At 80 K the spectra of such particles are usually magnetically split (Fig. 3.15) with average hyperfine fields of about 30–45 T (Table 3.7). Pedogenic maghemite occurs often in association with hematite [124]. The spectra of both phases possess a rather similar hyperfine field (in particular when the latter is in the WF state) and differ only in the quadrupole shift. This renders the distinction between both minerals rather difficult.

Al substitution is commonly found in soil-related maghemite. The spectra are somewhat similar to that of Al-free maghemite, but the sextet is often accompanied by a doublet (Fig. 3.16), the contribution of which decreases at lower temperatures. This typical superparamagnetic behavior is a consequence of a combined effect of particle size and diamagnetic substitution [125, 126].

Titanomaghemites originating from oxidation of lithogenic titanomagnetite show similar features in their Mössbauer spectra, i.e. asymmetrically broadened sextet lines accompanied by a central doublet [127, 128]. These maghemites seem to have vacancies on both sites [127, 129].

3.3.10 Green Rust Related Minerals

Green rusts are mixed-valence iron hydroxysalts that have been extensively studied for understanding the corrosion behavior of iron based materials and steels. They are obtained by oxidizing metallic iron or $Fe(OH)_2$ and are intermediate compounds comprising ferrous and ferric ions. Several types have been synthesized incorporating various anions such as CO_3^{2-}, Cl^- or SO_4^{2-} and were studied by MS [130–133]. They all belong to the double layered hydroxide family (HDL) where in the most stable one, i.e. with CO_3^{2-}, positive charged brucite-like layers $\left(Fe_4^{2+}Fe_2^{3+}(OH)_{12}\right)^{2+}$ alternate with $(CO_3.3H_2O)^{2-}$ interlayers, possessing a ferric molar fraction $x = Fe^{3+}/Fe_{total} = 1/3$. Green rusts dissolve usually by oxidation whereas a ferric oxyhydroxide precipitates such as ferrihydrite and/or goethite. However, rapid oxidation under alkaline conditions (in situ deprotonation) keeps the green rust structure essentially unchanged, leading finally to so-called "ferric green rust" which has in fact an orange color [134].

A green rust related mineral in soils was first discovered in a so-called gleysol at Fougères (Brittany, France) and has therefore been named fougèrite [135]. The Mössbauer spectrum, usually measured at lower temperatures (78 K) to avoid oxidation, is typical for green rust, consisting of one or two Fe^{2+} doublets with $\delta \approx 1.25$ mm/s and with $\Delta \approx 2.60$–2.90 mm/s and a Fe^{3+} doublet with $\delta \approx 0.48$ mm/s and with $\Delta \approx 0.50$–0.70 mm/s. The spectrum from the first measured sample is shown in Fig. 3.17a. For different fougèrite samples, the ferric molar fraction x turned out to be between 1/3 and 2/3 [136]. A study in depth with a miniaturized Mössbauer spectrometer showed that fougèrite is more ferric in the upper horizons and variations are consistent with the fluctuations in the water table and thus with aerobic and anaerobic conditions [137]. Those in situ experiments yield spectra and thus hyperfine values close to RT (283 K). They are $\delta \approx 1.05$–1.08 mm/s and $\Delta \approx 2.7$ mm/s, $\delta \approx 0.75$–1.00 mm/s and $\Delta \approx 2.2$–2.7 mm/s both for Fe^{2+}, and $\delta \approx 0.21$–0.30 mm/s and $\Delta \approx 0.6$–0.8 mm/s for Fe^{3+}.

In contrast to the permanently waterlogged soils of a continental aquifer with green rust related minerals with values of x within the (1/3, 2/3) range as firstly extracted in Fougères, similar minerals being still more ferric with x values within the (2/3, 1) range occurs in gley from the schorre of a maritime marsh, as was firstly extracted in Trébeurden (Brittany, France) [138]. The spectrum of a typical Trébeurden sample with $x \approx 0.75$ measured at 78 K and fitted with 3 doublets is shown in Fig. 3.17c. Similar spectra were recorded from samples extracted from

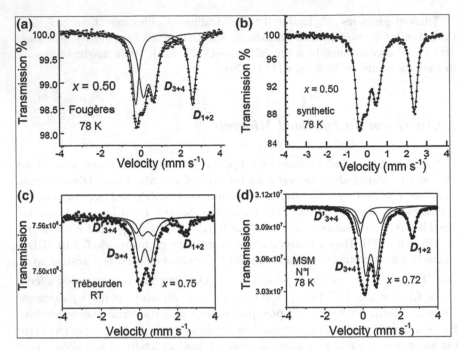

Fig. 3.17 Mössbauer spectrum of a Fougèrite sample at 78 K (**a**) compared with the spectrum of a synthetic oxidized green rust sample with the same Fe^{3+} molar fraction x (**b**) Mössbauer spectrum of a green rust related sample from Trébeurden (**c**) and the spectrum of a similar sample from the Mont Saint Michel (**d**) (adapted from Ref. [307])

the bay of Mont Saint Michel (Normandy, France) [138] as can be seen from the example shown in Fig. 3.17d.

Structural studies on synthetic samples [133] suggest that long ranger order appears for $x = 1/3$, $x = 2/3$ and $x = 1$ which is confirmed by magnetic measurements showing a ferrimagnetic behavior for $x = 1/3$, $x = 2/3$ with Néel temperatures of 5 and 20 K respectively and being ferromagnetic for $x = 1$ with a Curie temperature around 80 K [139]. From these findings together with the appearance of typical doublets in the Mössbauer spectra Génin et al. [138] suggest that the intermediate samples consist in fact of topotaxically mixed domains of the ordered definite compounds. Therefore the natural green rust related minerals must be considered as a mixture of three basic minerals:

$$Fe_4^{2+}Fe_2^{3+}(OH)_{12}CO_3.3H_2O \ (x = 1/3);$$
$$Fe_2^{2+}Fe_4^{3+}(OH)_{10}O_2CO3.3H_2O \ (x = 2/3) \text{ and}$$
$$Fe_6^{3+}(OH)_8O_4CO_3.3H_2O \ (x = 1).$$

They suggest to redefine the first one as fougérite, whereas in the case of $x = 2/3$ and $x = 1$, the new mineral names "trébeurdenite" and "mössbauerite" are respectively proposed [138].

3.4 Sulfides, Sulfates and Carbonates

3.4.1 Sulfides

The most abundant and widespread sulfides are pyrite and marcasite. They both have the formula FeS_2 but the former crystallizes in a cubic structure whereas the latter is orthorhombic. They both contain divalent iron which is in a non-magnetic low spin state (Fe^{II}). Consequently the Mössbauer spectra (Fig. 3.18.) consist of a doublet with low isomer shifts and moderate quadrupole splittings [130].

Although the hyperfine parameters of natural marcasites and pyrites vary somewhat from sample to sample, probably due to chemical impurities, they can be unambiguously distinguished from their spectra. However, more problematic is to recognize the presence of these minerals among Fe^{3+}-containing clay minerals because of the similar range of hyperfine parameter values. Other marcasites exist with S replaced by Se or Te. They show hyperfine parameters comparable to those of the sulfides (Table 3.8). Löllingites ($FeAs_2$ and $FeSb_2$) and mispickel FeAsS, on the other hand, exhibit larger quadrupole splittings. Both the presence of formal Fe^{IV} and the very distorted structure are responsible for the high quadrupole splitting observed.

Another class of iron sulfides is $Fe_{1-x}S$ which crystallize in a hexagonal structure. Stoichiometric FeS, called troilite, is an antiferromagnet with $T_N = 595$ K. At RT the spectrum shows a hyperfine field of about 30.8 T and a positive quadrupole shift of $2\varepsilon \approx 0.3$ mm/s [144].

Most of the monosulfides, however, are not stoichiometric and natural samples consist often of a mixture of compounds with compositions between troilite, FeS and FeS_2, namely $Fe_{11}S_{12}$ ($x = 0.083$), $Fe_{10}S_{11}$ ($x = 0.091$), and Fe_9S_{10}

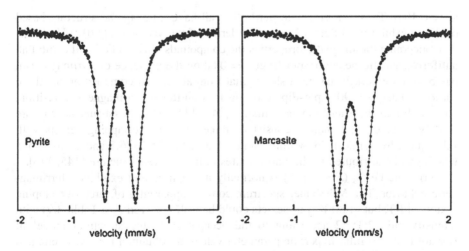

Fig. 3.18 RT spectrum of pyrite and marcasite (adapted from Evans et al. [130])

Table 3.8 RT hyperfine parameters of FeS_2 and related minerals

Mineral	Formula	δ_{Fe} (mm/s)	Δ (mm/s)
Pyrite	FeS_2	0.31	0.61
Marcasite	FeS_2	0.27	0.50
Marcasite (Se)	$FeSe_2$	0.40	0.58
Marcasite (Te)	$FeTe_2$	0.47	0.50
Löllingite (Sb)	$FeSb_2$	0.45	1.28
Löllingite (As)	$FeAs_2$	0.31	1.68
Mispickel	$FeAsS$	0.26	2.10

Table 3.9 Representative hyperfine parameters at RT of various sulfides

Mineral	Formula $Fe_{1-x}S$	B (T)	2ε or Δ (mm/s)	δ_{Fe} (mm/s)
Triolite	FeS	30–32	−0.3	0.7–0.9
Intermediate	$Fe_{11}S_{12}$; $Fe_{10}S_{11}$ Fe_9S_{10}	22.0–23.5	0.05	0.55
pyrrhotite	($x = 0.83$; 0.091; 0.100)	25.5–26.5	0.05	0.55
		27.5–31.5	0.10	0.55
Pyrrhotite	Fe_7S_8 ($x = 0.125$)	22.9	0.08	0.77
monoclinic		26.7	0.03	0.79
		31.1	0.15	0.79
		34.5	−0.09	0.81
Smythite	Fe_9S_{11} ($x = 0.182$)	22.4	0.33	0.58
		25.9	0.04	0.58
		30.3	0.17	0.47
Greigite	Fe_3S_4 ($x = 0.250$)	31.3	0.0	0.17
		31.3	−0.08	0.42
Mackinawite	$Fe_{1.01}S - Fe_{1.07}S$		(0.02)	(0)
Pentlandite	Fe_9S_8		0.36	0.30–0.37
			0.60–0.68	0

($x = 0.100$), These intermediate pyrrhotites all have a hexagonal structure based on NiAs exhibiting different kinds of ordering of the vacancies [110, 141, 142].

Concerning the magnetic properties the compounds close to FeS still exhibit an antiferromagnetic behavior, but from $x = 0.08$ on the presence of ferrimagnetism has been observed. It has been shown that from about that composition on, these sulfides undergo a 90° spin-flip resulting in a antiferro-ferrimagnetic transition, comparable to the transition in hematite [143, 144]. The Mössbauer spectra are usually complex and mainly consist of three to four overlapping sextets with subsequent hyperfine fields within ranges 22.0–23.5, 25.5–26.5 and 27.5–31.5 T which can be attributed to the various sites in the ordered structure [145, 146].

Pyrrhotite Fe_7S_8 ($x = 0.125$) is normally monoclinic and exhibits a ferrimagnetic behavior. The Mössbauer spectrum consists apparently of three overlapping sextets at RT, but at 80 K four sextets could be well resolved (see Table 3.9) with intensity ratio 2:1:2:2 according to the occupation in the structural model of Bertaut [139]. Similar hyperfine parameter values are obtained for a synthetic and a natural sample by Jeandey et al. [147].

Fig. 3.19 RT spectrum of a natural greigite-smythite sample

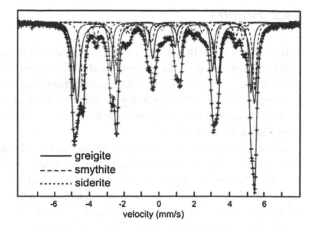

    greigite
    smythite
    siderite

Smythite is a hexagonal iron sulfide with composition close to Fe_9S_{11} ($x = 0.182$). There is still considerable doubt about the exact composition and whether it should be included as a separate phase in the iron-sulfur system. The Mössbauer spectra taken for some natural samples, containing other sulfides as well, reveal more or less three sextets [148] for the smythite phase, which are quite similar to three of the four sextets of pyrrhotite. Hoffmann et al. [149] also found three to four sextets but the one with the highest hyperfine field and nearly zero quadruple shift might be a sextet of greigite because the samples were from the same origin.

Greigite (Fe_3S_4, $x = 0.250$) is the sulfide analogue of magnetite and crystallizes in the cubic spinel structure. The RT spectrum exhibits an asymmetric sextet (Fig. 3.19) which is composed of two sextets: one arising from the A sites and one from the B sites with nearly equal hyperfine fields but different isomer shifts [148, 150] (Table 3.9). In contrast to magnetite, greigite does not show a Verwey transition [148, 151]. Because at higher temperature greigite transforms to pyrrhotite and FeS_2 the magnetic transition temperature cannot be determined, but lies above 480 K [148]. However, the magnetic properties are strongly dependent on the grain size [152].

Mackinawite, $Fe_{1+x}S$, with ($x = 0.01–0.07$) is a tetragonal iron sulfide with excess of iron. Morice et al. [153] reported a complex Mössbauer spectrum consisting of at least three sextets with hyperfine fields 29.8, 26.2 and 22.8 T and small quadrupole shifts of about 0.09, 0.06 and 0.09 mm/s respectively. On the other hand, only a singlet spectrum, even down to 4 K, has been observed by Vaughan and Ridout [154]. Probably the concentration of Co and Ni found to be present in the involved natural samples is decisive for the different magnetic behavior.

Pentlandite, $(Fe,Ni,Co)_9S_8$, has a face-centered cubic structure with iron and other metal atoms such as Ni and Co distributed among tetrahedral and octahedral sites. At RT the Mössbauer spectrum appears as an asymmetric doublet consisting of a quadrupole doublet with $\delta_{Fe} = 0.36$ and $\Delta = 0.30–0.37$ mm/s and an additional singlet with $\delta_{Fe} \approx 0.6$ mm/s, which is responsible for the asymmetry [155, 156].

3.4.2 Sulfates

Sulfates are known in nature as a series of minerals that are formed by oxidation of sulfides. Depending on the environment of formation they possess a different stage of hydration. Iron sulfates often accompany coal and are indicative of the weathering of the latter. Recently, a renewed interest for these minerals has been evoked in connection with the mineralogy on Mars, prospected by the Mars Exploration Rovers (MERs).

In recent years, the structural classification of the sulfates has been well established [157]. Sulfates consist predominantly of a polymerization of MO_6 octahedra, where M represents divalent or trivalent cations such as Fe^{2+}, Fe^{3+} and Mg^{2+}, Mn^{2+}, Zn^{2+}, Al^{3+}, etc.) and TO_4 tetrahedra where T stands for hexavalent or pentavalent cations such as S^{6+} and Mo^{6+}, P^{5+}, V^{5+}, etc.

The most common iron sulfate is szomolnokite, $FeSO_4.H_2O$. Other sulfates are the ferrous species rozenite, $FeSO_4.4H_2O$, and melanterite, $FeSO_4.7H_2O$, and the ferric species kornelite, $Fe_2(SO_4)_3.7H_2O$, and coquimbite $Fe_2(SO_4)_3.9H_2O$. Sulfates of mixed valence between ferrous melanterite and ferric kornelite are known as römerite. Another well-known mixed valence sulfate group is voltaite with formula $K_2Fe_5^{2+}Fe_3^{3+}Al(SO_4)_{12}.18H_2O$, but containing a large variety of substituting elements. A diversity of anhydrated sulfate minerals exists with general formula $MFe_3(SO_4)_3(OH)_6$ where $M = Na^+$, K^+, ... being catalogued as jarosites.

Distortions of polyhedra are a common phenomenon in sulfates and result in different stereographic environments. Moreover, natural samples contain a diversity of cations other than iron, which also influences the local environment of the latter. Therefore, the Mössbauer spectra of iron sulfates usually exhibit one or more predominant doublets and some additional smaller doublets. Such spectra can either be fitted with discrete doublets or by quadrupole distributions (Fig. 3.20).

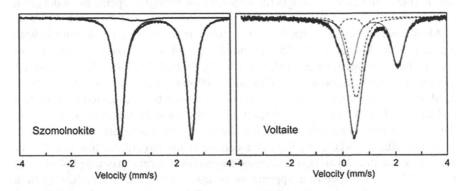

Fig. 3.20 RT spectra of some sulfates: szomolnokite (adapted from Van Alboom et al. 2009 [162]) and voltaite (adapted from Ertl et al. [161])

Table 3.10 Representative hyperfine parameters at RT of some iron sulfates

Mineral	Formula	Fe site	δ_{Fe} (mm/s)	Δ (mm/s)
Szomolnokite	$FeSO_4.H_2O$	Fe^{2+}oct	1.18–1.27	2.67–3.07
		(Fe^{2+} tetr)	0.23	0.69
		(Fe^{3+} oct)	0.55	0.38
Rozenite	$FeSO_4.4H_2O$	Fe^{2+} oct	1.27	3.33
		(Fe^{2+} oct)	0.37	1.15
		(Fe^{2+} tetr)	0.20	0.48
Melanterite	$FeSO_4.7H_2O$	Fe^{2+} oct	1.25–1.27	2.7–3.5
Kornelite	$Fe_2(SO_4)_3.7H_2O$	Fe^{3+} oct	0.47	0.45
		Fe^{2+} oct	1.18–1.32	1.57–1.62
		(Fe^{2+} tetr)	0.21	0.67
Coquimbite	$Fe_2(SO_4)_3.9H_2O$	Fe^{3+} oct	0.46–0.48	0.33–0.36
		(Fe^{3+} tetr)	0.1	0.67
Römerite	$FeSO_4.Fe_2(SO_4)_3.14H_2O$	Fe^{2+} oct	1.27–1.30	3.31
		Fe^{2+} oct	1.25–1.28	2.71–2.75
		Fe^{3+} oct	0.39–0.43	0.35–0.39
		(Fe^{3+} tetr)	0.1	0.66–0.84
Voltaite	$K_2Fe_8Al(SO_4)_{12}.18\ H_2O$	Fe^{2+} oct-M2	1.20	1.56–1.62
		Fe^{3+} oct-M1	0.50	0.13–0.18
		Fe^{3+} oct-M2	0.36	1.06–1.18
Jarosite	$KFe_3(SO_4)_3(OH)_6$	Fe^{3+} oct	0.38	1.23

Table 3.10 summarizes representative hyperfine parameters for a number of sulfate minerals. The data are taken from various authors as collected by Stevens et al. [158] and from others [159–162].

Apparently, four kinds of doublets may be encountered in sulfates: one or two Fe^{2+} doublets with hyperfine parameters in the range $\delta_{Fe} = 1.1$–1.3 mm/s and $\Delta = 2.6$–3.3 mm/s; one Fe^{2+} doublet with $\delta_{Fe} \approx 0.2$ and $\Delta = 0.65$–0.70 mm/s; one Fe^{3+} doublet with $\delta_{Fe} = 0.3$–0.5 and $\Delta = 0.4$–0.5 mm/s and one Fe^{3+} doublet with $\delta_{Fe} = 0$–0.1 and $\Delta = 0.6$–0.8 mm/s.

3.4.3 Carbonates

Siderite, $FeCO_3$, has a rhombohedral structure and contains divalent iron in the high spin state. The magnetic structure is antiferromagnetic with a low Néel temperature of 38 K. At room temperature as well as at 80 K a doublet is observed with hyperfine parameters $\delta_{Fe} = 1.2$ mm/s, $\Delta = 1.79$ mm/s at RT and $\delta_{Fe} = 1.36$ mm/s, $\Delta = 2.04$ mm/s at 80 K. In contrast to X-ray diffraction the Mössbauer spectra are not distinctively influenced by substitution of Mg and Mn for Fe [163], although there are indications that Ca^{2+} provokes an asymmetry in the lineshape, pointing to a second doublet with larger quadrupole splitting [66]. The siderite spectra are mostly asymmetric, namely, one line is deeper than the other one, whereas the widths are the

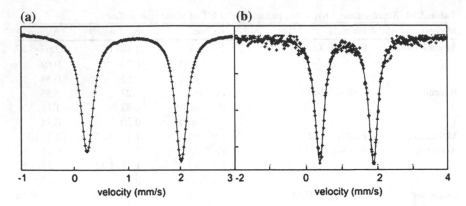

Fig. 3.21 RT Mössbauer spectra of siderite (**a**) and ankerite under the magic angle (**b**) (after De Grave and Vochten [166])

same (Fig. 3.21a). This feature was firstly explained as due to the so-called Goldanskii-Karayagin effect [164] being an anisotropy in the Mössbauer fraction. However, measurements under the so-called "magic angle", i.e. the absorber placed under an angle of 54 degrees with the gamma ray direction [165], yielded nearly equal intensities for both doublet lines pointing to texture effects as the reason for the asymmetry. At very low temperatures the magnetic transition is not sharply displayed in the Mössbauer spectra because of relaxation effects [166].

Ankerite, $CaFe(CO_3)_2$, has a similar rhombohedral structure as siderite but with calcium and iron in alternate positions. In natural ankerites the composition deviates always from the ideal one through the presence of considerable amounts of Mg and Mn. The magnetic transition temperature is much lower than for siderite and in a range of more than 10 K below the transition complicated spectra are observed due to spin–lattice relaxation [167, 168]. Anyway, a long-range order was not observed down to 1.7 K [169]. The spectra at RT and 80 K consist similarly to siderite of a single doublet (Fig. 3.21b), which is mostly asymmetric, most probably also as a result of preferential orientation of the crystallites in the absorber. The isomer shift is about the same as for siderite, but, the quadrupole splitting is somewhat smaller and amounts to 1.44–1.48 mm/s at RT, depending on the composition [170].

3.5 Silicates

3.5.1 Introduction

Silicate minerals are all structurally derived from the tetrahedral bonding of silicon to oxygen. For a relatively small group of these minerals, the structure consists of discrete orthosilicate anions SiO_4^{4-}, but, in the vast majority, the SiO_4 tetrahedra are joined by oxygen sharing into chains or sheets. Apart from the silicon, they all further contain monovalent, divalent and trivalent metal ions of which a substantial amount of Fe^{2+} and Fe^{3+} can be present. Therefore, Mössbauer studies in silicate mineralogy have already been started in the early days of the application of the Mössbauer effect. The power of the Mössbauer technique resides in the ability to determine qualitatively and quantitatively the iron in the different lattice sites with their specific valence and the distortion of their environments. This has been well demonstrated in the numerous publications of G.M. Bancroft, Roger.G. Burns, M. Darby Dyar, Stefan S. Haffner, Georg Amthauer, Victor A. Drits... and their respective co-workers. Dyar et al. [150] have composed a comprehensive list of the hyperfine parameters of a large variety of silicates.

Commonly, silicates are only magnetic at very low temperatures and hence their MS recorded at RT and at 80 K generally consist of doublets. Moreover, the divalent and trivalent iron cations are in the high spin state yielding comparable hyperfine parameters for a given Fe^{2+} or Fe^{3+} site in various minerals. This means that distinct doublets due to a particular valence, if present in a given spectrum, may strongly overlap and are sometimes difficult to resolve. Furthermore, absorbers from silicates that are crystallized in chains or sheets may be subject to texture effects resulting in a different intensity of the two lines of a given doublet component. All this makes MS not so powerful as far as direct identification of silicates is concerned and one has to rely on the results of other techniques such as X-ray diffraction. Nevertheless, a better resolution for the Fe^{2+} doublets in particular can often be obtained from measurements at several temperatures because of the divergent variation of the quadrupole splitting with temperature for the different ferrous sites. Further, asymmetry in the doublets can be avoided by a suitable absorber preparation eliminating texture effects to a large extent. If necessary, measurements under the so-called "magic angle", i.e. with the absorber at 54 degrees with respect to the direction of the γ-ray, may also be helpful in that respect.

From the spectral data of silicates some general rules can be put forward:

- the Fe^{2+} and Fe^{3+} oxidation states are usually easily distinguished
- the isomer shift for Fe^{2+} ions in silicates depends both on coordination number and symmetry in the following order: δ (square planar) $\ll \delta$ (tetrahedral) $\ll \delta$ (octahedral) $\ll \delta$ (dodecahedral); δ shows a linear relation with bond distance and bond strength
- the quadrupole splitting of octahedral Fe^{2+} is very sensitive to site symmetry and generally decreases with increasing distortion, due to the lattice contribution being opposite to the dominant valence contribution.

According to the way of stacking the SiO_4 tetrahedra, the silicates are usually divided into different classes (Fig. 3.22): nesosilicates (single tetrahedra), sorosilicates (double tetrahedra), cyclosilicates (tetrahedra joined to rings), inosilicates (tetrahedra joined to single or double chains), phyllosilicates (sheets of tetrahedra) and tektosilicates (three dimensional stacking of tetrahedra). This classification

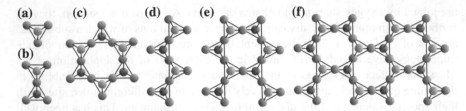

Fig. 3.22 SiO_4 stacking in silicates (only O atoms are shown): **a** Nesosilicates, **b** sorosilicates, **c** cyclosilicates, **d, e** inosilicates, **f** phyllosilicates

will henceforth be adopted for further discussion of the specific silicate minerals.

3.5.2 Nesosilicates

Nesosilicates is the class of silicate minerals in which the structure is built up by independent tetrahedra of SiO_4. This class contains some mineral groups like the olivines, the silicate garnets, the epidote group and some separate minerals.

3.5.2.1 Olivines

Olivines consist of independent SiO_4 tetrahedra surrounded by six-coordinated metal cations in two distinct sites M1 and M2. Mössbauer studies of the forsterite-fayalite (Mg_2SiO_4-Fe_2SiO_4) and fayalite-tephroite (Fe_2SiO_4–Mn_2SiO_4) series all revealed at RT a simple doublet with a quadrupole splitting Δ in the range 2.80–3.02 mm/s and an isomer shift $\delta_{Fe} = 1.16$–1.18 mm/s [171]. The doublet is often asymmetric with different line widths and depths indicating the presence of two doublets, one arising from M1 and the other one from M2 sites [172]. However, these doublets overlap to such a large extent that they can hardly be separated in the spectrum analysis. At high temperatures (1000 K) a separation of both doublets can be possible. Fayalite itself orders antiferromagnetically below 66 K, and becomes towards forsterite at a lower composition-dependent temperature a canted antiferromagnet [173–175]. At very low temperatures two distinct magnetic patterns are observed with B of 12.0 and 32.3 T [176, 177].

A mixed-valence iron olivine is laihunite (ferrifayalite) with ideal composition $Fe^{2+}Fe^{3+}(SiO_4)_2$. In this olivine species Fe^{2+} occupies M1 sites, alternated with vacancies, whereas Fe^{3+} occupies the M2 sites. Due to the difference in oxidation state two well resolvable doublets are observed with $\delta_{Fe} = 0.39$ mm/s and $\Delta = 0.91$ mm/s for Fe^{3+} and $\delta_{Fe} = 1.13$ mm/s and $\Delta = 2.75$ mm/s for Fe^{2+} [178]. Ferrifayalite often occurs as an intergrowth of fayalite and laihunite [179] (Table 3.11, 3.12, and 3.13).

Table 3.11 Representative hyperfine parameters at RT for some olivines

Mineral	Formula	Fe site	δ_{Fe} mm/s	Δ (mm/s)
Fayalite	$(Fe^{2+})_2SiO_4$	Fe^{2+} M1,M2	1.17	2.81
Fayalite-forsterite-tephroite	$(Fe^{2+},Mg,Mn)_2SiO_4$	Fe^{2+} M1,M2	1.16–1.18	2.80–3.02
Laihunite	$(Fe^{2+}, Fe^{3+},)_2SiO_4$	Fe^{2+} M1	1.13–1.16	2.75–2.83
		Fe^{3+} M2	0.37–0.43	0.85–0.91

Table 3.12 Representative hyperfine parameters at RT for some garnets

Mineral	Formula	T (K)	Fe site	δ_{Fe} (mm/s)	Δ (mm/s)
Almandine	$(Fe^{2+})_3Al_2(SiO_4)_3$	RT	Fe^{2+} dodec	1.28	3.51
		77		1.43	3.66
Pyrope-almandine	$(Mg,Fe^{2+})_3Al_2(SiO_4)_3$	RT	Fe^{2+} dodec	1.20–1.30	3.47–3.70
		77		1.33–1.44	3.47–3.70
Spessartine	$(Mn,Fe^{2+})_3Al_2(SiO_4)_3$	RT	Fe^{2+} dodec	1.28	3.52
		77		1.42	3.64
Andradite	$Ca_3(Al,Fe^{3+})_2(SiO_4)_3$	RT	Fe^{3+} oct	0.41	0.5–0.6
		77		0.50	0.5–0.6
Almandine-grossular	$(Ca,Fe^{2+})_3(Al,Fe^{3+})_2(SiO_4)_3$	RT	Fe^{2+} dodec	1.20–1.30	3.47–3.70
		77	Fe^{3+} oct	0.35–0.45	0.29–0.75
			Fe^{2+} dodec	1.33–1.44	3.47–3.70
			Fe^{3+} oct	0.42–0.52	0.26–0.64
Titano-andratite	$Ca_3(Fe^{3+},Ti,Si)_2(SiO_4)_3$	RT	Fe^{3+} tetr	0.20	1.15
		77	Fe^{3+} oct	0.40	0.75
			Fe^{3+} tetr	0.30	1.15
			Fe^{3+} oct	0.50	0.75

Table 3.13 Representative hyperfine parameters at RT for some other nesosilicates

Mineral	Formula	Fe site	δ_{Fe} (mm/s)	Δ (mm/s)
Epidote	$Ca_2(Fe^{3+},Al)_3O(OH)(Si_2O_7)(SiO_4)$	Fe^{3+} M3	0.34–0.36	1.9–2.1
Piemontite	$Ca_2(Fe^{3+},Mn,Al)_3O(OH)(Si_2O_7)(SiO_4)$	Fe^{3+} M3	0.34–0.36	2.05–2.10
		Fe^{3+} M1	0.29–0.33	0.9–1.1
Allanite	$(Ca,RE)_2(Fe^{3+},Al)_3O(OH)(Si_2O_7)(SiO_4)$	Fe^{3+} M3	0.37	1.97
		Fe^{3+} M1	0.29	1.33
		Fe^{2+} M3	1.07	1.66
		Fe^{2+} M1	1.24	1.93
Staurolite	$(Fe,Mg,Zn)_2Al_9(Si,Al)_4O_{20}(OH)_4$	Fe^{2+} tetr	0.98 (av)	2.45 (av)
		Fe^{2+} tetr	0.98 (av)	2.10 (av)
		Fe^{2+} tetr	0.97 (av)	1.6 (av)
		Fe^{3+} tetr	0.15 (av)	0.7 (av)

3.5.2.2 Garnets

The most important iron and silicon containing garnet minerals are the pyrope-almandine series $(Mg,Fe^{2+})_3Al_2(SiO_4)_3$ and andradite $Ca_3(Al,Fe^{3+})_2(SiO_4)_3$. When Mg is replaced by Mn the garnet is called spessartine $(Mn,Fe^{2+})_3Al_2(SiO_4)_3$. The garnet structures contain SiO_4 tetrahedra with commonly trivalent cations in 6-coordination and divalent cations in 8-coordination.

Members of the pyrope-almandine series exhibit a doublet at RT with an isomer shift $\delta_{Fe} = 1.2$–1.4 mm/s and a very large quadrupole splitting Δ in the range 3.47–3.70 mm/s [180]. An additional doublet of Fe^{3+} due to the six-fold coordinated Al site is observed with $\delta_{Fe} = 0.35$–0.45 mm/s and $\Delta = 0.29$–0.75 mm/s when going from almandine towards andradite. These intermediate minerals are often called grossular. Pure almandine orders magnetically around 10 and at 4 K an Fe^{2+} octet is observed with a hyperfine field of 23–24. 5 T [181]. The typical large quadrupole splitting in garnets is shown in the spectrum of spessartine in Fig. 3.23.

Andradite possesses at RT the typical Fe^{3+} hyperfine parameters, namely $\delta_{Fe} = 0.39$–0.41 mm/s and $\Delta = 0.5$–0.6 mm/s [180]. The latter is, however, strongly dependent on the composition. It orders magnetically below 11 and at 4 K the hyperfine field amounts to 51.9 T [182]. Actually, the spectrum is asymmetric and can be analyzed with two sextets having different quadrupole shifts $2\varepsilon = 0.32$ and -0.16 mm/s, due to a complex spin structure with different orientations with respect to the EFG principal axis. In some garnets such as titanium andradite, $Ca_3(Fe,Ti,Si)_5O_{12}$, Fe^{3+} also occupies partly the tetrahedral Si sites, exhibiting very low isomer shifts and rather high quadrupole splittings [183].

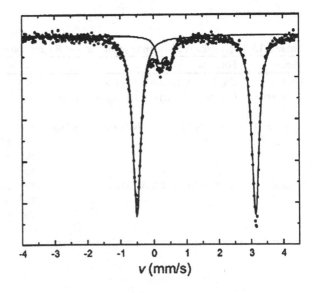

Fig. 3.23 RT spectrum of spessartine showing the predominant Fe^{2+} doublet with the typical large quadrupole splitting for garnets. The spectrum shows an additional small doublet from octahedral Fe^{3+} (after Eeckhout et al. [308])

3.5.2.3 Epidote Group

Epidote has the chemical formula $Ca_2(Fe,Al)_3O(OH)(Si_2O_7)(SiO_4)$. The structure is mainly monoclinic and consists of Al oxide and hydroxide chains linked by single-tetrahedra SiO_4 and double-tetrahedra Si_2O_7. Because of the presence of the latter the epidotes can also be placed under the class of sorosilicates. The Fe^{3+} is mainly located in a very irregular polyhedron called M3 site and hence, unusually large quadrupole splittings in the range 1.9–2.1 mm/s are observed for Fe^{3+}, whereas the isomer shifts 0.34–0.36 mm/s clearly reflects the trivalent state [184–187]. Epidotes have a large variety in chemical composition and trivalent iron has also been found to be present in the more regular octahedral M1 and M2 sites. Also, Ca can partly be substituted by trivalent rare earth elements (RE), such as Ce^{3+}, with simultaneously Fe^{2+} substituting for Al^{3+} to make up the charge balance. Such minerals are called allanites. In the latter some part of Fe^{3+} occupies the M1 sites, whereas also Fe^{2+} is present on both sites yielding four doublets [185].

3.5.2.4 Staurolite

Staurolite belongs to the so-called subsaturate group of nesosilicates because it has in its structure additional oxygen ions to those forming the SiO_4 tetrahedra. The chemical composition is rather complex and can be written as $(Fe,Mg,Zn)_2Al_9$ $(Si,Al)_4O_{20}(OH)_4$. It crystallizes in a monoclinic structure with Fe^{2+} located in different tetrahedral coordinations, having average hyperfine parameters: $\delta_{Fe} \approx 0.98$ mm/s and $\Delta \approx 2.3$ mm/s [184, 188]. Fe^{3+} was found to be present with $\delta_{Fe} \approx 0.20$ mm/s and $\Delta \approx 0.60$ mm/s [188]. However, a more thorough analysis revealed the appearance of up to six doublets in the spectra [189].

3.5.3 Sorosilicates

Sorosilicates consist of double tetrahedral Si_2O_7 units and are relatively uncommon. Representative minerals in this class are the melilites and ilvaite.

3.5.3.1 Melilites

Melilites with chemical formula $(Ca,Na)_2(Mg,Fe^{2+},Fe^{3+},Zn,Al)(Si,Al)_2O_7$ possess a tetragonal structure. It consists of tetrahedral layers with cations (Ca, Na) halfway between adjacent layers. The layer modulation is incommensurate (Seifert et al. 1987). Well-known examples are gehlenite ($Ca_2Al(AlSiO_7)$), åkermanite ($Ca_2Mg(Si_2O_7)$) and hardystonite ($Ca_2Zn(Si_2O_7)$). Calcium occupies a dodecahedral site, whereas the other cations are distributed among two kinds of tetrahedral sites T1 and T2, the latter being the normal silicon site. Fe^{3+} enters the melilites mainly the T1 site. The hyperfine parameters are $\delta_{Fe} \approx 0.2$ mm/s, $\Delta \approx 0.9$ mm/s for Fe^{3+} and $\delta_{Fe} \approx 1.00$ mm/s, $\Delta \approx 2.3$ mm/s for Fe^{2+}. Although it was suggested that a small amount of Fe^{3+} enters also the T2, it is finally found in ferric melilites that Fe^{3+} actually occupies two differently distorted T1 sites with $\delta_{Fe} \approx 0.19$ mm/s, $\Delta = 0.75$ and 1.04 mm/s [190].

3.5.3.2 Ilvaite

Ilvaite with chemical formula $CaFe_2^{2+}Fe^{3+}(Si_2O_7)O(OH)$ has an orthorhombic structure. Technically spoken, it can be considered as a subsaturated sorosilicate in view of the additional O^{2-} and OH^- groups. The structure can be described as ribbons of double rows of edge-shared octahedra consisting of five oxygens and one hydroxyl anion and designated as M1(8d) sites. Above and below the ribbons are octahedral M2(4c) sites with solely oxygen surrounding. It was later found that low-impurity ilvaite possesses a slightly monoclinic structure with two M1 sites, M11 and M12, and becoming orthorhombic with increasing substitution (e.g. by Mn^{2+}) and temperature [191]. Iron enters all the octahedral sites. The Mössbauer spectra show globally three absorption peaks (Fig. 3.24) which can essentially be analyzed with 3 doublets, one for Fe^{2+} in M11 site, one for Fe^{3+} in M12 site and one for Fe^{2+} in M2 [192, 193]. Representative hyperfine parameters are listed in Table 3.14.

On the other hand, two additional doublets are sometimes introduced to account for Fe^{2+}–Fe^{3+} electronic relaxations [194, 195], in particular for Mn-rich ilvaites [196]. The appearance of only two doublets in Mn-rich ilvaites at higher temperatures indicates a collapse to $Fe^{2.5+}$ in the M1 site, so that in the intermediate temperature region a relaxation process occurs which significantly broadens the central absorption peak [197].

Fig. 3.24 Typical RT spectrum of ilvaite (adapted from Dotson and Evans [193])

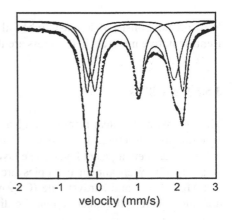

velocity (mm/s)

Table 3.14 Representative hyperfine parameters at RT for some sorosilicates

Mineral	Formula	Fe site	δ_{Fe} (mm/s)	Δ (mm/s)
Melilite	$(Ca,Na)_2(Mg,Fe^{2+},Fe^{3+},Zn,Al)(Si,Al)_2O_7$	Fe^{2+} T1,T2	≈ 1.00	2.3
	$(Ca,Na)_2(Mg,Fe^{3+})(Si_2O_7)$	Fe^{3+} T1$_1$	0.19	0.75
		Fe^{3+} T1$_2$	0.19	1.04
Ilvaite	$CaFe_2^{2+}Fe^{3+}(Si_2O_7)O(OH)$	Fe^{2+} M2	≈ 1.05	2.30–2.50
		Fe^{2+} M11	≈ 1	2.10–2.20
		Fe^{3+} M12	0.55	1.1–1.3

3.5.4 Cyclosilicates

The most common structural motif in the group of cyclosilicates is the six-membered Si_6O_{18} ring, as been found in beryl and tourmaline. The Mössbauer spectra of these minerals present a number of features that have remained incompletely understood for a long time.

Beryl has a hexagonal structure composed of a stacking of rings forming channels parallel to the c-axis. These columns are linked by Al^{3+} and Be^{2+} in six- and four-fold coordination, respectively. The ideal formula is $Be_3Al_2Si_6O_{18}$. The iron contents in these minerals are small and usually less than 1 wt %. In general, the Mössbauer spectrum consists of a broad peak near zero velocity and a sharp line near 2.5 mm/s. This particular spectral behavior has been attributed to an unspecified relaxation process. At low temperatures two well resolved ferrous doublet spectra are observed. At 4 K the hyperfine parameters are $\delta_{Fe} = 1.3$ mm/s, $\Delta = 2.7$ mm/s and $\delta_{Fe} = 1.0$ mm/s, $\Delta = 1.5$ mm/s, respectively [198]. The first doublet has been assigned to octahedral Fe^{2+} whereas the second less intense one represents substitutional Fe^{2+} in the highly distorted Be^{2+} tetrahedral sites. Additionally, small amounts of Fe^{3+} are located in the octahedral Al^{3+} sites. In a sample of deep-bleu beryl a fourth doublet has been resolved and assigned to Fe^{2+} in the channels [199].

Cordierite with ideal formula $Al_3(Fe^{2+})_2(AlSi_5)O_{18}$ is structurally similar to beryl. The principal octahedral cations are Fe^{2+} and some Mg instead of Al. The spectra of both magnesium and iron cordeirites exhibit spectra consisting of a predominant ferrous doublet with $\delta_{Fe} = 1.15$ mm/s and $\Delta = 2.3$ mm/s. A second, much weaker, ferrous component is observed which was attributed to channel iron [200]. However, this second has also been interpreted as being due to Fe^{2+} replacing Al, whereas Na enters the center of the rings [201].

Tourmalines have a complex structure which includes Si_6O_{18} rings, BO_3 groups, spiral chains of $AlO_5(OH)$ octahedral (C) sites and edge-sharing clusters of three $MgO_4(OH)_2$ octahedral (B) sites. The ideal formula would be $X(Y_3Z_6)(Si_6O_{18})(BO_3)_3(OH)_3(OH,F)$ with X = Na, but also K, Ca or vacancies, Y = Mg, but also Fe, Mn, Al, Fe, Cr, Ti and Z = Al, but also Cr, and V. Due to the large variety of cations in this structure a wide variety of spectra have been reported. The most common tourmalines are within the elbaite-schorl series

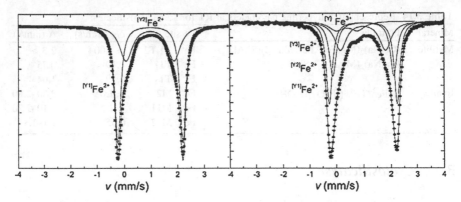

Fig. 3.25 RT spectra of some tourmalines from the English Lake District which needed to be fitted with 2 to 4 doublets (adapted from Eeckhout et al. [207])

Table 3.15 Representative hyperfine parameters at RT for cyclosilicates

Mineral	Formula	Fe site	δ_{Fe} (mm/s)	Δ (mm/s)
Beryl	$Be_3Al_2Si_6O_{18}$	Fe^{2+} oct	1.16	2.70
		$(Fe^{3+}$ oct)	0.59	0.86
Cordierite	$Al_3(Fe^{2+})_2(AlSi_5)O_{18}$	Fe^{2+}	1.15	2.3
Tourmaline	$Ca(Fe^{2+})_3]Al6(Si_6O_{18})(BO_3)_3OH)_3(OH)$	Fe^{2+} Y1	1.08–1.10	2.50
(schorl)		Fe^{2+} Y2	1.07–1.10	2.20–2.35
		Fe^{2+} Y3	1.07–1.10	1.40–1.70
		+ other D		

$[Na(Li_{0.5}Al_{0.5})_3\text{-}Ca(Fe^{2+})_3]Al_6(Si_6O_{18})(BO_3)_3(OH)_3(OH)$- and the dravite-schorl series $[Na(Mg)_3\text{-} Ca(Fe^{2+})_3]Al_6(Si_6O_{18})(BO_3)_3(OH)_3(OH)$.

The spectrum of tourmaline consists basically of a quadrupole doublet with $\delta_{Fe} = 1.1$ mm/s and $\Delta = 2.46$–2.48 mm/s originating from Fe^{2+} on the B (Y) sites. Although this doublet is dominant in most of the spectra in tourmalines, other doublets are clearly observed which have been attributed to Fe^{2+}, Fe^{3+} and intermediate Fe^{n+} (from electron exchange processes). Additional iron in C (Z) sites and a cis–trans isomerism in B (Y) sites may render the spectra rather complicated leading up to five or even seven components to be used in the fits [202–207]. RT spectra of some tourmalines are displayed in Fig. 3.25 (Table 3.15).

3.5.5 Inosilicates

The inosilicate minerals are built up by chains of SiO_4 tetrahedra. They can further be divided into two subclasses: the pyroxenes with single-stranded chains, having an overall composition SiO_3^{2-} and the amphiboles with double-stranded chains of $Si_4O_{11}^{6-}$ stoichiometry.

3.5.5.1 Pyroxenes

Pyroxenes have the general formula $X(Y)(SiO_3)_2$. They crystallize in a monoclinic (clinopyroxenes) or orthorhombic (orthopyroxenes) structure. The Y cations are located in the M1 sites which are moderately distorted octahedra whereas X represents cations in highly distorted sites with six-, seven- or eight-fold coordination according to the kind of mineral. Generally, the structure is orthorhombic when X is Mg or Fe^{2+} and is monoclinic when X is Ca or Na.

Orthopyroxenes have compositions close to the enstatite-ferrosilite $MgSi_2O_6$–$FeSi_2O_6$ tie line. The spectra consist of two ferrous doublets, which can be clearly resolved (Fig. 3.25). The one with the small quadrupole splitting (see Table 3.16) can be assigned to the distorted M2 sites [208]. In the $(Mg,Fe)Si_2O_6$ series this doublet remains rather intense evoking the preference of iron for the M2 sites. [209].

A larger variety of crystal chemistry is possible in the clinopyroxenes. Besides the compositions on and close to the diopside-hedenbergite $CaMgSi_2O_6$-$CaFeSi_2O_6$ tie line (augites) there exist sodium pyroxenes such as jadeite $NaAlSi_2O_6$, aegirine $NaFe^{3+}Si_2O_6$ (also called acmite), kosmochlor $NaCrSi_2O_6$ and jervisite $NaScSi_2O_6$, lithium pyroxenes as spodumene $LiAlSi_2O_6$, monoclinic forms of enstatite and ferrosilite (clinoenstatite and clinoferrosilite) and of course all compositions between them of which some received mineral names like omphacites $(Ca,Na)(Fe^{2+},Fe^{3+},Mg,Al)Si_2O_6$ and pigeonites $(Ca,Fe^{2+},Mg)(Fe,Mg)Si_2O_6$. Ca-free clinopyroxenes along the $MgSiO_3$–$FeSiO_3$ tie line only exist under high pressure in the earth crust [210] (Fig. 3.26).

The Mössbauer spectra of the diopside-hedenbergite series are straightforward and predominantly consist of a ferrous doublet (Fig. 3.27a) for which the quadrupole splitting Δ varies between 2.30 and 1.85 mm/s (Table 3.16), decreasing on going from hedenbergite to diopside. [211–213]. Deviations from stoichiometry introduce a small amount of an Fe^{2+} doublet from M2 sites and a weak Fe^{3+} doublet. Because the quadrupole splitting of the M1 doublet is strongly temperature dependent whereas the one of the M2 doublet remains invariant at about 2.0 mm/s, measurements at various temperatures are usually necessary to resolve both doublets. Trivalent iron can also be introduced to some extent in diopside, thus substituting for Mg and Si.

Table 3.16 Table Representative hyperfine parameters at RT for some pyroxenes

Mineral	Formula	Fe site	δ_{Fe} (mm/s)	Δ (mm/s)
Ferrosilite	$Fe_2Si_2O_6$	Fe^{2+} M1	1.17	2.48
		Fe^{2+} M2	1.13	1.93
Enstatite- ferrosilite	$(Mg,Fe^{2+},Mn)_2Si_2O_6$	Fe^{2+} M1	1.15–1.18	2.35–2.69
		Fe^{2+} M2	1.12–1.16	1.91–2.13
Hedenbergite	$CaFe^{2+}Si_2O_6$	Fe^{2+} M1	1.19	2.20
		$(Fe^{3+}$M1)	0.34	0.68
Diopside-hedenbergite	$Ca(Mg,Fe^{2+})Si_2O_6$	Fe^{2+} M1	1.19	1.85–2.30
Aegerine	$(Na,Li)Fe^{3+}Si_2O_6$	Fe^{3+} M1	0.39	0.30

Fig. 3.26 RT Mössbauer
spectra of synthetic samples
of the enstatite-ferrosilite
series showing a gradual
increase of Fe^{2+} on M1 sites
(from Dyar et al. [209])

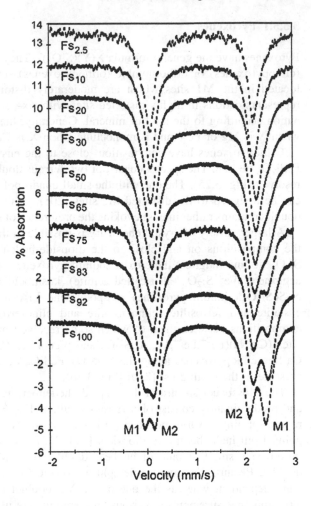

Ca–Fe pyroxenes with compositions between hedenbergite and clinoferrosilite $(Ca,Fe^{2+})(Fe^{2+})Si_2O_6$ show structure in their spectra which requires more than two ferrous doublets for an adequate fit [214]. The observed subspectra may arise from the different numbers of iron and calcium neighbors on the three M3 sites. Even a more distinct structure is observed in the omphacites in which the distribution of the charges among the neighboring M2 sites (Ca^{2+} and Na^+) has a large effect on the Fe^{2+} located in the M1 sites [215].

Aegirine $NaFe^{3+}Si_2O_6$ shows a ferric doublet with $\delta_{Fe} = 0.39$ mm/s and $\Delta = 0.30$ mm/s (Fig. 3.27b). Replacing Na by Li causes a change in coordination number from eight to six for M2 sites, but has no effect on the hyperfine parameters. Systematic studies of solid solutions of aegirine with diopside, hedenbergite, kosmochlor and $LiFeSi_2O_6$ showed that neither δ nor Δ changes markedly as long as the valences of the neighboring cations of Fe remain unaltered. Whenever ions of different valences are mixed on the same site electron

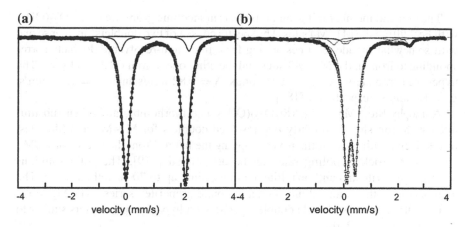

Fig. 3.27 RT spectra of hedenbergite (a) and aegirine (b)

hopping occurs [197, 216] and consequently the quadrupole splitting depends strongly on the composition.

Beside the clino- and orthopyroxenes there exists a group of minerals with similar chemical compositions of pyroxenes called pyroxenoids. There structures are based on twisted chains of silica tetrahedra. An example is rhodonite $(Mn,X)SiO_3$ ($X = Fe$, Ca, Mg), which contains five different sites: three moderately and one strongly distorted octahedral sites and one seven-fold coordinated site. Spectra of rhodonite and fowlerite (Zn-rich rhodonite) needed to be adjusted with five doublets showing Fe^{2+} to be present in all the available sites [217, 218].

Other minerals, which are structurally similar to pyroxenes except that the octahedral strips form zigzags, are carpholites, $(Mn, Fe^{2+})Al_2(Si_2O_6)(OH)_4$. In these mineral species the ferrous iron is found to be in a M2 position that is virtually undistorted. Consequently a very large quadrupole splitting of 3.20 mm/s is observed [209, 210].

3.5.5.2 Amphiboles

The amphiboles consist structurally of double Si_4O_{11} chains parallel to the orthorhombic or monoclinic c axis. The general formula of the amphiboles is $W_xX_2(Y_5)(Si_4O_{11})_2(OH)_2$ in which Y represents the octahedral M1, M2 and M3 sites, X denotes the cation in the irregular M4 site, similar to the M2 site in pyroxenes, and W is the cation in the ten- to 12-fold coordinated A site which resembles the interlayer sites in micas. The A site is often empty. Similarly to the pyroxenes the M4 site can accommodate Ca, Na, Li, Mg and Fe^{2+} and the M1–M3 sites are occupied by Fe^{2+}, Mg, Fe^{3+}, Al, Mn,... The A site, if occupied, accommodates large monovalent ions such as Na or K. The relative abundances of the various sites M1:M2:M3:M4:A is 2:2:1:2:x with $0 \ll x \ll 1$.

The iron end member of the amphiboles is monoclinic grunerite $Fe_7Si_8O_{22}(OH)_2$ with Fe^{2+} filling the M1–M4 sites. The spectra consist of two doublets, an intense one with somewhat broadened lines arising from three non-resolvable doublets corresponding to iron on the M1–M3 sites and the other one to iron on the M4 sites. The hyperfine parameters are $\delta_{Fe} = 1.16$ mm/s, $\Delta = 2.82$ mm/s and $\delta_{Fe} = 1.10$ mm/s, $\Delta = 1.8$ mm/s, respectively [184].

Anthophyllite, $(Mg_2)[Mg_7](Si_8O_{22})(OH)_2$, is orthorhombic and when substituted with iron shows similarly two resolved doublets for M1–M3 and M4 sites, respectively, with most of the iron occupying the latter. From the degree of Fe/Mg disorder the rock's cooling rate can be determined [219]. The solid solutions between grunerite and anthophyllite are monoclinic up to 70 % anthophyllite. The members of the monoclinic cummingtonite-grunerite series, $(Fe,Mg,Mn)_7$ $Si_8O_{22}(OH)_2$, all show double doublet spectra with hyperfine parameters similar to those of grunerite [220].

Gedrite is an orthorhombic amphibole obtained from anthophyllite by substitution of some Mg and Si by Na and Al. The ideal formula is $Na_{0.5}(Mg_{5.5},Al_{1.5})(Si_6Al_2)O_{22}(OH)_2$ with Na entering the A sites. The spectrum of ferrous iron in gedrites exhibits a broadened doublet because the M1, M2 and M3 sites have no longer similar distortions from octahedral symmetry [221, 222]. The spectra can be fitted with three Fe^{2+} components and a weak Fe^{3+} component, but it remains difficult to assign unambiguously the subspectra to the different sites. Although there is a miscibility gap between anthophyllite and gedrite [223], some Al may substitute in the former and the main effect is to reduce Δ and increase the linewidth of the M1–M3 doublet with increasing Al content.

Holmquistite is another orthorhombic amphibole obtained from anthophyllite by substitution of some Mg by Li and Al, the lithium entering the M4 site. In a sample containing 1.44 iron atoms per formula unit two ferrous doublets could be resolved with $\delta_{Fe} = 1.13$ mm/s, $\Delta = 2.8$ mm/s and $\delta_{Fe} = 1.11$ mm/s, $\Delta = 2.0$ mm/s which were attributed to Fe^{2+} in M1 and M3 sites, respectively. Fe^{3+} and Al enter the M2 sites [224]. Holmquisite has also a monoclinic variant, which has a similar spectrum with comparable hyperfine parameters.

The calcic amphiboles, actinolites, are close to the tremolite-ferroactinolite, $Ca_2Mg_5Si_8O_{22}(OH)_2$–$Ca_2Fe_5Si_8O_{22}(OH)_2$, tie line. Ferrous iron is mainly present in the M1 and M3 sites yielding $\Delta = 2.8$–2.9 mm/s, but is also found in the M2 sites with $\Delta = 1.7$–1.9 mm/s [225]. However, the latter doublet with low Δ value has been assigned by Goldman [226] to Fe^{2+} at the M4 sites, whereas this author introduced an additional doublet with an intermediate $\Delta = 2.0 - 2.4$ mm/s for Fe^{2+} in the M2 sites. Ferric iron, if present, enters also the M2 sites.

Similarly to actinolites a solid solution series of sodic amphiboles exists along the line from glaucophane, $Na_2Mg_3Al_2Si_8O_{22}(OH)_2$, to riebeckite, $Na_2(Fe^{2+})_3$ $(Fe^{3+})_2Si_8O_{22}(OH)_2$. The spectra are rather different from those of the actinolites in that the M1 and M3 subspectra are better separated and that a substantial contribution of an Fe^{3+} doublet is observed. The involved Fe^{3+} is believed to be mainly present in the M2 sites [227, 228] (Table 3.17).

Table 3.17 Representative hyperfine parameters at RT for some amphiboles

Mineral	Formula	Fe site	δ_{Fe} (mm/s)	Δ (mm/s)
Grunerite	$Fe_7Si_8O_{22}(OH)_2$	Fe^{2+} M1	1.16	2.82
		Fe^{2+} M4	1.10	1.8
Cummingtonite-grunerite	$(Mg,Fe^{2+},Mn)_7Si_8O_{22}(OH)_2$	Fe^{2+} M1–M3	1.16	2.81
		Fe^{2+} M4	1.10	1.5–1.8
Anthophyllite	$(Mg,Fe^{2+})_7Si_8O_{22}(OH)_2$	Fe^{2+} M1–M3	1.12	2.6
		Fe^{2+} M4	1.10	1.8
Riebeckite	$Na(Fe^{2+})_3(Fe^{3+})_2Si_8O_{22}(OH)_2$	Fe^{2+} M1	1.14	2.83
		Fe^{2+} M3	1.11	2.32
		Fe^{3+} M2	0.38	0.43
Holmquistite	$(Li,Fe^{3+},Mg,Fe^{2+})_7Si_8O_{22}(OH)_2$	Fe^{2+} M1	1.13	2.8
		Fe^{2+} M3	1.1	2.0
		Fe^{3+} M2	0.38	0.3
Ferroactinolite	$Ca_2Fe_5Si_8O_{22}(OH)_2$	Fe^{2+} M1,M3	1.15	2.81
		Fe^{2+} M2	1.14	1.85–2.1
		$(Fe^{2+}$ M4)	1.10	<1.8

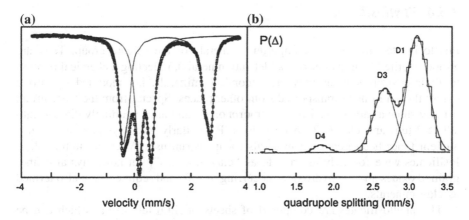

Fig. 3.28 Spectrum at 80 K of a natural riebeckite sample, fitted with quadrupole distributions (**a**); resulting quadrupole distribution for Fe^{2+} showing three peaks (adapted from Van Alboom and De Grave [221])

It is worth to mention that the analysis of the spectra of pyroxenes has not always been successful by using discrete doublets. Already in the early days of Mössbauer investigations of pyroxenes, Bancroft [3] argued that the shortcomings of the until-then commonly applied fitting procedures are the result of the non-uniform chemical environment for both iron sites requiring rather a set of doublets for each site. The use of shape-independent quadrupole distributions seem to be more appropriate in that case [229] and has indeed been successfully applied to the analyses of the spectra of pyroxene minerals such as aluminium diopsides [230] and magnesian hedenbergites [211] and of amphiboles such as riebeckites [231] (Fig. 3.28).

3.5.5.3 Other Chain Silicates

There exist a number of other chain silicates for which the chains are not linked in the way as they are in pyroxenes or amphiboles. A first example is babingtonite in which two chains of SiO_4 tetrahedra are joined by groups of four edge-sharing octahedra. The ideal formula of the iron member is $Ca_2Fe^{2+}Fe^{3+}Si_5O_{14}(OH)$ in which Fe^{2+} and Fe^{3+} each occupy a distinct site. Calcium occupies two types of large interstices. The spectra are consequently quite simple, consisting of narrow ferric and ferrous doublets [232]. Another particular chain silicate is deerite which has a structure based on a hybrid single-double Si_6O_{17} chain. There is a strip of edge-sharing octahedra with nine crystallographically distinct sites, which are classified into three groups of each three sites. The ideal formula is $(Fe^{2+})_6(Fe^{3+})_3O_3Si_6O_{17}(OH)_5$. The spectra are complicated by the effects of electron hopping [233]. There are signs of a preference of Fe^{3+} for M1 and M3 positions and of Fe^{2+} for the others.

3.5.6 Phyllosilicates

Phyllosilicates are the most important minerals of the silicate group. They are either inherited from parent rocks (detrital minerals), reflecting a chemical relation to their environment, or they are secondary minerals. i.e. modified by strong external conditions or transported from other places. Species from the latter group usually show a rather small-particle morphology and are accordingly divided into silt (>2 µm) and clay (<2 µm) fractions. Particularly the clays have ever since ancient times been important minerals for industrial uses. Microcrystalline phyllosilicates were formerly referred to as "clay minerals", but nowadays also fine-grained oxides and oxyhydroxides occurring in soils and sediments are also termed as clay minerals.

The phyllosilicates are composed of sheets of SiO_4 tetrahedra, which can be divided into two groups: the 1:1 layer and the 2:1 layer minerals. The 1:1 layer silicates are composed of alternating tetrahedral Si_4O_{10} and octahedral $Al_4(OH)_{12}$ layers. The octahedra are formed by two oxygen and four hydroxyl anions (Fig. 3.29). The layers themselves are electrically neutral and the stacks are held together by van der Waals or hydrogen bonding. The repeat distance is about 0.7 nm. The general formula is $(M_2^{3+} \text{ or } M_3^{2+})Si_2O_5(OH)_4$. According to the filling of the octahedra the silicate can either be dioctahedral with 2/3 of the octahedral sites filled or trioctahedral in which all octahedra are filled.

The basic block of a 2:1 layer silicate is a sandwich layer of one octahedral sheet between two tetrahedral ones. The octahedral sheet contains now two apices of two sheets of tetrehedra, so four of the ligands are now oxygen and two are hydroxyl ions (Fig. 3.29). The hydroxyls may be at opposite or adjacent corners of the octahedron, giving trans and cis coordination, respectively, mostly denoted as

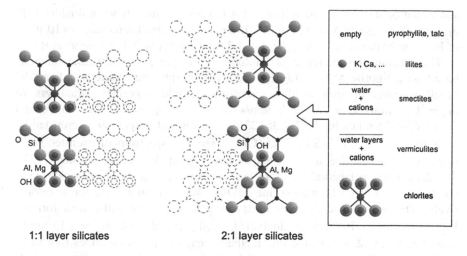

Fig. 3.29 Stacking of octahedra and tetrahedra in phyllosilicates

M1 and M2. The basic blocks may either electrically neutral such as in pyrophyllite or talc, but more frequently there is a charge deficiency which must be compensated. In the micas the substitution of Al^{3+} for a quarter of the Si^{4+} is compensated by a 12-fold coordinated large K^+ ion in the interlayer position. In the so-called brittle micas two silicons are replaced by aluminum and the interlayer cation is divalent. The repeat distance of the layer is close to 1 nm. The smectites and vermiculites contain water or water layers with cations in solution. In the former the repeat distance is variable according to the water content. Because vermiculites are usually better crystallized the charge deficit is higher which makes them difficult to expand. Their normal state is a double layer of water giving the characteristic 1.4 nm spacing. In chlorites the interlayer is occupied by a brucite $(Mg,Al)_3(OH)_6$ sheet which compensates for the charges. The ionic bonding between the brucite sheet and the talc layers also prevents the chlorites to expand. Similar to 1:1 layer silicates, 2:1 layer species can be dioctahedral or trioctahedral.

3.5.6.1 1:1 Layer Silicates

Kaolinite is the basic dioctahedral 1:1 layer silicate. The ideal formula is $Al_2Si_2O_5(OH)_4$. It tends to contain very little iron and it has been shown that ferric iron can substitute for Al to a very small amount. Murad and Wagner [234] summarized the published Mössbauer data giving on the average $\delta_{Fe} = 0.34$ mm/s and $\Delta = 0.52$ mm/s for the ferric doublet. These hyperfine parameters are also representative for superparamagnetic iron oxyhydroxides and oxides which are commonly associated with kaolinite. Therefore spectra at very low temperatures (below the blocking temperatures of the oxides) are necessary to distinguish the associated oxides from lattice iron. At 4 K kaolinite still shows a doublet with $\delta_{Fe} = 0.48$ mm/s and $\Delta = 0.52$ mm/s. In some cases a small percentage (<10 at %) of Fe^{2+} can be detected with $\delta_{Fe} = 1.1$–1.3 mm/s and $\Delta = 2.4$–2.9 mm/s at RT.

The trioctahedral analogue of kaolinite is the group of serpentine minerals based on antigorite $Mg_3Si_2O_5(OH)_4$ with polymorphs chrystolite and lizardite. Small amounts of iron in those serpentines yield hyperfine parameters of $\delta_{Fe} = 1.14$ mm/s, $\Delta = 2.68$–2.76 mm/s for Fe^{2+} and $\delta_{Fe} = 0.37$–0.42 mm/s, and $\Delta = 0.65$–0.85 mm/s for Fe^{3+} [235, 236]. The iron-rich serpentine is cronstedtite with formula $Fe_2^{2+}Fe^{3+}(SiFe^{3+})O_5(OH)_4$. The RT spectrum of cronstedtite is rather complicated due to the electron hopping on the octahedral sites [237]. Tetrahedral and octahedral Fe^{3+} are best distinguished in the magnetically split spectrum at 4 K for which hyperfine fields of 40.6 and 46.7 T are found respectively. The ferrous end member of the serpentines is greenalite with formula $Fe_3^{2+}Si_2O_5(OH)_4$. Its ferrous doublet is well defined with $\delta_{Fe} = 1.15$ mm/s, $\Delta = 2.75$ mm/s [238]. For the intermediate members, the berthierines (not to be confused with chamosite, which is a 2:1 silicate), the quadrupole splitting Δ falls within the range 2.62–2.68 mm/s [239].

3.5.6.2 2:1 Layer Silicates

The basic 2:1 dioctahedral silicate is pyrophyllite with formula $Al_2Si_4O_{10}(OH)_2$. In muscovite, $KAl_2(Si_3Al)O_{10}(OH)_2$, interlayer K compensates electrically for the Al replacement for Si. Illites have a more variable composition with a general formula $(H_30,K)_xAl_2(Si_{4-x}Al_x)O_{10}(OH)_2$. Nontronite is the iron-rich dioctahedral silicate with general formula $M^+Fe_2^{3+}(Si_{4-x}Al_x)O_{10}(OH)_2$.

Although, Mössbauer spectra of the various silicates turned out so far to be reasonably analyzable in a relatively unambiguous way, this is surely not the case for 2:1 layer silicates. In all these silicates Al^{3+} is partly replaced by Fe^{3+} resulting in one or two ferric doublets, which can be interpreted by the Fe-for-Al substitution on trans and cis sites. This is the basic spectrum for all the 2:1 dioctahedral silicates. The quadrupole splitting of the inner doublet (cis) is found to increase in the sequence ferripyrophyllite, nontronite, glauconite, montmorillonite, illite, muscovite (Table 3.18), which in some sense corresponds to increasing distortion of the octahedral [240]. The presence of small amounts of Fe^{2+} possibly introduces two additional doublets similarly attributed to cis and trans arrangements. A third Fe^{3+} doublet with very small quadrupole splitting can be assigned to iron in the tetrahedral sites. (see Johnston and Cardile [241] and references therein). However, in general there is still a lack of common approach for the analysis of the spectra. This is clearly illustrated in the published Mössbauer results for illites and nontronites [242]. Hence, only a fit with two broad-lined doublets or by quadrupole splitting distributions, one for each Fe^{2+} and Fe^{3+}, can provide some straightforward insight in the iron behavior in the structure of these minerals [243]. The iron-rich variants of illite are glauconite and celadonite. In glauconite there is

Table 3.18 Representative hyperfine parameters at RT of some phyllosilicates

Mineral	Site	δ_{Fe} (mm/s)	Δ (mm/s)
Mineral	Site	δ_{Fe} (mm/s)	Δ (mm/s)
1:1 dioctahedral			
Kaolinite	Fe^{3+}	0.48	0.52
	Fe^{2+}	1.1–1.3	2.4–2.9
1:1 trioctahedral			
Antigorite	Fe^{2+}	1.14	2.68–2.76
	Fe^{3+}	0.37–0.42	0.65–0.85
Greenalite	Fe^{2+}	1.15	2.75
Berthierine	Fe^{2+}	1.15	2.62
2:1 dioctahedral			
Muscovite	Fe^{3+}(M2)	0.40	0.72
	Fe^{2+}(M2)	1.13	3.00
	Fe^{2+}(M1)	1.12	2.20
Illite	Fe^{3+}(M2)	0.33	0.59–0.73
	Fe^{3+}(M1)	0.38	1.21
	Fe^{2+}(M2)	1.14	2.75
Nontronite	Fe^{3+}(M2)	0.36–0.39	0.24–0.27
	Fe^{3+}(M1)	0.37–0.40	0.59–0.68
Glauconite	Fe^{3+}(M2)	0.34	0.33
	Fe^{3+}(M1)	0.32	0.69
	Fe^{2+}(M2)	1.1	2.68
	Fe^{2+}(M1)	1.1	1.7–1.9
	Fe^{3+}(interlayer)	0.50	0.5–0.8
Celadonite	Fe^{3+}(M2)	0.35	0.39
	Fe^{2+}(M2)	1.12	2.64
	Fe^{2+}(M1)	1.12	1.76
	Fe^{3+}(DSS)	0.39	1.16
2:1 trioctahedral			
Talc	Fe^{2+}(M2)	1.13	2.6
Biotite	Fe^{2+}(M2)	1.12–1.14	2.57–2.63
	Fe^{2+}(M1)	1.07–1.15	2.12–2.22
	Fe^{3+}(M2)	0.39–0.44	0.37–0.66
	Fe^{3+}(M1)	0.39–0.49	1.16–1.24
Vermiculite	Fe^{3+}(M2)	0.29	0.57
	Fe^{3+}(M1)	0.37	1.10
	Fe^{2+}(M2)	1.11	2.58
Chlorite	Fe^{2+}(M2)	1.12	2.68
	Fe^{2+}(M1)	1.12	2.3–2.4
	Fe^{2+}(brucite)	1.16–1.20	2.67–2.70
	Fe^{3+}(?)	0.38–0.46	0.5–0.75

a much more significant replacement of Si by Al or Fe^{3+} in the tetrahedra than in celadonite, for which the substitution is considered to be 20 at % or less. The Mössbauer spectra of glauconites can be consistently fitted with two ferrous doublets (cis and trans), two ferric doublets (cis and trans) and a third ferric

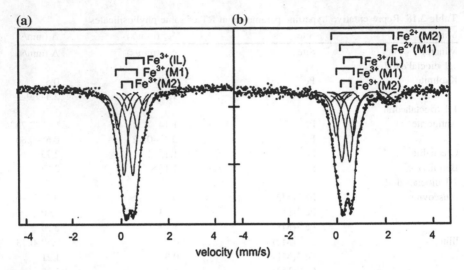

Fig. 3.30 RT spectra of two natural glauconiet samples fitted with three (**a**) or five doublets (**b**) (adapted from De Grave and Geets [309])

Fig. 3.31 RT spectrum of a natural celadonite sample fitted with four doublets (adapted from Bowen et al. [245])

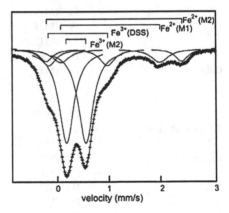

doublet (Fig. 3.30). De Grave et al. [244] assigned the latter to interlayer Fe^{3+}, whereas Johnson and Cardile [241] analyzed with a third ferric doublet with rather small Δ which they attributed to tetrahedral Fe^{3+}.

The spectrum of celadonite can be adjusted with two ferrous doublets and one ferric doublet, the latter assigned to M2 sites (cis) [245]. A second weak Fe^{3+} doublet with relatively large quadrupole splitting ($\Delta \approx$ 1.1–1.2 mm/s) (Fig. 3.31) has been ascribed to iron in dehydroxylated surface sites (DSS) [246, 247].

The basic mineral of the 2:1 trioctahedral silicates is talc with formula $Mg_3Si_4O_{10}(OH)_2$. It is obvious that for this silicate Fe^{2+} will be the most abundant valence of iron and consequently the spectra consist predominantly of ferrous doublets [246, 248]. Similarly, this is also the basic spectral appearance for all trioctahedral silicates. The spectra of biotites show consistent peak positions for Fe^{2+} in M2 sites and for Fe^{2+} in M1 sites with broader variation in hyperfine parameters for the latter [247, 249]. Additional Fe^{3+} components that appear in the spectra are resolved as two distinct doublets with hyperfine parameters representing the M2 and M1 sites [249, 250].

Chlorites exhibit the main features of biotites. Well-known representatives of the chlorite group are clinochlore, $(Mg,Fe^{2+})_5Al(Si_3Al)O_{10}(OH)_8$ and its iron-rich variant chamosite. The major Fe^{2+} doublet in the spectra of chlinichlore and chamosite has the typical values $\delta_{Fe} = 1.13$ mm/s and $\Delta = 2.70$ mm/s whereas a minor Fe^{3+} doublet gives $\delta_{Fe} = 0.39$ mm/s and $\Delta = 0.67$ mm/s [243]. A third doublet is observed at lower temperatures, but has nearly the same hyperfine parameters as the M2 ferrous iron at RT. This doublet might be attributed to substitutions in the brucite layer [251]. The hyperfine parameters of trioctahedral micas obtained in about fifty studies of these silicates are summarized in a review by Dyar [250].

Finally, Table 3.18 presents a survey of δ_{Fe} and Δ values that have been reported so far in the literature for a number of selected phyllosilicate species. It is obvious from these data that the interpretation of the Mössbauer spectra of micas is not straightforward, the more so nature has provided us with such silicates with an enormous variety in chemical composition.

3.6 Phosphates

3.6.1 Introduction

A large variety of phosphate minerals are found on Earth. Among these, apatite, $Ca_5(PO_4)_3(F,OH,Cl)$, is the most abundant one and consequently serves as the major source from which other naturally occurring phosphates originate by diverse transformational processes. The structure of apatite can accommodate numerous metal cations that substitute for Ca. By far most of the phosphates exhibit a structure that is based on a polymerization of (PO_4) tetrahedra and $(M\phi_6)$ octahedra, with $\phi = O^{2-}$ or OH^-. As such one distinguishes structures with finite clusters of (PO_4) tetrahedra and $(M\phi_6)$ octahedra (e.g. anapaite), structures with infinite chains of tetrahedra and octahedra (eosphorite), structures with infinite sheets of tetrahedra and octahedra (strunzite, vivianite), and finally structures with infinite frameworks of tetrahedra and octahedra (triphylite, heterosite, leucophosphite). Iron, both as a divalent or trivalent cation, is a common substituent and even complete solid solutions may be encountered. As a particular iron phosphate generally occurs in low concentrations and as fine-grained material in association with other mineral species, including other phosphates that may or may not contain iron, it is often a pain-staking task to collect a sufficient amount of uncontaminated sample to derive a sound and diagnostic Mössbauer-spectroscopic fingerprint for that given Fe phosphate. This drawback is possibly one of the reasons why in the past Fe-bearing phosphates, compared to silicates, have less frequently been studied by the Mössbauer effect. In what follows a few Fe phosphates are selected and their Mössbauer characteristics as observed from their paramagnetic spectra (mostly room temperature) are presented.

3.6.2 Anapaite

The ideal chemical formula of anapaite is $Ca_2Fe^{2+}(PO_4)_2 \cdot 4H_2O$. It has the triclinic structure. The Ca and Fe cations exhibit an eight- and a six-fold co-ordination respectively. The iron is bound to four basal water molecules and two apical oxygen atoms, which belong to a phosphate tetrahedron. The Fe octahedron shares two edges with the Ca dodecahedron. Hence, there is only one type of octahedral site available for the Fe^{2+} cations with coordination $O_2(OH_2)_4$. Consequently, the Mössbauer spectrum (see Fig. 3.32a) consists of a sharp quadrupole doublet with parameters as indicated in Table 3.19. The mineral remains paramagnetic down to a temperature as low as 4.2 K. From the temperature dependence of the quadru-pole splitting, combined with external-field spectra, it is concluded that the Fe octahedron is subjected to a tetragonal compression [252]. Oxidation treatments in solutions with various H_2O_2 concentrations up to 20 % and subsequent Mössbauer experiments at room temperature, have revealed that the anapaite structure is unusually highly resistant against oxidation since eventually only a small amount of Fe^{2+} (~ 6.5 %) is converted into Fe^{3+} (see Fig. 3.32b). The Fe^{2+} doublet parameters are not affected by this partial oxidation.

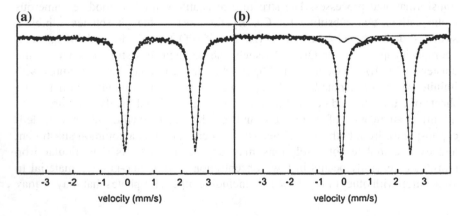

Fig. 3.32 Mössbauer spectra of anapaite at 11 K (**a**) and of oxidized anapaite at 80 K (**b**)

Table 3.19 Representative hyperfine parameters at RT of some iron-bearing phosphates

Mineral	Formula	Fe site	δ_{Fe} (mm/s)	Δ (mm/s)
Anapaite	$Ca_2Fe^{2+}(PO_4)_2.4H_2O$	$Fe^{2+}O_2(OH_2)_4$	1.19	2.49
Eosphorite	$Mn^{2+}AlPO_4(OH)_2.H_2O$	$Fe^{2+}O_4(OH)_2$	1.25	1.74
Manganostrunzite	$Mn^{2+}Fe_2^{3+}(PO4)_2(OH)_2 \cdot 6(H_2O)$	$Fe^{3+}O_2(OH_2)_4$	0.42	0.61
		$Fe^{3+}O_3(OH_2)(OH)_2$ [Fe(1)]	0.41	0.79
			0.41	0.63
		$Fe^{3+}O_3(OH_2)(OH)_2$ [Fe(2)]		
Ferristrunzite	$Fe_3^{3+}(PO4)_2(OH)_2 \cdot [(H_2O)5(OH)]$	$Fe^{3+}O_2(OH_2)_3(OH_2)$	0.40	0.64
		$Fe^{3+}O_3(OH_2)(OH)_2$ [Fe(1)]	0.39	0.89
		$Fe^{3+}O_3(OH_2)(OH)_2$ [Fe(2)]	0.39	0.74
Ferrostrunzite[a]	$Fe^{2+}Fe_2^{3+}(PO4)_2(OH)_2 \cdot 6(H_2O)$	$Fe^{2+}O_2(OH_2)_4$	1.33	2.65
		$Fe^{2+}O_2(OH_2)_4$	1.33	3.06
		$Fe^{3+}O_3(OH_2)(OH)_2$ [Fe(1)]	0.52	0.85
			0.52	0.64
		$Fe^{3+}O_3(OH_2)(OH)_2$ [Fe(2)]		
Vivianite	$Fe_3^{2+}(PO_4)_2.8H_2O$	$Fe^{2+}O_2(OH_2)_4$ [Fe(1)]	1.19	2.48
		$Fe^{2+}O_4(OH_2)_2$ [Fe(2)]	1.22	2.96
Ox. vivianite	$Fe_{3-q}^{2+}Fe_q^{3+}(PO_4)_2 \cdot (8-q)H_2Oq(OH)$	$Fe^{2+}O_2(OH_2)_4$ [Fe(1)]	1.23	2.48–2.39
		$Fe^{2+}O_4(OH_2)_2$ [Fe(2)]	1.22	3.00
		$Fe^{3+}O_2(OH_2)_4$ [Fe(1)]	0.41	0.91–1.01
		$Fe^{3+}O_4(OH_2)_2$ [Fe(2)]	0.42	0.44
Triphylite	$LiFePO_4$	$Fe^{2+}O_6$ [M2]	1.22	2.96
Heterosite	$Fe^{3+}PO_4$	$Fe^{3+}O_6$ [M2]	0.41	1.63
Leucophosphite	$KFe_2(PO_4)_2(OH).2H_2O$	$Fe^{3+}O_5(OH)$ [Fe(1)]	0.41	0.84
		$Fe^{3+}O_4(OH)_2$ [Fe(2)]	0.41	0.55

[a] data at 50 K

3.6.3 Eosphorite

The manganese phosphate eosphorite, $Mn^{2+}AlPO_4(OH)_2.H_2O$, has the orthorhombic symmetry. The structure consists of alternating chains of $MnO_4(OH)_2$ octahedra and $AlO_2(OH)_2(H_2O)_2$ octahedra that run parallel to the c-axis. The strongly distorted

Mn octahedra share opposite O–O edges, while the more regular Al octahedra share opposite H_2O corners. The two types of chains are linked to one another by sharing their OH corners, thus forming parallel sheets that are held together by phosphorous species in a tetrahedral O_4 co-ordination. Eosphorite is isomorphous with childrenite, $FeAlPO_4(OH)_2.H_2O$, and the two mentioned minerals form the end members of a complete solid-solution series. Naturally occurring eosphotites usually exhibit a substantial Fe-for-Mn substitution. It is generally accepted in the literature that only Fe^{2+} is present in the structure of iron-substituted eosphorite, $(Mn,Fe)AlPO_4(OH)_2.H_2O$. This belief is corroborated by its Mössbauer spectrum, which consists of a relatively narrow ferrous doublet at temperatures as low as ~ 35 K. At RT the relevant hyperfine parameters are as listed in Table 3.19 [253].

3.6.4 Strunzite

Three distinct variants of strunzite are known to occur in nature: manganostrunzite, $Mn^{2+}Fe_2^{3+}(PO_4)_2(OH)_2 \cdot 6(H_2O)$, ferristrunzite, $Fe_3^{3+}(PO_4)_2(OH)_2 \cdot [(H_2O)_5(OH)]$, and ferrostrunzite $Fe^{2+}Fe_2^{3+}(PO_4)_2(OH)_2 \cdot 6(H_2O)$, respectively. Ferristrunzite may be regarded as the fully oxidized form of ferrostrunzite. The strunzite structure is triclinic and consists of infinite chains of octahedral ferric sites along the c axis which are linked one another by sharing hydroxyl groups and by PO_4 tetrahedra. The latter also bind adjacent chains, thus forming slabs that are connected to each other by Mn octahedra between remaining PO_4 vertices. Within the chains two ferric sites Fe(1) and Fe(2) alternate, both having an octahedral $O_3(OH)_2(OH_2)$ coordination. The Fe(1) is somewhat more distorted than Fe(2) as indicated by a slight difference in average Fe–O bond length and average O–Fe–O bond angle. The crystallographic unit cell contains two iron octahedra of each type and two Mn octahedra with coordination $O_2(OH_2)_4$. In ferristrunzite the manganese is substituted by Fe^{3+}, in ferrostrunzite by Fe^{2+}. In the first case the charge balance is re-established by substitution of OH by H_2O at a non-bridging vertex of the Mn octahedron. Mössbauer spectra (MS), both for these three mineral species have been reported [254–256]. The spectrum of manganostrunzite recorded at ~ 60 K is reproduced in Fig. 3.33a. It was decomposed into three ferric quadrupole doublets. This model was imposed by the results at 4.2 K at which temperature the sample is magnetically ordered and clearly three distinct sextet components are recognized. Obviously, there is no indication whatsoever that Fe^{2+} would be present in the structure of this manganostrunzite species. The doublet hyperfine parameters observed at RT are included in Table 3.19. The presence of three spectral components implies that the manganese sublattice is partly substituted by Fe^{3+} and from the relative spectral area of the corresponding doublet ($\Delta = 0.61$ mm/s), i.e. ~ 12 %, it is inferred that the Fe-for-Mn substitution is around 25 %, which is in accordance with the results of the chemical analysis [255]. The two other ferric doublet components are assigned to the Fe(1) and Fe(2) sites on the basis that the former sites exhibit a somewhat higher distortion and hence are expected to produce the larger quadrupole splitting. The

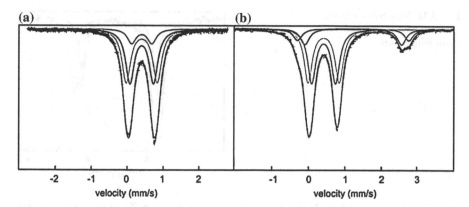

Fig. 3.33 Mössbauer spectra of manganostrunzite at 60 K (**a**) and of ferrostrunzite at 50 K (**b**)

paramagnetic spectra of the ferristrunzite variant are very similar to those of manganostrunzite, the major distinction being the higher relative contribution of the first doublet ($\Delta = 0.64$ mm/s), i.e. ~ 30 %, which is due to Fe^{3+} cations on the Mn sites. The ideal value for a complete Fe-for-Mn substitution would be 33 %. The small deviation is ascribed to the presence of Al cations in the structure to an amount of approximately 0.1 p.f.u., as derived from the chemical analysis. A spectrum of ferrostrunzite is shown in Fig. 3.33b and refers to a temperature of ~ 60 K. Clearly two ferrous components are present in addition to a dominant ferric doublet. The latter was decomposed into two contributions which were assigned to Fe(1) and Fe(2) sites, respectively (see Table 3.19). Rationally, the two ferrous doublets were attributed to the Mn sites. However, the reason why two distinct Mn sites appear has remained unexplained. Moreover, the ferrous-to-ferric concentration range was found to be 0.36, which is significantly smaller than the value 0.5 for ideal ferrostrunzite.

3.6.5 Vivianite

Ideally, the chemical formula of vivianite is $Fe_3^{2+}(PO_4)_2.8H_2O$. As such the mineral is colorless. In soils vivianite species are commonly fine-grained and easily oxidize in the air, the color becoming pale blue. There are no structural changes observed upon oxidation and oxidized vivianite can be represented as $Fe_{(3-q)}^{2+}Fe_q^{3+}(PO_4)_2.[(8-q)H_2Oq(OH)]$. The unit cell of (oxidized) vivianite is monoclinic. The structure consists of single and paired octahedra bound together by PO_4 tetrahedra, thus forming infinite sheets parallel to the (010) plane. The single octahedron has a $O_2(OH_2)_4$ coordination with the oxygens in *trans* position, and is denoted as the Fe(1) site. The paired octahedra form the group $O_6(OH_2)_4$ and share two O^{2-} anions along their common edge. For both octahedra of the

Fig. 3.34 Mössbauer spectra
of non-oxidized vivianite at
room temperature

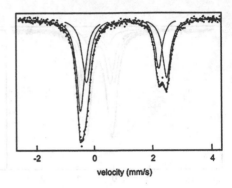

velocity (mm/s)

paired unit the two other O^{2-} anions are in a *trans* position. The two octahedra are identical and are henceforward referred to as the Fe(2) sites. There are twice as many Fe(2) sites in the vivianite structure than there are Fe(1) sites.

MS for a single crystal of non-oxidized vivianite were reported by [257–259]. The spectrum at RT as presented in the latter work is reproduced in Fig. 3.34 and concerns a species from Anloua, Cameroon. It was recorded under a geometry whereby the incident γ-ray was perpendicular to the crystallographic *ac* plane and had therefore to be analyzed by a superposition of two asymmetric quadrupole doublets arising from Fe^{2+} at the Fe(1) and Fe(2) sites respectively. The asymmetry is due to texture effects as a result of the EFG's principal axis being non-randomly oriented with respect to the incident γ-ray direction. Such a situation leads to a quadrupole doublet of which the two composing absorption lines have unequal spectral areas. Fitting the spectrum of Fig. 3.34 yielded a line-area asymmetry of 0.64 for both doublets, which is consistent with the conclusion of Forsyth (1970) that for both Fe sites the EFG's principal axis is lying in the *ac* plane. The hyperfine parameters are given in Table 3.19. Both Gonser et al. [257] and De Grave [259] observed a significant deviation of the area ratio of the two subspectra from the ideal value of 1:2 and ascribed this feature to the pronounced effective thickness of the single-crystal absorber.

Partly oxidized vivianites, commonly as powders, have been studied by Mössbauer spectroscopy by De Grave et al. [260], McCammon and Burns [261], and Dormann et al. [262]. Various oxidation degrees q have been considered. Two room temperature spectra are reproduced in Fig. 3.35 referring to $q = 0.31$ and $q = 0.14$, respectively, as derived from the analyses of the spectra using two ferrous and two ferric doublets. The assignment of the various doublet components to the Fe(1) and Fe(2) sites is specified in Table 3.19, together with their respective δ_{Fe} and Δ values. It was found that the quadrupole splitting for the Fe^{2+} and the Fe^{3+} cations at the Fe(2) sites is not affected by the oxidation degree q. On the other hand, for the Fe(1) sites both the ferrous and ferric Δ decrease with increasing q. It was further noticed that the oxidation of Fe^{2+} to Fe^{3+} preferably takes place at the Fe(1) octahedra so that the fraction of total iron as Fe^{3+} at the Fe(2) sites does not exceed 0.05 in the range q $\leq$ 0.32.

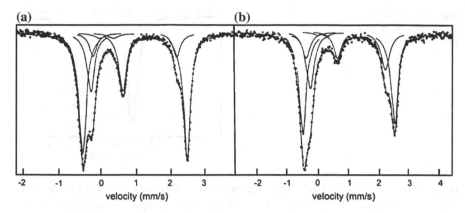

Fig. 3.35 Mössbauer spectra at room temperature of oxidized vivianites, $Fe^{2+}_{3-q}Fe^{3+}_q(PO_4)_2 \cdot (8 - q)H_2Oq(OH)$, with (a) $q = 0.31$ and (b) $q = 0.14$

3.6.6 Triphylite

Lithium-iron phosphate, $LiFePO_4$, has received ample attention in recent years because of its potential application as electrode active material for rechargeable lithium batteries (see [263] and references therein). It occurs in nature and as such is known as the mineral triphylite. It has an olivine- type crystallographic structure in which the ferrous cations occupy strongly distorted corner-sharing octahedral M2 sites, which form zig-zag chains running parallel to the c-axis. A second type of edge-sharing octahedral sites, M1, forms linear chains that are also directed along the c-axis and are occupied by lithium cations. Each $M1O_6$ octahedron shares edges with two adjacent $M1O_6$ octahedra, with two $M2O_6$ octahedra and with two PO_4 tetrahedra. The $M2O_6$ octahedron has common edges with two $M1O_6$ octahedra and one PO_4 tetrahedron. Mössbauer spectra of both naturally occurring and synthetic triphylites have been reported by several authors (Van Alboom et al. [264] and references therein). Below ~ 52 K the material is antiferromagnetically ordered. At higher temperatures the Mössbauer spectrum consists of a narrow doublet with a relatively high quadrupole splitting (see Table 3.19).

3.6.7 Heterosite

This mineral, ideally $Fe^{3+}PO_4$, possesses the same structure as triphylite, however, with all M1 sites being vacant. Actually, the two mentioned minerals are the end members of a complete solid solution. Heterosite also forms a solid solution with purpurite, $MnPO_4$. The Mössbauer spectrum recorded at 80 K for a natural Mn-substituted heterosite species (Buranga, Rwanda) is reproduced in Fig. 3.36 and very similar spectra were observed for all temperature higher than ~ 60 K

Fig. 3.36 Mössbauer
spectrum at 80 K of a natural
Mn-substituted heterosite
species

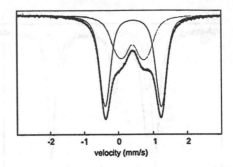

(unpublished results). Clearly the spectra consist of an outer doublet component
and an inner much broader doublet, and are most adequately analyzed with a
superposition of two quadrupole-splitting distributions. The maximum-probability
Δ values at RT were calculated to be 1.63 mm/s and 0.40 mm/s, respectively. On
the basis of the isomer shifts both doublets are attributable to ferric cations.

A value of 1.63 mm/s for the quadrupole splitting is unusually high for Fe^{3+}
species and is ascribed to a strong deformation of the M2 octahedra as a result of
the presence of Mn^{3+} cations, which are Jahn–Teller active. The appearance of two
distinct spectral components is not consistent with the availability of only one type
of site for the cations. Yamada and Chung [265] reported MS at RT for synthetic
$(Mn_yFe_{1-y})PO_4$ compounds, showing indeed only one doublet with Δ increasing
with increasing Mn content from 1.53 mm/s for $y = 0$, to 1.65 mm/s for $y = 0.5$.
Therefore, it is likely that the inner doublet in the MS of the Buranga heterosite is
due to a second Fe-bearing phase. This suggestion is corroborated by the observation that the inner doublet remains to be present at temperatures as low as 15 K,
at which the heterosite is magnetically ordered and produces a well-resolved sextet
(unpublished results).

Considering the substantial fraction of total iron that is present in this second
phase (~ 40 %), it is puzzling that this phase was not detected in the X-ray
diffractogram. A possible explanation could be that the involved impurity concerns
an amorphous iron phosphate. It should be noted that recently Fehr et al. [266]
reported the Mössbauer spectrum of a naturally occurring Mn-substituted heterosite
(Sandamab, Namibia). In addition to the heterosite doublet with $\Delta = 1.63$ mm/s, the
authors found an inner doublet having $\Delta = 0.69$ mm/s. They concluded that this
inner doublet is due to an associated iron-phosphate mineral.

3.6.8 Leucophosphite

Leucophosphite is an iron-potassium phosphate with ideal composition given as
$KFe_2(PO_4)_2(OH).2H_2O$. It is often found in guano deposits, in pegmatite deposits
and in lateritic crusts. The crystal structure of leucophosphite is triclinic. The
atomic arrangement is based on an octahedral tetramer involving an edge-sharing

Fig. 3.37 Mössbauer
spectrum at room temperature
of synthetic leucophosphite

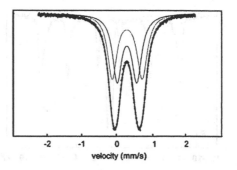

dimer (hereafter called Fe(2) sites) which further links by corner sharing to two other octahedrons (Fe(1) sites). The edge-sharing octahedrons possess two OH^- groups in *cis* position along the shared edge. The two other octahedrons each have a water molecule at the corner that is in *cis* position with respect to the shared corner. The K ions and one water molecule per formula unit are located in channels that run parallel to the [010] and [001] crystallographic axes.

The Mössbauer spectrum for a synthetic leucophosphite at RT is reproduced in Fig. 3.37 [267]. It is adequately fitted with two ferric quadrupole doublets, which are assigned to the Fe(1) and Fe(2) sites (see Table 3.19) on the basis that the Fe(1) are expected to exhibit the largest ΔE_Q value since the ferric species in these sites experience the largest charge asymmetry for their closest co-ordination shell. Within error limits the area ratio of the two doublets is 1:1, but the Fe(1) doublet has a slightly broader linewidth.

3.7 Examples

3.7.1 Goethite-Hematite-Ferrihydrite Associations

Most weathered mineral or soil-related samples contain simultaneously goethite, hematite and ferrihydrite. As a general rule, hematite occurs more frequently in places below 40° latitude whereas ferrihydrite is mostly found above that latitude [124]. In such composite samples, hematite exhibits nearly always a sextet with asymmetric lines at RT and remains predominantly weakly ferromagnetic down to 80 K. On the other hand, the spectrum of goethite is strongly dependent on the crystallinity and at RT can vary from a doublet for poorly crystallized goethite over a largely collapsed spectrum to a sextet with strongly asymmetric lines for relatively well-crystallized goethite. Due to the asymmetry of the absorption lines in the latter case, the involved subspectrum was often fitted with a few sextets with consecutively decreasing intensity. However, an adequate fit is only obtained if a distribution of hyperfine fields is considered.

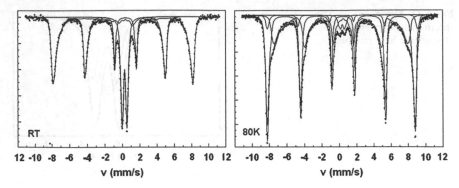

Fig. 3.38 Mössbauer spectrum at RT and at 80 K of a hematite-rich sample containing hematite and goethite

A first example concerns a sample from a weathered boulder in a granite outcrop in Sungai Ringit (Malaysia), which is relatively rich in hematite. The spectra at RT and at 80 K are displayed in Fig. 3.38.

The RT spectrum is composed of a sextet with asymmetric lines having a relatively high hyperfine field and a negative quadrupole shift, which is typical for hematite. The hyperfine parameters are collected in Table 3.19. The center of the spectrum shows a doublet that corresponds to Fe^{3+}, but can in principle not directly be assigned to a particular mineral. Further, close inspection learns that the spectrum exhibits a slightly curved bag-like shape, which has been accounted for by including in the fit a B-distributed sextet component in the low hyperfine-field region (5–35 T). Although the latter adjustment is not always theoretically correct because of the comparable strength of the quadrupole and the magnetic interaction [268], it adequately describes the overall spectral shape. This collapsed spectrum can usually be considered as originating from that fraction of the goethite particles that exhibit the highest degree of crystallinity.

At 80 K the doublet has almost completely disappeared in favor of a sextet with strongly asymmetrical line shape. The hyperfine parameters of this sextet point to goethite. This change is clearly the effect of super paramagnetism in which the sextet increases at the expense of the doublet by lowering the temperature as a result of a distribution of particle sizes. Because the hyperfine fields in the derived goethite field distribution are tending to low values, it can be expected that the remaining doublet at 80 K is due to that part of the goethite with the poorest crystallinity. The moderate quadruple splitting of that doublet (0.57 mm/s) points indeed in that direction because ferrihydrite, the presence of which cannot be excluded a priori, has usually $\Delta \geq 0.6$ mm/s. The sextet lines of hematite at 80 K are sharp and only slightly asymmetric. However, the spectrum itself is asymmetric in the sense that the first line is significantly deeper than the sixth one, whereas the it is opposite is noticed for lines 2 and 4. This feature is typical for the presence of the two magnetic hematite phases (cfr. Fig. 3.9). Introducing a weak AF hematite component indeed improved the fit significantly.

Table 3.20 Hyperfine parameters for the goethite-hematite sample of example 1

T	B_{av} (T)	B_p (T)	2ε or Δ (mm/s)	δ_{Fe} (mm/s)	RA (%)	Assignment
RT	48.3	49.7	−0.20	0.36	63	Hematite (WF)
	(21)	–	−0.2	0.36	10	Goethite (coll.)
	–	–	0.57	0.35	27	Goethite (SP)
80 K	52.4	52.9	−0.20	0.47	52	Hematite (WF)
	53.5	53.9	0.20	0.47	6	Hematite (AF)
	39.7	48.0	−0.24	0.48	39	Goethite (AF)
	–	–	0.56	0.46	3	Goethite (SP)

The relative spectral areas RA (see Table 3.20) provide an indication of the amount of Fe species in the different components. However, the relative area of hematite derived from the RT spectrum (63 %) is not the same as that observed at 80 K (52 % + 6 %). This quite strong reduction (about 8 %) cannot be attributed to the relative change in the Mössbauer fraction of hematite with respect to that of goethite on decreasing the temperature from RT to 80 K. This difference is rather an artifact of the distribution fitting applied to sextets where there is strong overlap of the broad goethite sextet with the low-field tail of the hematite sextet. So, the relative area is most accurately determined from the RT spectrum in which the goethite appears as a doublet and consequently the sextet of hematite is well resolved.

In a second example it is demonstrated that Mössbauer spectroscopy is helpful in determining the relative change in the amounts of the different Fe phases present in samples taken at different depths from a so-called Griffin Farmhill soil profile in South Africa [269]. The RT spectra show a dominant doublet together with a hematite sextet of which the intensity increases in the sequence C1 to C6, i.e. with increasing depth along the profile (Fig. 3.39). The 80 K spectra reveal a similar increase of goethite, leaving only a small amount of doublet for C6. It was found that the remaining doublet at 80 K for sample C1 possessed a large quadrupole splitting of 0.68 mm/s (Table 3.21) pointing to ferrihydrite. In order to verify the latter, spectra of sample C1 have been taken at lower temperatures.

The spectrum at 4 K shows three sextets: one of hematite, one of goethite and one of ferrihydrite (Fig. 3.40). However, there is still a doublet present at 4 K. At 15 K the spectrum shows that the contribution of this doublet has increased at the expense of the ferrihydrite sextet. This means that the doublet arises from ferrihydrite species behaves superparamagnetically at temperatures as low as 4 K. This feature can be explained by the presence of so-called DOM ferrihydrite [73] in view of the high carbon content in the topsoil samples.

The evolution of the relative spectral areas for the different iron-bearing phases as a function of depth is represented in Fig. 3.41, which thus provides an idea of the transformation of ferrihydrite to goethite and hematite in the different horizons. From this picture is clear that in the deeper layers ferrihydrite is nearly completely transformed to goethite and hematite.

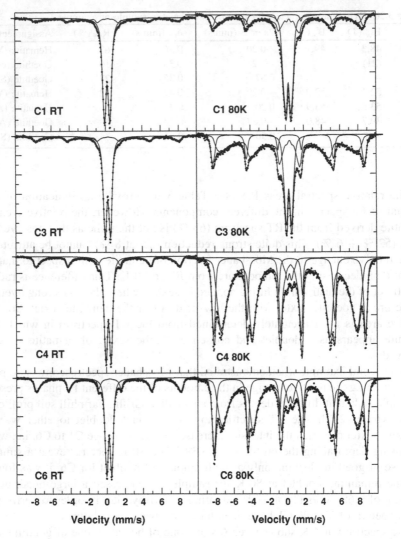

Fig. 3.39 Mössbauer spectra at RT and 80 K of samples from a Griffin Farmhill soil profile

3.7.2 Hidden Doublets

Relevant doublets of weak intensity are often largely hidden under other, more intense doublets or inner lines of sextet components and their presence will generally not be recognized although they have hyperfine parameters that deviate from the predominant doublets. In order to overcome this shortcoming to some extent, it is always advisable to throw a closer look at the central part of the spectrum.

Table 3.21 Hyperfine parameters of sample C1 at different temperatures

T (K)	B_{av} (T)	B_p (T)	$2\varepsilon/\Delta$ (mm/s)	δ_{Fe} (mm/s)	RA (%)	Assignment
300	37.0	49.1	−0.17	0.35	22	Hematite
	−	−	0.62	0.34	78	Goethite + Ferrihydrite
15	52.4	52.5	−0.17	0.47	23	Hematite (WF)
	47.9	49.1	−0.20	0.47	32	Goethite
	29.7	37.7	−0.06	0.46	19	Ferrihydrite
	−	−	0.69	0.44	26	Fe^{3+} doublet (Fh)
4	52.5	52.5	−0.17	0.47	27	Hematite (WF)
	48.6	49.3	−0.19	0.47	34	Goethite
	37.2	42.7	−0.06	0.42	32	Ferrihydrite
	−	−	0.67	0.41	7	Fe^{3+} doublet (Fh)

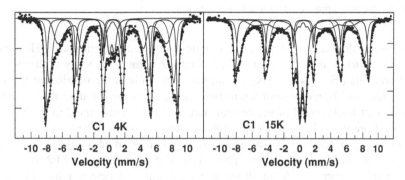

Fig. 3.40 Mössbauer spectra of C1 at 4 and 15 K

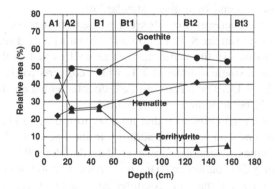

Fig. 3.41 Evolution of the relative spectral areas representing Fe in the different iron-bearing components as a function of depth

Figure 3.42a represents the full RT spectrum of one of the soil samples from Cameroun, which were investigated by MS in the framework of a soil mapping project. The spectrum shows the typical hematite sextet together with the

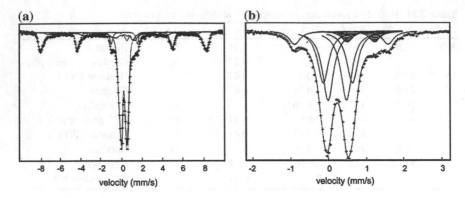

Fig. 3.42 RT Mössbauer spectrum of Cameroun soil sample (**a**) full spectrum, **b** central part—the shaded doublet represents the ilmenite subspectrum

non-resolved doublets of superparamagnetic goethite and ferrihydrite. However, the central part of that spectrum has been analyzed more carefully with conventional doublets (Fig. 3.42b). The overlapping long tails of the distributed lines of the goethite and hematite sextets can be considered as contributing to a constant non-resonant background in the central part. It was thus attempted to fit this part with two Lorentzian lines to account for the inner lines of the hematite sextet and two doublets for goethite. It was found, however, that an additional doublet was needed to obtain a reasonable fit. This extra doublet with $\delta_{Fe} = 1.02$ mm/s and $\Delta = 0.8$ mm/s corresponds to ilmenite which was not detected by any other technique. So, it clear that a closer inspection of the central part of the spectra might reveal the presence of some minor amounts of Fe-bearing minerals which would not have been found by a quick standard fit.

Another example concerns the Mössbauer measurement of iron-containing nodules in a planosol from the S–W Ethiopian highlands [270]. The RT Mössbauer spectrum of the nodules in the vertic horizon as well as that of the accumulated nodules consists of a single doublet at RT, which converts partly into a goethite sextet at 80 K (Fig. 3.43). The remaining Fe^{3+} doublet with a quadrupole splitting of 0.60 mm/s might be attributed to ferrihydrite.

At this point, one might conclude that no other iron-bearing phases are present unless another spectral component in addition to the one of goethite/ferrihydrite would be hidden in the Fe^{3+} doublet. However, if the doublet at RT is analyzed with a quadrupole-splitting distribution an extra maximum is found around 1.1 mm/s in addition to the expected one for goethite/ferrihydrite at about 0.5–0.6 mm/s (Fig. 3.44). The second sample showed the same features. The RT spectra were then accordingly reanalyzed with two doublets. The second doublet with $\delta_{Fe} = 0.37$ mm/s and $\Delta = 1.1$ mm/s appeared indeed in the fit (Fig. 3.44) and could be identified as the main doublet of bixbyite (FeMnO$_3$) [271, 272], the presence of which was detected by XRD. The other bixbyite doublet with a lower quadrupole splitting of 0.6 mm/s coincides with the main goethite/ferrihydrite doublet.

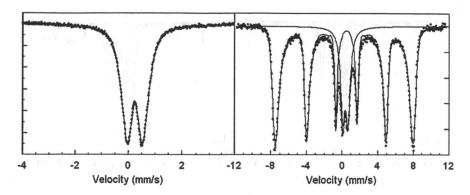

Fig. 3.43 Mössbauer spectrum at RT (*left*) and 80 K (*right*) of an iron nodule in a planosol

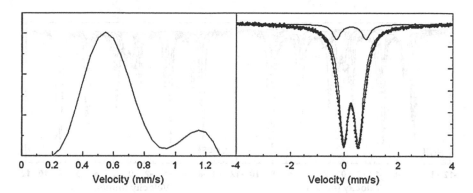

Fig. 3.44 Quadrupole distribution (*left*) from the RT spectrum of the nodule and the spectrum refitted with 2 doublets (*right*)

In conclusion, also the use of quadrupole-splitting distributions may reveal the presence of minor doublet fractions in multi-phase Mössbauer spectra.

3.7.3 Magnetic Mineralogy

Many soil or rock samples contain magnetic minerals other than hematite, such as magnetite, titanomagnetite and maghemite, which are often only present in small amounts. These magnetic compounds appear then as vague sextets in the Mössbauer spectra and as-such can hardly be analyzed. Magnetic separation is then the pre-eminently solution to bring this interesting magnetic mineralogy on the foreground. A strong hand magnet can already be used for that purpose, but a more sophisticated magnetic separator is obviously more efficient.

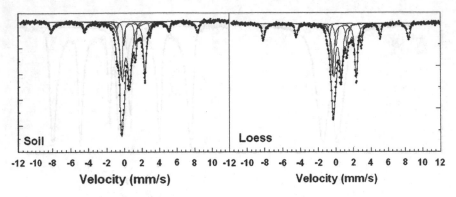

Fig. 3.45 RT spectra of topsoil and underlying loess

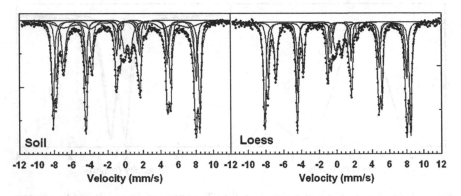

Fig. 3.46 RT spectra of magnetically separated topsoil and underlying loess

As a first example, a sample from the topsoil and one from an underlying loess of a paleosol-loess sequence at Xiadong in the Jixian loess section (China), will be considered. The RT Mössbauer spectra of soil and loess show a series of central doublets, which can be attributed to illite. Apart from these doublets the sextet typical for soil hematite is observed (Fig. 3.45).

However, it is well known that these soils, which possess an even higher magnetic susceptibility than the underlying loess, contain other magnetic components such as magnetite and maghemite. Apparently, their spectra fall below the detection or analyzing limit of MS. After the sample was submitted to a wet magnetic separation in a Franz isodynamic separator a totally different picture appeared in the Mössbauer spectra (Fig. 3.46). The sextets of magnetite, maghemite and of a more course-grained hematite could be adjusted to the spectrum of the soil, whereas for the loess sample only the sextets for magnetite and hematite were resolved. This was the first time that the magnetic components of loess and paleosol were visualized in a direct way [273, 274].

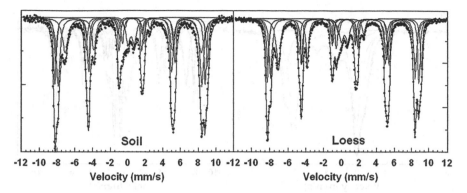

Fig. 3.47 Spectra at 130 K of magnetically separated topsoil and underlying loess

The samples were also measured at low temperature i.e. at 130 K in order to remain above the Verwey transition of magnetite (Fig. 3.47). In these spectra the AF and WF phase of hematite were simultaneously present demonstrating a better crystallinity for this hematite phase than for the hematite observed in the total soil and loess sample, for which at 80 K only the WF state was revealed.

This procedure of separation has also been applied to a series of samples from a paleosol-loess sequence section in Huangling (China). From the corresponding Mössbauer spectra and particularly from the relative spectral area ratios, interesting conclusions could be drawn with respect to the origin of the enhanced magnetic susceptibility of the soils in comparison with that of the loess [275]. A similar method was applied and concurrent results were obtained in a series of samples from a section in the north-eastern area of the Buenos Aires province [276].

A second example of magnetic separation concerns a study of fresh dolerite samples from Berg en Dal, 90 km north of Paramaribo in Suriname, in which MS was utilized as an additional characterization technique of the samples [277]. Apart from several doublets, the RT spectrum shows two weak sextets that can be assigned to magnetite. In order to study more carefully the magnetic components the sample has been subjected to a magnetic separation yielding a magnetic and a non-magnetic fraction. The RT spectra of both fractions are depicted in Fig. 3.48.

The spectrum of the magnetic fraction shows clearly the two sextets of magnetite and no other sextets were visible at first sight. In order to fit the spectrum adequately in the complex central part of the spectrum, those hyperfine parameters were used that were obtained from the fit of the spectrum of the non-magnetic fraction, which is not disturbed by inner sextet lines. The ratio of the relative areas for magnetite, $S(Fe^{2.5+})/S(Fe^{3+})$, turned out to be 2.2 which is by far larger than 1.8 as expected for pure magnetite. Because oxidation or substitution lowers the amount of $Fe^{2.5+}$ pairs and thus the relative area of the $Fe^{2.5+}$ sextet, there must be another sextet component present, which overlaps to some extent with the latter. This can hardly be another Fe^{3+} oxide component because this low field would correspond to a small particle morphology, which would lead to a rather bag-like

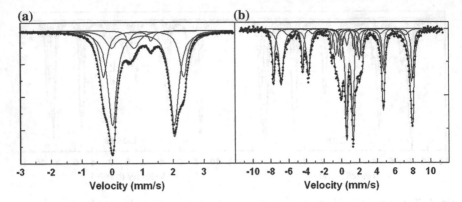

Fig. 3.48 RT spectrum of non-magnetic (**a**) and magnetic (**b**) fraction of a dolerite sample

shape of the spectrum. A sextet with low hyperfine field ($B = 42$ T) and high isomer shift ($\delta_{Fe} = 0.71$ mm/s) could be added pointing to a Fe^{2+} sextet as expected for a titanomagnetite (Fig. 3.48b).

3.7.4 The Strong Means: External Magnetic Fields

The question often arises to what extent one can resolve sextets in a Mössbauer spectrum. Of course there is no general answer. If sextets strongly overlap, it will be very difficult to resolve them in a correct way. Moreover, if they possess a hyperfine-field distribution, it will be practically impossible to discern the different components, even if other hyperfine parameters like the quadrupole shift or the isomer shift are significantly different. Only when the lines exhibit distinct shoulders one can expect another sextet component to be present because strong anomalies in hyperfine field distributions for the same iron species are rare.

In some cases, Mössbauer measurements with the sample in a high external field might be a welcome solution to improve the resolution of sextets. Ferrimagnetic compounds have the advantage that their magnetic moments align in a more or less strong external magnetic field. Consequently, the external field is added to the hyperfine field of the site with the smallest magnetic moment and subtracted from that of the site with the strongest magnetic moment because the hyperfine field is opposite to the magnetic moment.

A typical example is maghemite with its strongly overlapping sextets of A and B sites. This spectrum can be simply fitted with 2 sextets, but the resulting hyperfine parameters will depend on the initial values used in the fit. This ill-posed problem can only be solved by performing measurements in an external magnetic field, where the outer sextet from the A sites is well separated from that of the B sites (Fig. 3.49), yielding accurate values for both isomer shifts and hyperfine fields.

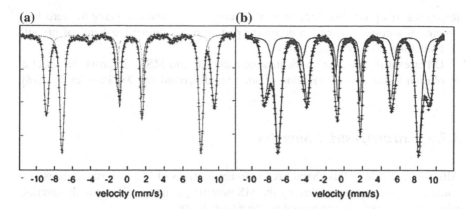

Fig. 3.49 External-field (6 T) spectra at RT of well-crystallized (**a**) and poorly crystallized (**b**) maghemite. Note the presence of lines 2 and 5 as an indication of spin canting

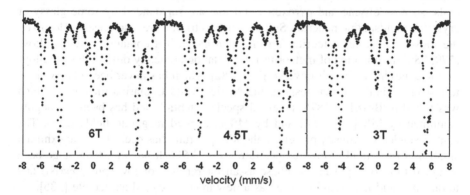

Fig. 3.50 Mössbauer spectra of the greigite-smythite sample at 80 K in different external magnetic fields showing the separation of the greigite lines

A mineral-related example was the use of external field measurements in the study of greigite [148]. The RT spectrum of a greigite-smythite sediment sample has been reproduced in Sect. 3.4.1 (Fig. 3.19). This spectrum has been fitted with two sextets from greigite, three sextets from smythite and a siderite doublet. As shown, the spectrum fit seems at first sight to be straightforward, but that was not at all the case. At that moment, the hyperfine fields on tetrahedral and octahedral sites of greigite were not well known and the purpose of that study was to determine the temperature behavior of the hyperfine fields. The only way to tackle the problem consisted of measuring the sample at 80 K in applied fields of various strengths. Because greigite is ferrimagnetic the external field separates its A- en B-site spectrum (Fig. 3.50), whereas the lines of smythite, being ferromagnetic shift more to center of the spectrum. Extrapolation for the different fields yielded the values of the hyperfine fields at 80 K and the respective isomer shifts.

Repeating this procedure for a few other temperatures and analyzing the zero-field spectra in a consistent way a complete picture of the hyperfine fields of greigite and smythite could be determined.

Other examples of the application of external-field MS to improve the spectral resolution in the characterization of soils are described by De Grave et al. [278].

3.7.5 Extraterrestrial Samples

Although the title of this chapter refers to applications of MS in earth sciences, one cannot exclude to mention briefly the MS investigations of soils and rock materials which are not strictly connected to the planet Earth.

First of all, a lot of extraterrestrial material has reached the Earth by means of meteorites. These meteorites are in 85 % of the cases so-called chondrites, which contain predominantly particles or chondrules that are composed of silicate minerals, such as olivine and pyroxene, and that are surrounded by some glassy or crystalline feldspathic material. Sulfides such as triolite are also present. However, the most interesting meteorites are the iron meteorites containing Fe–Ni alloys [279, 280]. In this kind of meteorite 1:1 crystallographically ordered Fe–Ni alloys (tetrataenite) have been discovered [281, 282], which has never been observed for terrestrial variants. In that respect, the Ni-rich Santa Catharina meteorite, which was found in Brazil in 1875, received special attention and has been thoroughly examined by MS [283, 284] and by MS in applied magnetic fields [285]. The central singlet, manifestly present in the RT spectrum, has been shown to exhibit a small hyperfine field at low temperatures (Fig. 3.51). It has in a later stage been assigned to a low-spin Fe–Ni phase as an intergrowth of low-spin γ-Fe(Ni) and tetrataenite and has been proposed as a new mineral, called antitaenite [286].

Secondly, different space programs for the exploration of the Moon were initiated in the sixties of the former century. During the American manned Apollo

Fig. 3.51 RT spectrum of a sample of the Santa Catharina meteorite showing the sextets of ordered 50–50 Fe–Ni (*dash*) and disordered 50–50 Fe–Ni (*dots*) and the singlet of the paramagnetic 28 % Ni phase (adapted from [284])

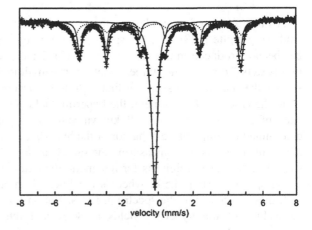

Fig. 3.52 RT spectra for impact-derived regolith samples at different lunar locations. Peak positions for the individual phases are indicated by the stick diagram (after Morris et al. [291])

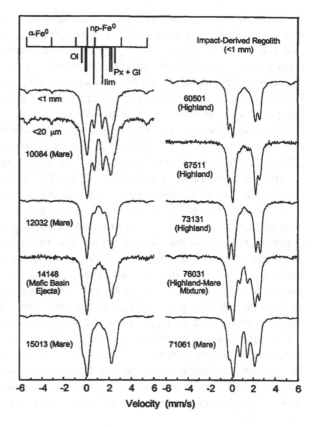

lunar missions between 1969 and 1972 almost 400 kg of material have been brought to earth. Also, the Russian Luna program succeeded in 1970 in returning lunar samples to the Earth. All those regolith samples have been investigated by MS, showing the spectra of Fe-bearing phases, such as olivine, glass, pyroxene, ilmenite, triolite, iron metal and Fe–Ni alloys [287–291]. Figure 3.52 shows a collection of RT spectra of impact-derived regolith samples, i.e. surface dust less than 1 mm depth, collected at different lunar locations, illustrating ilmenite and pyroxene to be the dominant crystalline phases in mare samples, whereas pyroxene and olivine are the dominant crystalline phases in highland samples.

Two Mars exploration rovers (MER) both equipped with a miniaturized Mössbauer spectrometer (MIMOS), developed by Gustar Klingelhöfer and coworkers [292, 293], were launched in 2003. The first Mössbauer spectrum was transmitted to Earth in January 2004 and was undoubtedly the most exciting event for every Mössbauer spectroscopist. Since then both spectrometers have collected more than thousand spectra during 5 years. The spectrometer on the Opportunity (MER-B) is still operating in 2011, although the collecting time has increased considerably due to the weak source, whereas since March 2010 the other rover, the Spirit (MER-A), is not responding anymore.

All the Mössbauer spectra recorded on Mars have been publically released so that every research group could dispose of them freely. Together with each sampled spectrum, a spectrum of a reference absorber was collected and also the recorded error signal of the driver is provided. As expected, the spectra are not linear in the velocity scale, which causes some calibration difficulties. Agresti et al. [284] calibrated the spectra by considering the α-Fe lines in the reference target, combined with the error signal of the drive.

The many Mössbauer investigations of the spectra taken by the MERs on soils and rocks along their pathways showed olivines and pyroxenes as the main silicates [294–296]. Most interesting was the obvious observation of a jarosite spectrum, inferring the former presence of water on Mars [297]. This was also corroborated by the identification of goethite in the spectra from some locations [298].

Concerning the magnetic spectra, hematite and magnetite were found as the predominant iron oxides. Unfortunately, those spectra with relatively high hyperfine fields were not so well defined due to significant deviations from linearity experienced in the highest velocities, which are not covered by the sextet lines of metallic Fe in the calibration spectra. The direct use of the hematite and magnetite in the reference target for calibration was initially not simple because along with the strongly overlapping spectral lines, hematite showed the two magnetic phases in the temperature windows of the measurements. By determining the hyperfine parameters of the five sextets of Fe metal, hematite and magnetite in a laboratory sample consisting of the same reference target material measured at the appropriate temperature windows, the spectra of the reference target could be calibrated [299]. With this calibration method, several magnetic spectra recorded on Mars at different temperature windows could be accurately analyzed [299, 300] (Fig. 3.53).

For instance, magnetite could be fitted without parameter constraints. From the fit there was also a strong indication that goethite exhibits two maxima in the hyperfine field distribution inferring two different goethite formations. It was further demonstrated that the Mars hematite measured in different temperature windows between 210 and 260 K showed both the AF and WF phase giving an idea about its morphology. Some spectra are shown in Fig. 3.53. Similar results were recently obtained by Agresti et al. [301], using a simfit method consisting of the simultaneous analysis of 60 spectra with multi-spectrum constraints. However, the latter results were interpreted by the presence of two kinds of hematites, i.e. one showing the Morin transition and another remaining in the WF phase.

Finally, it is worth to mention that the Opportunity came across several meteorites on the surface of Mars [302]. The first one was found in Meridiani Planum [303]. The recorded Mössbauer spectra showed this meteorite to consist mainly of kamacite, having about 7 wt % Ni [304, 319]. No taenite spectrum was observed indicating that this iron-nickel phase probably occurs in the meteorite only in a very small amount below the detection limit of the spectrometer.

Further extraterrestrial Mössbauer measurements are planned using an improved version of the MIMOS [319]. Phobos, one of the moons of Mars, is the next target for Mössbauer investigations during the Phobos-Grunt mission by the

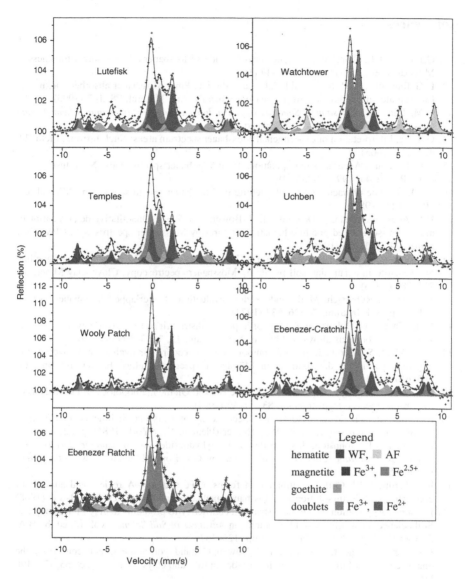

Fig. 3.53 Analyzed Mössbauer spectra taken on Mars by the Spirit at different spots in the temperature window 210–220 K (m 5) (adapted from Van Cromphaut et al. [299, 300])

Russian Space Agency. A joint ESA-NASA project aims to send two new rovers, both again equipped with a MIMOS, to Mars in 2018.

Acknowledgments We are grateful to Darby Dyar, Enver Murad and Jean-Marie Génin for providing precious information about some specific items. Darby Dyar, Richard Morris and Jean-Marie Génin are also acknowledged for giving the authorization of using some of their figures.

References

1. G.J. Long, T.E. Cranshaw, G. Longworth, The ideal Mössbauer effect absorber thicknesses. Mössbauer Eff. Ref. Data J. **6**, 42–49 (1983)
2. D.G. Rancourt, A.M. McDonald, A.E. Lalonde, J.Y. Ping, Mössbauer absorber thicknesses for accurate site populations in Fe-bearing minerals. Am. Mineral. **78**, 1–7 (1993)
3. G.M. Bancroft, *Mössbauer Spectroscopy. An Introduction for Inorganic Chemists and Geochemists* (McGraw-Hill, Maidenhead, 1973)
4. G. Lang, Interpretation of experimental Mössbauer spectrum areas. Nucl. Instrum. Meth. **24**, 425–428 (1963)
5. D.G. Rancourt, Accurate site populations from Mössbauer spectroscopy. Nucl. Instr. Meth. Phys. Res. B **44**, 199–210 (1989)
6. E. Murad, The characterization of goethite by Mössbauer spectroscopy. Am. Mineral. **67**, 1007–1011 (1982)
7. D.D. Amasiriwardena, E. De Grave, L.H. Bowen, S.B. Weed, Quantitative determination of aluminium-substituted goethite-hematite mixtures by Mössbauer spectroscopy. Clays Clay Miner. **34**, 250–256 (1986)
8. R.E. Vandenberghe, E. De Grave, G. De Geyter, C. Landuydt, Characterization of goethite and hematite in a Tunisian soil profile by Mössbauer spectroscopy. Clays Clay Miner. **34**, 275–280 (1986)
9. J. Hesse, A. Rübartsch, Model independent evaluation of overlapped Mössbauer spectra. J. Phys. E. Sci. Instrum. **7**, 526–532 (1974)
10. G. Le Caer, J.M. Dubois, Evaluation of hyperfine distributions from overlapped Mössbauer spectra of amorphous alloys. J. Phys. E. Sci. Instrum. **12**, 1083–1090 (1979)
11. C.O. Wivel, S. Morup, Improved compotutional procedure for evaluation of overlapping hyperfine parameter distributions in Mössbauer spectra. J. Phys. E. Sci. Instrum. **14**, 605–610 (1981)
12. R.E. Vandenberghe, E. De Grave, P.M.A. de Bakker, On the methodology of the analysis of Mössbauer spectra. Hyp. Interact. **83**, 29–49 (1994)
13. L.H. Bowen, S.B. Weed, Mössbauer spectroscopy of soils and sediments, in *Chemical Mössbauer Spectroscopy*, ed. by R.H. Herber (Plenum, New York, 1984), pp. 217–242
14. E. Murad, J.H. Johnston, Iron oxides and oxyhydroxides, in *Mössbauer Spectroscopy Applied to Inorganic Chemistry*, vol. 2, ed. by G.J. Long (Plenum, New York, 1987), pp. 507–582
15. R.L. Parfitt, C.W. Childs, Estimation of forms of Fe and Al—A review and analysis of contrasting soils by dissolution and Mössbauer methods. Aust. J. Soil Res. **26**, 121–144 (1988)
16. E. Murad, Application of ^{57}Fe Mössbauer spectroscopy to problems in clay mineralogy and Soil Science: possibilities and limitations, in *Advances in Soil Science*, vol. 12, ed. by B.A. Stewart (Springer, New York, 1990), pp. 125–157
17. R.E. Vandenberghe, E. De Grave, L.H. Bowen, C. Landuydt, Some aspects concerning the characterization of iron oxides and hydroxides in soils and clays. Hyp. Interact. **53**, 175–196 (1990)
18. L.H. Bowen, E. De Grave, R.E. Vandenberghe, Mössbauer effect studies of magnetic soils and sediments, in *Mössbauer Spectroscopy Applied to Magnetism and Material Science*, vol. 1, ed. by G.J. Long, F. Grandjean (Plenum, New York, 1993), pp. 115–159
19. E. Murad, Mössbauer spectroscopy of clays, soils and their mineral constituents. Clay Miner. **45**, 413–430 (2010)
20. J.B. Forsyth, I.G. Hedley, C.E. Johnson, The magnetic structure and hyperfine field of goethite (α-FeOOH). J. Phys. Chem. C **1**, 179–188 (1968)
21. E. De Grave, R.E. Vandenberghe, 57Fe Mössbauer effect study of well crystallized goethite (α-FeOOH). Hyp. Interact. **28**, 643 (1986)
22. D.C. Golden, L.H. Bowen, S.B. Weed, J.M. Bigham, Mössbauer studies of synthetic and soil-occuring aluminium-substituted goethites. Soil Sci. Soc. Am. J. **43**, 802–808 (1979)

23. B.A. Goodman, D.G. Lewis, Mössbauer spectra of aluminous goethites (α—FeOOH). J. Soil Sci. **32**, 351–364 (1981)
24. S.A. Fysh, P.E. Clark, Aluminium goethite: A Mössbauer study. Phys. Chem. Miner. **8**, 180–187 (1982)
25. E. Murad, U. Schwertmann, The influence of aluminum substitution and crystallinity on the Mössbauer spectra of goethite. Clay Miner. **18**, 301–312 (1983)
26. J. Fleisch, R. Grimm, J. Grübler, P. Gütlich, Determination of the aluminum content in natural and synthetic alumo-goethites using Mössbauer spectroscopy. J. Phys. Colloq. C1(41), 169–170 (1980)
27. T. Ericsson, A. Krishnamurthy, B. Srivastava, Morin-transition in Ti-substituted hematite: A Mössbauer study. Phys. Scr. **33**, 88–90 (1986)
28. J.H. Johnston, K. Norrish, [57]Fe Mössbauer spectroscopic study of a selection of Australian and other goethites. Aust. J. Soil Res. **19**, 231–237 (1981)
29. J. Friedl, U. Schwertmann, Aluminium influence on iron oxides: XVIII. The effect of Al substitution and crystal size on magnetic hyperfine fields of natural goethites. Clay Miner. **31**, 455–464 (1996)
30. D.G. Schulze, U. Schwertmann, The influence of aluminum on iron oxides. X. Properties of Al substituted goethites. Clay Miner. **19**, 521–539 (1984)
31. E. Wolska, U. Schwertmann, The mechanism of solid solution formation between goethite and diaspore. N Jb. Miner. Mh. **5**, 213–223 (1993)
32. C.A. Barrero, R.E. Vandenberghe, E. De Grave, G.M. Da Costa, A qualitative analysis of the Mössbauer spectra of aluminous goethites based on existing models, in ed. by I. Ortali. Conference Proceedings, vol 50 "ICAME 95" (Editrice Compositoir, Bologna, 1996)
33. C.A. Barrero, R.E. Vandenberghe, E. De Grave, The electrical hyperfine parameters in synthetic aluminogoethites. Czech J. Phys. **47**, 533–536 (1997)
34. C.A. Barrero, R.E. Vandenberghe, E. De Grave, A.L. Morales, The influence of the sample properties on the electrical hyperfine parameters in synthetic aluminogoethites. Hyp. Interact. **C2**, 209–212 (1997)
35. R.E. Vandenberghe, C.A. Barrero, G.M. Da Costa, E. Van San, E. De Grave, Mössbauer characterization of iron oxides and (oxy) hydroxides: The present state of the art. Hyp. Interact. **126**, 247–259 (2000)
36. C.A. Barrero, R.E. Vandenberghe, E. De Grave, A.L. Morales, H. Perez, The experimental nuclear quadrupole interaction in synthetic Al-goethites of various crystallinity. Hyp. Interact. **148/149**, 337–344 (2003)
37. J.A.M. Gómez, V.G. de Resende, J. Antonissen, E. De Grave, Influence of Mn-for-Fe substitution on structural properties of synthetic goethite. Hyp. Interact. **189**, 143–149 (2009)
38. R.E. Vandenberghe, A.E. Verbeeck, E. De Grave, W. Stiers, [57]Fe Mössbauer effect Study of Mn-substituted goethite and hematite. Hyperfine Interact. **29**, 1157–1160 (1986)
39. R.M. Cornell, R. Giovanoli, P.W. Schindler, Clays Clay Miner. **35**, 21–28 (1987)
40. T.G. Quin, G.J. Long, C.G. Benson, S. Mann, R.J. Williams, Influence of silicon and phosphorus on structural and magnetic properties of synthetic goethite and related oxides. Clays Clay Miner. **36**, 165–175 (1988)
41. S.K. Kwon, K. Kimijima, K. Kanie, S. Suzuki, A. Muramatsu, M. Saito, K. Shinoda, Y. Waseda, Influence of silicate ions on the formation of goethite from green rust in aqueous solution. Corros. Sci. **49**, 2946–2961 (2007)
42. D. Chambaere, E. De Grave, On the Néel temperature of β-FeOOH: structural dependence and its implications. J. Magn. Magn. Mater. **42**, 263–268 (1984)
43. C.W. Childs, B.A. Goodman, E. Paterson, F.W.D. Woodhams, Nature of iron in akaganeite (9-FeOOH). Aust. J. Chem. **33**, 15–26 (1980)
44. D. Chambaere, E. De Grave, R.L. Vanleerberghe, R.E. Vandenberghe, The electric field gradient at the iron sites in β-FeOOH. Hyp. Interact. **20**, 249–262 (1984)
45. D. Chambaere, E. De Grave, On the influence of the double iron co-ordination on the hyperfine field in β-FeOOH. J. Magn. Magn. Mater. **44**, 349–352 (1984)

46. C.E. Johnson, Antiferromagnetism of γ-FeOOH: A Mössbauer effect study. J. Phys. C: Solid State Phys. **2**, 1996–2002 (1969)

47. E. Murad, U. Schwertmann, The influence of crystallinity on the Mössbauer spectrum of lepidocrocite. Mineral. Mag. **48**, 507–511 (1984)

48. E. De Grave, R.M. Persoons, D.G. Chambaere, R.E. Vandenberghe, L.H. Bowen, An ^{57}Fe Mössbauer effect study of poorly crystalline γ-FeOOH. Phys. Chem. Miner. **13**, 61–67 (1986)

49. U. Schwertmann, E. Wolska, The influence of aluminum on iron oxides. XV. Al-for-Fe substitution in synthetic lepidocrocite. Clays Clay Miner. **38**, 209–212 (1990)

50. E. De Grave, G.M. da Costa, L.H. Bowen, U. Schwertmann, R.E. Vandenberghe, ^{57}Fe Mössbauer-effect study of Al-substituted lepidocrocites. Clays Clays Miner. **44**, 214–219 (1996)

51. F.V. Chukhrov, B.B. Zvyagin, A.I. Gorshkov, L.P. Ermilova, V.V. Korovushkin, E.S. Rudnitskaya, N. Yu Yakubovskaya, Ferroxyhyte, a new modification of FeOOH. Izvest. Akad. Nauk. SSSR. Ser. Geol. **5**, 5–24 (1976)

52. Y.N. Vodyanitskii, A.V. Sivtsov, Formation of ferrihydrite, ferroxyhyte, and vernadite in soil. Eurasian Soil Sci. **37**, 863–875 (2004)

53. L. Carlson, U. Schwertmann, Natural occurrence of feroxyhite (δ'-FeOOH). Clays Clay Miner. **28**, 272–280 (1980)

54. M.B. Madsen, S. Mørup, C. Bender Koch, A study of microcrystals of synthetic feroxyhite (Ω'-FeOOH). Surf. Sci. **156**, 328–334 (1985)

55. K.M. Towe, W.F. Bradley, Mineralogical constitution of colloidal 'hydrous ferric oxides'. J. Colloid Interface Sci. **24**, 384–392 (1967)

56. R.A. Eggleton, R.W. Fitzpatrick, New data and a revised structural model for ferrihydrite. Clays Clay Miner. **36**, 111–124 (1988)

57. C.W. Childs, Ferrihydrite: A review of structure, properties and occurrence in relation to soils. Z. Pflanzenernähr. Bodenk. **155**, 441–448 (1992)

58. R.L. Parfitt, C.W. Childs, A structural model for natural siliceous ferrihydrite. Clays Clay Miner. **40**, 675–681 (1992)

59. V.A. Drits, B.A. Sakhorov, A.L. Salyn, A. Manceau, Structural model for ferrihydrite. Clay Miner. **28**, 185–207 (1993)

60. E. Jansen, A. Kyek, W. Schäfer, U. Schwertmann, The structure of six-line ferrihydrite. Appl. Phys. A **74**, 04–06 (2002)

61. F.M. Michel, L. Ehm, S.M. Antao, P.L. Lee, P.J. Chupas, L. Gang, D.R. Strongin, M.A.A. Schoonen, B.L. Phillips, J.B. Parise, The structure of ferrihydrite, a nanocrystalline material. Science **316**, 1726–1729 (2007)

62. R. Harrington, M. Michel, J. Parise, D. Hausner, D. Strongin, Powder neutron diffraction studies of ferrihydrite, a nanocrystalline material. Geochim. Cosmochim. Acta **74**(Suppl 1), A383–A383 (2010)

63. E. Murad, L.H. Bowen, G.J. Long, G.J. Quin, The influence of crystallinity on magnetic ordering in natural ferrihidrites. Clay Miner. **23**, 161–173 (1988)

64. C. Gilles, P. Bonville, K.K.W. Wong, S. Mann, Non-Langevin behaviour of the uncompensated magnetization in nanoparticles of artificial ferritin. Eur. Phys. J. B **17**, 417–427 (2000)

65. Y. Guyodo, S.K. Banerjee, R.L. Penn, D. Burleson, T.S. Berquó, T. Seda, P. Solheid, Magnetic properties of synthetic six-line ferrihydrite nanoparticles. Phys. Earth Plant. Int. **157**, 222–233 (2006)

66. G. De Geyter, R.E. Vandenberghe, L. Verdonck, G. Stoops, Mineralogy of Holocene bog iron ore in northern Belgium. Neues Jahrb. Miner. Abh. **153**, 1–17 (1985)

67. U. Schwertmann, J. Friedl, A. Kyek, Formation and properties of a crystallinity series of synthetic ferrihydrites (2- to 6-line) and their relation to FeOOH forms. Clays Clay Miner. **52**, 221–226 (2004)

68. E. Murad, The Mössbauer spectrum of "well"-crystallized ferrihydrite. J. Magn. Magn. Mater. **74**, 153–157 (1988)

69. J. Chadwick, D.H. Jones, M.F. Thomas, G.J. Tatlock, R.W. Devenish, A Mössbauer study of ferrihydrite and aluminium substituted ferrihydrites. J. Magn. Magn. Mater. **61**, 88–100 (1986)

70. L. Carlson, U. Schwertmann, Natural ferrihydrites in surface deposits from Finland and their association with silica: Geochim. Cosmochim. Acta **45**, 421–429 (1981)

71. A.S. Campbell, U. Schwertmann, H. Stanjek, J. Friedl, A. Kyek, P.A. Campbell, Si incorporation into hematite by heating Si-ferrihydrite. Langmuir **18**, 7804–7809 (2002)

72. T.S. Berquó, S.K. Banerjee, R.G. Ford, R.L. Pichler, T. Penn, High crystallinity Si-ferrihydrite: An insight into its Néel temperature and size dependence of magnetic properties. J. Geophys. Res. **112**, B02102 (2007). doi:10.1029/2006JB004583

73. U. Schwertmann, F. Wagner, H. Knicker, Ferrihydrite–humic associations: magnetic hyperfine interactions. Soil Sci. Soc. Am. J. **69**, 1009–1015 (2005)

74. U. Schwertmann, Differenzierung der Eisenoxide des Bodens durch Extraktion mit Ammoniumoxalat-Lösung Z. Pflanzenernähr. Düng. Bodenk. **195**, 194–202 (1964)

75. U. Schwertmann, D.G. Schulze, E. Murad, Identification of ferrihydrite in soils by dissolution kinetics, differential X-ray-diffraction, and Mössbauer-spectroscopy. Soil Sci. Am. J. **46**, 869–875 (1982)

76. J.A. McKeague, J.H. Day, Dithionite- and oxalate-extractable Fe and Al as aids in differentiating various classes of soils. Can. J. Soil Sci. **46**, 13–22 (1966)

77. U. Schwertmann, Use of oxalate for Fe extraction from soils. Can. J. Soil Sci. **53**, 244–246 (1973)

78. A.L. Walker, The effects of magnetite on oxalate-and dithionite-extractable iron. Soil Sci. Soc. Am. J. **47**, 1022–1026 (1983)

79. A.S. Campbell, U. Schwertmann, Evaluation of selective dissolution extractants in soil chemistry and mineralogy by differential X-ray diffraction. Clay Miner. **20**, 515–519 (1985)

80. J. Arocena, G. De Geyter, C. Landuydt, U. Schwertmann, Dissolution of soil iron oxides with ammonium oxalate: Comparison between bulk samples and thin sections. Pedologie **XXXIX-3**, 275–297 (1989)

81. F. van der Woude, Mössbauer Effect in α-Fe₂O₃. Phys. Status Solidi **17**, 417–432 (1966)

82. W. Kündig, H. Bömmel, G. Constabaris, R.H. Linquist, Some properties of supported small α-Fe₂O₃ particles determined with the Mössbauer effect. Phys. Rev. **142**, 327–333 (1966)

83. A.M. van der Kraan, Mössbauer effect studies of surface ions of ultrafine α-Fe₂O₃ particles. Phys. Status Solidi (a) **18**, 215–226 (1973)

84. T. Shinjo, M. Kiyama, N. Sugita, K. Watanabe, K. Takada, Surface magnetism of α- Fe₂O₃ by the Mössbauer spectroscopy. J. Magn. Magn. Mater. **35**, 133–135 (1983)

85. T. Yang, A. Krishnan, N. Benczer-Koller, G. Bayreuther, Surface magnetic hyperfine interactions in Fe₂O₃ determined by energy-resolved conversion-electron. Phys. Rev. Lett. **48**, 1292–1295 (1982)

86. C. Van Cromphaut, V.G. de Resende, E. De Grave, R.E. Vandenberghe, Surface effects in α-Fe₂O₃ nanoparticles studied by ILEEMS and TMS. Hyp. Interact. **191**, 167–171 (2009)

87. D.G. Rancourt, S.R. Julian, J.M. Daniels, Mössbauer characterization of very small superparamagnetic particles: Application to intra-zeolitic Fe₂O₃. J. Magn. Magn. **49**, 305–316 (1985)

88. R.C. Nininger Jr, D. Schroeer, Mössbauer studies of the Morin transition in bulk and microcrystalline α-Fe₂O₃. J. Phys. Chem. Solids **39**, 137–144 (1978)

89. A.E. Verbeeck, E. De Grave, R.E. Vandenberghe, Effect of the particle morphology on the Mössbauer effect in α-Fe₂O₃. Hyp. Interact. **28**, 639–642 (1986)

90. E. De Grave, R.E. Vandenberghe, Mössbauer effect study of the spin structure in natural hematites. Phys. Chem. Miner. **17**, 344–352 (1990)

91. N. Amin, S. Arajs, Morin temperature of annealed submicronic α-F₂O₃ particles. Phys. Rev. B **35**, 4810–4811 (1987)

92. E. De Grave, L.H. Bowen, D.D. Amarasiriwardena, R.E. Vandenberghe, ⁵⁷Fe Mössbauer effect study of highly substituted aluminum hematites: determination of the magnetic hyperfine field distributions. J. Magn. Magn. Mater. **72**, 129–140 (1988)

93. M.-Z. Dang, D.G. Rancourt, J.E. Dutrizac, G. Lamarche, R. Provencher, Interplay of surface conditions, particle size, stoichiometry, cell parameters, and magnetism in synthetic hematite-like materials. Hyp. Interact. **117**, 271–319 (1998)

94. J.F. Bengoa, M.S. Moreno, S.G. Marchetti, R.E. Vandenberghe, R.C. Mercader, Study of the Morin transition in pseudocubic α-Fe_2O_3 particles. Hyp. Interact. **161**, 177–183 (2005)

95. P.M.A. de Bakker, E. Grave, R.E. Vandenberghe, L.H. Bowen, R.J. Pollard, R.M. Persoons, Mössbauer study of the thermal decomposition of lepidocrocite and characterization of the decomposition products. Phys. Chem. Miner. **18**, 131–143 (1991)

96. E. Van San, E. De Grave, R.E. Vandenberghe, H.O. Desseyn, L. Datas, V. Barrón, A. Rousset, Study of Al-substituted hematites, prepared from thermal treatment of lepidocrocite. Phys. Chem. Miner. **28**, 488–497 (2001)

97. R.E. Vandenberghe, E. Van San, E. De Grave, G.M. Da Costa, About the Morin transition in hematite in relation with particle size and aluminium substitution. Czech J. Phys. **51**, 663–675 (2001)

98. U. Schwertmann, R.W. Fitzpatrick, R.M. Taylor, D.G. Lewis, The influence of aluminum on iron oxides: II. Preparation and properties of aluminum-substituted hematites. Clays Clay Miner. **27**, 105–112 (1979)

99. S.A. Fysh, P.E. Clark, Aluminium hematite: a Mössbauer study. Phys. Chem. Miner. **8**, 257–267 (1982)

100. E. De Grave, L.H. Bowen, S.B. Weed, Mössbauer study of aluminum-substituted hematites. J. Magn. Magn. Mater. **27**, 98–108 (1982)

101. E. De Grave, D. Chambaere, L.H. Bowen, Nature of the Morin transition in Al-substituted hematite. J. Magn. Magn. Mater. **30**, 349–354 (1983)

102. E. Murad, U. Schwertmann, Influence of Al substitution and crystal size on the room-temperature Mössbauer spectra of hematite. Clays Clay Miner. **34**, 1–6 (1986)

103. G.M. da Costa, E. Van San, E. De Grave, R.E. Vandenberghe, V. Barrón, L. Datas, Al hematites prepared by homogeneous precipitation of oxinates: material characterization and determination of the Morin transition. Phys. Chem. Miner. **29**, 122–131 (2002)

104. V. Baron, J. Gutzmer, H. Rundlof, R. Tellgren, Neutron powder diffraction study of Mn-bearing hematite, α-$Fe_2 - {}_xMn_xO_3$, in the range $0 \leq x \leq 0.176$. Solid State Sci. **7**, 753–759 (2005)

105. R.M. Cornell, R. Giovanoli, Effect of manganese on the transformation of ferrihydrite into goethite and jacobsite in alkaline media. Clays Clay Miner. **35**, 11–20 (1987)

106. R.E. Vandenberghe, A.E. Verbeeck, E. De Grave, On the Morin transition in Mn-substituted hematite. J. Magn. Magn. Mater. **54–57**, 898–900 (1986)

107. G. Shirane, D.E. Cox, W.J. Takei, S.L. Ruby, A study of the magnetic properties of the $FeTiO_3$-αFe_2O_3 system by neutron diffraction and the Mössbauer effect. J. Phys. Soc. Jpn. **17**, 1598–1611 (1962)

108. A.H. Muir Jr, R.M. Housley, R.W. Grant, M. Abdel-Gawad, M. Blander, Mössbauer spectroscopy of Moon samples. Science **167**, 688–690 (1970)

109. R.W. Grant, R.M. Housley, S. Geller, Hyperfine interactions of Fe^{2+} in ilmenite. Phys. Rev. B **5**, 1700–1703 (1972)

110. W. Kim, I.I.J. Park, C.S. Kim, Mössbauer study of magnetic structure of cation-deficient iron sulfide $Fe_{0.92}S$. J. Appl. Phys. **105**, 07D535–07D535-3 (2009)

111. W.Q. Guo, S. Malus, D.H. Ryan, Z. Altounian, Crystal structure and cation distributions in the $FeTi_2O_5$-Fe_2TiO_5 solid solution series. J. Phys. Condens. Matter. **11**, 6337–6346 (1999)

112. L. Häggström, H. Annersten, T. Ericsson, R. Wäppling, W. Karner, S. Bjarman, Magnetic dipolar and electric quadrupolar effects on the Mössbauer spectra of magnetite above the Verwey transition. Hyp. Interact. **5**, 201–214 (1978)

113. H. Annersten, S.S. Hafner, Vacancy distribution in synthetic spinels of the series Fe_3O_4-γ-Fe_2O_3. Z. Kristallogr. **137**, 321–340 (1973)

114. A. Ramdani, J. Steinmetz, C. Gleitzer, J.M.D. Coey, J.M. Friedt, Perturbation de l'échange electronique rapide par des lacunes cationiques dans $Fe_{3-x}O_4$ ($x \leq 0.09$). J. Phys. Chem. Solids **48**, 217–228 (1987)

115. C.I. Pearce, C.M.B. Henderson, N.D. Teiling, R.A.D. Pattrick, D.J. Vaughan, J.M. Charnock, E. Arenholz, F. Tuna, V.S. Coker, G. van der Laan, Iron site occupancies in magnetite-ulvöspinel solid solution: a new approach using X-ray magnetic circular dichroism. Am. Mineral. **95**, 425–439 (2010)

116. H. Tanaka, M. Kono, Mössbauer spectra of titanomagnetite: A reappraisal. J. Geomag. Geoelectr. **39**, 463–475 (1987)

117. H.H. Hamdeh, K. Barghout, J.O. Ho, P.M. Shand, L.L. Miller, A Mössbauer evaluation of cation distribution in titanomagnetites. J. Magn. Magn. Mater. **191**, 72–78 (1999)

118. R.S. Hargrove, W. Kündig, Mössbauer measurements of magnetite below the Verwey transition. Solid State Commun. **8**, 303–308 (1970)

119. Y. Miyahara, Impurity effects on the transition temperature of magnetite. J Phys. Soc. Jpn. **32**, 629–634 (1972)

120. V.A.M. Brabers, F. Waltz, H. Kronmuller, Impurity effects upon the Verwey transition in magnetite. Phys. Rev. B **58**, 14163–14166 (1998)

121. M.M. Hanzlik, N. Petersen, R. Keller, E. Schmidbauer, Electron microscopy and ^{57}Fe Mössbauer spectra of 10 nm particles, intermediate in composition between Fe_3O_4–γ-Fe_2O_3, produced by bacteria. Geophys. Res. Lett. **23**, 479–482 (1996)

122. G.M. da Costa, E. De Grave, L.H. Bowen, P.M.A. de Bakker, R.E. Vandenberghe, The center shift in Mössbauer spectra of maghemite and aluminum maghemites. Clays Clay Miner. **42**, 628–633 (1994)

123. J.M.D. Coey, D. Khalafalla, Superparamagnetic γ-Fe_2O_3. Phys. Status Solidi (a) **11**, 229–242 (1972)

124. U. Schwertmann, Occurrence and formation of iron in various pedenvironments, in *Iron in Soils and Clay Minerals*, vol. 217, NATO ASI Series, Series C: Math. and Phys. Sci., ed. by J.W. Stucki, B.A. Goodman, U. Schwertmann (D.Reidel Publication, Dordrecht, 1988), pp. 267–308

125. G.M. da Costa, C.H. Laurent, E. De Grave, R.E. Vandenberghe, A comprehensive Mössbauer study of highly-substituted aluminium maghemite. eds. by M.D. Dyar, C. McCammon, M.W. Schaefer, Mineral Spectroscopy: A Tribute to Roger G. Burns, The Geochemical Society (Special Publication 5, 1996) pp. 93–104

126. G.M. da Costa, E. De Grave, R.E. Vandenberghe, Mössbauer studies of maghemites and Al-substituted maghemites. Hyp. Interact. **117**, 207–243 (1998)

127. J.E.M. Allan, J.M.D. Coey, I.S. Sanders, U. Schwertmann, G. Friedrich, A. Wiechowski, An occurrence of a fully-oxidized natural titanomaghemite in basalt. Miner. Mag. **53**, 299–304 (1989)

128. W. Xu, D.R. Peacor, W.A. Dollase, R. Van Der Voo, R. Beaubouef, Transformation of titanomagnetite to titanomaghemite: A slow-step oxidation ordering process in MORB. Am. Mineral. **82**, 1101–1110 (1997)

129. S. Collyer, N.W. Grimes, D.J. Vaughan, G. Longworth, Studies of the crystal structure and crystal chemistry of titanomaghemite. Am. Mineral. **73**, 153–160 (1988)

130. E. Murad, R.M. Taylor, The Mössbauer spectra of hydroxycarbonate green rusts. Clay Miner. **19**, 77–83 (1984)

131. J.M.R. Génin, Ph Bauer, A.A. Olowe, D. Rézel, Mössbauer study of the kinetics of simulated corrosion process of iron in chlorinated aqueous solution around room temperature: the hyperfine structure of ferrous hydroxides and green rust I. Hyp. Interact. **29**, 1355–1360 (1986)

132. A.A. Olowe, J.M.R. Génin, Ph Bauer, Hyperfine interactions and structures of ferrous hydroxide and green rust II in sulfated aqueous media. Hyp. Interact. **41**, 501–504 (1988)

133. S.H. Drissi, Ph Refait, M. Abdelmoula, J.-M.R. Génin, Preparation and thermodynamic properties of Fe(II)-Fe(III) hydroxycarbonate (green rust 1), Pourbaix diagram of iron in carbonate-containing aqueous media. Corros. Sci. **37**, 2025–2041 (1995)

134. C. Ruby, M. Abdelmoula, S. Naille, A. Renard, V. Khare, G. Ona-Nguema, G. Morin, J.M.R. Génin, Oxidation modes and thermodynamics of $Fe^{II–III}$ oxyhydroxycarbonate green

rust: dissolution-precipitation versus in situ deprotonation. Geochim. Cosmochim. Acta **74**, 953–966 (2009)

135. F. Trolard, J.M.R. Génin, M. Abdelmoula, G. Bourrié, B. Humbert, A.J. Herbillon, Identification of a green rust mineral in a reductomorphic soil by Mössbauer and Raman spectroscopies. Geochim. Cosmochim. Acta **61**, 1107 (1997)

136. M. Abdelmoula, F. Trolard, G. Bourrié, J.M.R. Génin, Evidence of Fe(II)-Fe(III) Green rust "Fougerite" mineral occurrence in hydromorphic soil and its transformation with depth. Hyp. Interact. **112**, 235–238 (1998)

137. F. Féder, F. Trolard, G. Klingelhöfer, G. Bourrié, In situ Mössbauer spectroscopy evidence for green rust (fougerite) in a gleysol and its mineralogical transformations with time and depth. Geochim. Cosmochim. Acta **69**, 4463–4483 (2005)

138. J.-M.R. Génin, O. Guérin, A.J. Herbillon, E. Kuzmann, S.J. Mills, G. Morin, G. Ona-Nguema, C. Ruby, C. Upadhyay, Redox topotactic reactions in Fe^{II-III} oxyhydroxycarbonate new minerals related to fougèrite in gleysols: "trébeurdenite and mössbauerite". Hyp. Interact. **204**(1–3), 71–81 (2012)

139. B. Rusch, J.M.R. Génin, C. Ruby, M. Abdelmoula, P. Bonville, Ferrimagnetic properties of Fe(II-III) (oxy)hydoxycarbonate green rust. Solid State Sci. **10**, 40 (2008)

140. S. Hafner, M. Kalvius, The Mössbauer resonance of Fe in troilite and pyrrhotite. Z. Krist. **123**, 443–458 (1966)

141. L.F. Power, H.A. Fine, The iron-sulphur system. Part 1. The structures and physical properties of the compounds of the low-temperature phase fields. Miner. Sci. Eng. **8**, 106–128 (1976)

142. A.D. Elliot, Structure of pyrrhotite $5C$ (Fe_9S_{10}). Acta Cryst. **B66**, 271–279 (2010)

143. R. Gosselin, M.G. Townsend, R.J. Tremblay, A.H. Webster, Mössbauer effect in single-crystal $Fe_{1-x}S$. J. Solid State Chem. **17**, 43–48 (1976)

144. M.G. Townsend, A.H. Webster, J.L. Harwood, H. Roux-Buisson, Ferrimagnetic transition in $Fe_{0.9}S$—Magnetic, thermodynamic and kinetic aspects. J. Phys. Chem. Solids **40**, 183–189 (1979)

145. V.P. Gupta, A.K. Singh, K. Chandra, S.K. Jaireth, Investigations of pyrrhotites of Indian ore deposits. in *ED Proceedings of the Indian Science Academy*, International Conferences on the Application of the Mössbauer Effect, Jaipur 1981, (Indian Nat Science Academy, New Delhi, 1982), pp. 863–865

146. M. Saporoschenko, C.C. Hinckley, H. Twardowska, G.V. Smith, O. Zahraa, R.H. Shiley, K.L. Konopka, Mössbauer study of synthetic pyrrhotite. in *ED Proceedings of the Indian Science Academy, International Conference on the Application of the Mössbauer Effect, Jaipur 1981*, (Indian National Science Academy, New Delhi, 1982), pp. 869–871

147. C. Jeandey, J.L. Oddou, J.L. Mattei, G. Fillion, Mössbauer investigation of the pyrrhotite at low temperature. Solid State Commun. **78**, 195–198 (1991)

148. R.E. Vandenberghe, E. De Grave, P.M.A. de Bakker, M. Krs, J.J. Hus, Mössbauer study of natural greigite. Hyp. Interact. **68**, 319–322 (1991)

149. V. Hoffmann, H. Stanjek, E. Murad, Mineralogical, magnetic and Mössbauer data of smythite (Fe_9S_{11}). Studia Geophys. Geod. **37**, 366–380 (1993)

150. J.M.D. Coey, M.R. Spender, A.H. Morrish, Magnetic structure of spinel Fe_3S_4. Solid State Commun. **8**, 1605–1608 (1970)

151. L. Chang, B.D. Rainford, J.R. Stewart, C. Ritter, A.P. Roberts, Y. Tang, Q. Chen, Magnetic structure of greigite (Fe_3S_4) probed by neutron powder diffraction and polarized neutron diffraction. J. Geophys. Res. **114**, B07101 (2009). doi:10.1029/2008JB006260

152. A.P. Roberts, L. Chang, C.J. Rowan, C.-S. Horng, F. Florindo, Magnetic properties of sedimentary greigite (Fe_3S_4): An update. Rev. Geophys. **49**, RG1002 (2011). doi:10.1029/2010RG000336

153. J.A. Morice, L.V.C. Rees, D.T. Rickard, Mössbauer studies of iron sulphides. J. Inorg. Nucl. Chem. **31**, 3797–3802 (1969)

154. D.J. Vaughan, M.S. Ridout, Mössbauer studies of some sulphide minerals. J. Inorg. Nucl. Chem. **33**, 741–746 (1971)

155. O. Knop, C.-H. Huang, F.W.D. Woodhams, Chalcogenides of the transition elements. VII. A Mössbauer study of pentlandite. Am. Mineral. **55**, 115–1130 (1970)
156. P.L. Wincott, D.J. Vaughan, Spectroscopic studies of sulfides. Rev. Mineral. **61**, 181–229 (2006)
157. F.C. Hawthorne, S.V. Krivovichev, P.C. Burns, The crystal chemistry of sulfate minerals. Rev. Mineral. Geochem. **40**, 1–112 (2000)
158. J.G. Stevens, A.M. Khasanov, J.W. Miller, H. Pollak, Z. Li (eds.), *Mössbauer Minerals Handbook* (Mössbauer Effect Data Center, Asheville, 1998)
159. V.A. O'Connor, Comparative crystal chemistry of hydrous iron sulfates from different terrestrial environments. Ph.D thesis, Mount Holyoke College, South Hadley, (2005)
160. M.D. Dyar, D.G. Agresti, M.W. Schaefer, C.A. Grant, E.C. Sklute, Mössbauer spectroscopy of earth and planetary materials. An. Rev. Earth Planetary Sci. **34**, 83–125 (2006)
161. A. Ertl, M.D. Dyar, J.M. Hughes, F. Brandstätter, M.E.M. Gunther, R.C. Peterson, Pertlikite, a new tetragonal Mg-rich member of the voltaite group from Madeni Zakh. Iran. Can. Mineral. **46**, 661–669 (2008)
162. A. Van Alboom, V.G. De Resende, E. De Grave, J.A.M. Gómez, Hyperfine interactions in szomolnokite (FeSO$_4$·H$_2$O). J. Molec. Struct. **924–926**, 448–456 (2009)
163. P.P. Gil, A. Pesquera, F. Velasco, X-ray diffraction, infrared and Mössbauer studies of Fe-rich carbonates. Eur. J. Miner. **4**, 521–526 (1992)
164. V.I. Goldanskii, E.F. Makarov, I.P. Suzdalev, I.A. Vinogradov, Quantitative test of the vibrational anisotropy origin of the asymmetry of quadrupole Mössbauer doublets. Phys. Rev. Lett. **20**, 137–140 (1968)
165. T. Ericsson, R. Wäppling, Texture effects in 3/2-1/2 Mössbauer spectra. J. Phys. **C6(37)**, 719–726 (1976)
166. K.K.P. Srivastava, A Mössbauer study of slow spin relaxation of paramagnetic Fe^{2+} in MgCO$_3$. J. Phys. C. Solid State Phys. **16**, 1137–1139 (1983)
167. E. De Grave, R. Vochten, An ^{57}Fe mössbauer effect study of ankerite. Phys. Chem. Miner. **12**, 108–113 (1985)
168. E. De Grave, ^{57}Fe Mössbauer effect in ankerite: Study of the electronic relaxation. Solid State Commun. **60**, 541–544 (1986)
169. G. Hilscher, P. Rogl, J. Zemann, T. Ntaflos, Low-temperature magnetic investigation of ankerite. Eur. J. Miner. **17**, 103–105 (2005)
170. R.J. Reeder, W.A. Dollase, Structural variation in the dolomite-ankerite solid-solution series: An X-ray, Mössbauer, and TEM study. Am Miner. **74**, 1159–1167 (1989)
171. M.W. Schaefer, Measurements of iron(Ill)-rich fayalites. Nature **303**, 325–327 (1983)
172. J.F. Duncan, J.H. Johnston, The determination of the cation distribution in olivine from single crystal Mössbauer studies. Aust. J. Chem. **26**, 231–239 (1973)
173. R. Santoro, R. Newnham, S. Nomura, Magnetic properties of Mn$_2$SiO$_4$ and Fe$_2$SiO$_4$. J. Phys. Chem. Solids **27**, 655–666 (1966)
174. W. Lottermoser, K. Forcher, G. Amthauer, H. Fuess, Powder- and single crystal Mössbauer spectroscopy on synthetic fayalite. J. Phys. Chem. Miner. **22**, 259–267 (1995)
175. F. Belley, E.C. Ferré, F. Martín-Hernández, M.J. Jackson, M.D. Dyar, E.J. Catlos, The magnetic properties of natural and synthetic (Fe$_x$, Mg$_{1-x}$)$_2$SiO$_4$ olivines. Earth and Planet. Sci. Lett. **284**, 516–526 (2009)
176. W. Kündig, J.A. Cape, R.H. Lindquist, G. Constabaris, Some magnetic properties of Fe$_2$SiO$_4$ from 4 K to 300 K. J. Appl. Phys. **38**, 947–948 (1967)
177. S.S. Hafner, J. Stanek, M. Stanek, ^{57}Fe hyperfine interactions in the magnetic phase of fayalite, Fe$_2$SiO$_4$. J. Phys. Chem. Solids **51**, 203–208 (1990)
178. X. Kan, J.M.D. Coey, Mössbauer spectra, magnetic and electrical properties of laihunite, a mixed-valence iron olivine mineral. Am. Mineral. **70**, 567–580 (1985)
179. M.W. Schaefer, Site occupancy and two-phase character of "ferrifayalite". Am. Mineral. **70**, 729–736 (1985)
180. G. Amthauer, H. Annersten, S.S. Hafner, The Mössbauer spectrum of ^{57}Fe in silicate garnets. Zeit. Kristallogr. **143**, 14–55 (1976)

181. E. Murad, F.E. Wagner, The Mössbauer spectrum of almandine. Phys. Chem. Miner. **14**, 264–269 (1987)
182. E. Murad, Magnetic ordering in andradite. Am. Mineral. **69**, 722–724 (1984)
183. K.B. Schwartz, D.A. Nolet, R.G. Burns, Mössbauer spectroscopy and crystal chemistry of natural Fe-Ti garnets. Am. Mineral. **65**, 142–153 (1980)
184. G.M. Bancroft, A.G. Maddock, R.G. Burns, Application of the Mössbauer effect of silicate mineralogy: I. Iron silicates of known crystal structure. Geochim. Cosmochim. Acta **31**, 831–834 (1967)
185. W.A. Dollase, Mössbauer spectra and iron distribution in the epidote-group minerals. Z. Krist. **138**, 41–63 (1973)
186. K.T. Fehr, S. Heuss-Assbichler, Intracrystalline equilibria and immiscibility along the join clinozoisite-epidote. An experimental and ^{57}Fe Mössbauer study. N. Jb. Min. Abh. **172**, 43–67 (1997)
187. M. Grodzicki, S. Heuss-Assbichler, G. Amthauer, Mössbauer investigations and molecular orbital calculations. Phys. Chem. Miner. **28**, 675–681 (2001)
188. A.K. Dzhemai, Distribution of cations in structures of iron magnesia silicates. Staurolites. V.A. Glebovitskii, ed. by in *Raspred Kationov Termodin Zhelezo-Magrez Tverd Rastvorov Silik.* (Izv. Nauka, Leningrad Old, Leningrad 1978), pp. 136–152
189. M.D. Dyar, C.L. Perry, C.R. Rebbert, B.L. Dutrow, M.J. Holdway, H.M. Lang, Mössbauer spectroscopy of synthetic and naturally occurring staurolite. Am. Mineral. **76**, 27–41 (1991)
190. M. Akasaka, M. Nagashima, K. Makino, H. Ohashi, Distribution of Fe$_3$ + in a synthetic (Ca, Na)$_2$(Mg, Fe^{3+})Si$_2$O$_7$–melilite: ^{57}Fe Mössbauer and X-ray Rietveld studies. J. Mineral. Petrol. Sci. **100**, 229–236 (2005)
191. B. Ghazi-Bayat, M. Behruzi, F.J. Litterst, W. Lottermoser, G. Amthauer, Crystallographic phase transition and valence fluctuation in synthetic Mn-bearing ilvaite CaFe2+2–xMnxFe3+[Si2O7/O/(OH)]. Phys. Chem. Miner. **18** 491–496 (1992)
192. F.J. Litterst, G. Amthauer, Electron delocalization in ilvaite, a reinterpretation of its ^{57}Fe Mössbauer spectrum. Phys. Chem. Miner. **10**, 250–255 (1984)
193. C.R. Dotson, B.J. Evans, The effects of chemical composition on electron delocalization and magnetic ordering in ilvaite, Ca[Fe^{2+},Fe^{3+}][Fe^{2+}]Si$_2$O$_7$O(OH). J. Appl. Phys. **85**, 5235–5236 (1999)
194. D.A. Nolet, Electron delocalization observed in the Mössbauer spectrum of ilvaite. Solid State Commun. **28**, 719–722 (1978)
195. D.A. Nolet, R.G. Burns, Ilvaite: A study of temperature dependent electron delocalization by the Mössbauer effect. Phys. Chem. Miner. **4**, 221–234 (1979)
196. N. Zotov, W. Kockelman, S.D. Jacobsen, I. Mitov, D. Paneva, R.D. Vassileva, I.K. Bonev, Structure and cation ordering in manganilvaite: a combined X-ray diffraction, neutron diffraction, and Mössbauer study. Can. Mineral. **43**, 1043–1053 (2005)
197. G. Amthauer, W. Lottermoser, G. Redhammer, G. Tippelt, Mössbauer studies of selected synthetic silicates. Hyp. Interact. **113**, 219–248 (1998)
198. D.C. Price, E.R. Vance, G. Smith, A. Edgar, B.L. Dickson, Mössbauer effect studies on beryl. J. Phys. **C6**(37), 811–816 (1976)
199. R.R. Viana, G.M. da Costa, E. De Grave, H. Jordt-Evangelista, W.B. Stern, Characterization of beryl (aquamarine variety) by Mössbauer spectroscopy. Phys. Chem. Miner. **29**, 78–86 (2002)
200. J.F. Duncan, J.H. Johnston, Single crystal ^{57}Fe Mössbauer studies of the site positions in cordierite. Aust. J. Chem. **27**, 249–258 (1974)
201. C.A. Geiger, T. Armbruster, V. Khomenko, S. Quartieri, Cordierite I: The coordination of Fe^{2+}. Am. Mineral. **85**, 1255–1264 (2000)
202. R.G. Burns, Mixed valencies and site occupancies of iron in silicate minerals from Mössbauer spectroscopy. Can. J. Spectr. **17**, 51–59 (1972)
203. Y. Fuchs, M. Lagache, J. Linares, R. Maury, F. Varret, Mössbauer and optical spectrometry of selected schorl-dravite tourmalines. Hyperfine Interact. **96**, 245–258 (1995)

204. A. Pieczka, J. Kraczka, W. Zabinski, Mössbauer spectra of Fe^{3+}-poor schorls: reinterpretation on the basis of the ordered structure model. J. Czech Geol. Soc. **43**, 69–74 (1998)

205. G.M. da Costa, C. Casteneda, N.S. Gomes, N.S. Pedrosa-Soares, C.M. Santana, On the analysis of the Mössbauer spectra of tourmalines. Hyp. Interact. **2**, 29–34 (1997)

206. M.D. Dyar, M.E. Taylor, T.M. Lutz, C.A. Francis, C.V. Guidotti, M. Wise, Inclusive chemical characterization of tourmaline: Mössbauer study of Fe valence site occupancy. Am. Mineral. **83**, 848–864 (1998)

207. S.G. Eeckhout, C. Corteel, E. Van Coster, E. De Grave, P. De Paepe, Crystal-chemical characterization of tourmalines from the English Lake District: Electron-microprobe analyses and Mössbauer spectroscopy. Am. Mineral. **89**, 1743–1751 (2004)

208. B.J. Evans, S. Ghose, S.S. Hafner, Hyperfine splitting of ^{57}Fe and Mg-Fe order-disorder in orthopyroxenes ($MgSiO_3$–$FeSiO_3$ solid solution). J. Geol. **75**, 306–322 (1967)

209. M.D. Dyar, R.L. Klima, D. Lindsley, C.M. Pieters, Effects of differential recoil-free fraction on ordering and site occupancies in Mössbauer spectroscopy of orthopyroxenes. Am. Mineral. **92**, 424–428 (2007)

210. S.G. Eeckhout, E. De Grave, C.A. McCammon, R. Vochten, Temperature dependence of the hyperfine parameters in synthetic $P21/c$ Mg-Fe pyroxenes along the $MgSiO_3$-$FeSiO_3$ join. Am. Mineral. **85**, 943–952 (2000)

211. G.M. Bancroft, P.G.L. Williams, R.G. Burns, Mössbauer spectra of minerals along the diopside—hedenbergite tie line. Am. Mineral. **56**, 1617–1625 (1971)

212. S.G. Eeckhout, E. De Grave, ^{57}Fe Mössbauer-effect studies of Ca-rich, Fe-bearing clinopyroxenes: Part I. Paramagnetic spectra of magnesian hedenbergite. Am. Mineral. **88**, 1128–1137 (2003)

213. E. De Grave, S.G. Eeckhout, ^{57}Fe Mössbauer-effect studies of Ca-rich, Fe-bearing clinopyroxenes: Part III Diopside. Am. Mineral. **88**, 1145–1152 (2003)

214. E. Dowty, D.H. Lindslay, Mössbauer spectra of synthetic hedenbergite-ferrosilite pyroxenes. Am. Mineral. **58**, 850–868 (1973)

215. L.P. Aldridge, G.M. Bancroft, M.E. Fleet, C.T. Herzberg, Omphacite studies; II, Mössbauer spectra of C2/c and P2/n omphacites. Am. Mineral. **63**, 1107–1115 (1978)

216. E. De Grave, A. Van Alboom, S.G. Eeckhout, Electronic and magnetic properties of a natural aegirine as observed from its Mössbauer spectra. Phys. Chem. Miner. **25**, 378–388 (1998)

217. W.R. Nelson, D.T. Griffen, Crystal chemistry of Zn-rich rhodonite ("fowlerite"). Am. Mineral. **90**, 969–983 (2005)

218. D.T. Griffen, W.R. Nelson, Mössbauer spectroscopy of Zn-poor and Zn-rich rhodonite. Am. Mineral. **92**, 1486–1491 (2007)

219. F.A. Seifert, D. Virgo, Kinetics of the Fe^{2+}-Mg, order-disorder reaction in anthophyllites: quantitative cooling rates. Science **188**, 1107–1109 (1975)

220. G.M. Bancroft, R.G. Burns, A.G. Maddock, Determination of cation distribution in the cummingtonite-grunerite series by Mössbauer spectra. Am. Mineral. **52**, 1009–1026 (1967)

221. G.M. Bancroft, A.G. Maddock, Cation distribution in anthophyllite from Mössbauer and infra-red spectroscopy. Nature **212**, 913–915 (1966)

222. M. Schindler, E. Sokolova, Y. Abdu, F.C. Hawthorne, B.W. Evans, K. Ishida, The crystal chemistry of the gedrite-group amphiboles. I. Crystal structure and site populations. Miner. Mag. **72**, 703–730 (2008)

223. F.S. Spears, The gedrite-anthophyllite solvus and the composition limits of orthoamphibole from the Post Pond Volcanics, Vermont. Am. Mineral. **65**, 1103–1118 (1980)

224. A.D. Law, E.J.W. Whittaker, Studies of the orthoamphiboles.1. The Mössbauer and infrared spectra of holmquistite. Bull. Mineral. **104**, 381–386 (1981)

225. R.G. Burns, C. Greaves, Correlations of infrared and Mössbauer site population measurements of actinolites. Am. Mineral. **56**, 2010–2033 (1971)

226. D.S. Goldman, A reevaluation of the Mössbauer spectroscopy of calcic amphiboles. Am. Mineral. **64**, 109–118 (1979)

227. G.M. Bancroft, R.G. Burns, A.J. Stone, Applications of the Mössbauer effect to silicate mineralogy. II. Iron silicates of unknown and complex crystal structures. Geochim. Cosmochim. Acta **32**, 547–559 (1968)

228. G.M. Bancroft, R.G. Burns, Mössbauer and absorption spectral study of alkali amphiboles. Mineral. Soc. Am. Spec. Pap. **2**, 137–148 (1969)

229. D.G. Rancourt, Mössbauer spectroscopy of minerals: I. Inadequacy of Lorentzian-line doublets in fitting spectra arising from quadrupole distributions. Phys. Chem. Miner. **21**, 244–249 (1994)

230. J. De Grave, P. De Paepe, E. De Grave, R. Vochten, S.G. Eeckhout, Mineralogical and Mössbauer spectroscopic study of a diopside occurring in the marbles of Andranondamo, southern Madagascar. Am. Mineral. **87**, 132–141 (2002)

231. A. Van Alboom, E. De Grave, Temperature dependence of the ^{57}Fe Mössbauer parameters in riebeckite. Phys. Chem. Miner. **23**, 377–386 (1996)

232. R.G. Burns, M.D. Dyar, Crystal chemistry and Mössbauer spectra of babingtonite. Am. Mineral. **76**, 892–899 (1991)

233. G. Amthauer, K. Langer, M. Schliestedt, Thermally activated electron delocalization in deerite. Phys. Chem. Miner. **6**, 19–30 (1980)

234. E. Murad, U. Wagner, Mössbauer spectra of kaolinite, halloysite and the firing products of kaolinite: new results and a reappraisal of published work. N. Jb Miner. Abh. **162**, 281–309 (1991)

235. I. Rozenson, E.R. Bauminger, L. Heller-Kallai, Mössbauer spectra of iron in 1:1 phyllosilicates. Am. Mineral. **64**, 893–901 (1979)

236. D.S. O'Hanley, M.D. Dyar, The composition of lizardite 1 T and the formation of magnetite in serpentinite. Am. Mineral. **78**, 391–404 (1993)

237. J.M.D. Coey, A. Moukarika, C.M. McDonagh, Electron hopping in cronstedtite. Solid State Commun. **41**, 797–800 (1982)

238. O. Ballet, J.M.D. Coey, Greenalite—A clay showing two-dimensional magnetic order. J. Phys. **C6**(39), 765–766 (1978)

239. K.J.D. Mackenzie, R.M. Berezowski, Thermal and Mössbauer studies of iron-containing hydrous silicates. V. Berthierine. Thermochimica Acta **74**, 291–312 (1984)

240. J.M.D. Coey, Mössbauer spectroscopy of silicate minerals, in *Mössbauer Spectroscopy Applied to Inorganic Chemistry*, vol. 1, ed. by G.J. Long (Plenum, New York, 1984), pp. 443–509

241. J.H. Johnston, C.M. Cardile, Iron substitution in montmorillonite, illite and glauconite by ^{57}Fe Mössbauer spectroscopy. Clays Clay Miner. **35**, 170–176 (1987)

242. E. Murad, J. Cashion, *Mössbauer spectroscopy of environmental materials and their industrial utilization* (Kluwer, Boston, 2004)

243. E. Murad, U. Wagner, Mössbauer spectrum of illite. Clay Miner. **29**, 1–10 (1994)

244. E. De Grave, J. Vandenbruwaene, E. Elewaut, An ^{57}Fe Mössbauer effect study on glauconites from different locations in Belgium and northern France. Clay Miner. **20**, 171–179 (1985)

245. L.H. Bowen, E. De Grave, D.A. Reid, R.C. Graham, S.B. Edinger, Mössbauer study of a California desert celadonite and its pedogenically-related smectite. Phys. Chem. Miner. **16**, 697–703 (1989)

246. J.M.D. Coey, T. Bakas, S. Guggenheim, Mössbauer spectra of minnesotaite and ferrous talc. Am. Mineral. **76**, 1905–1909 (1991)

247. O. Ballet, J.M.D. Coey, Magnetic properties of sheet silicates; 2:1layer minerals. Phys. Chem. Miner. **8**, 218–229 (1982)

248. C. Blaauw, G. Stroink, W. Leiper, Mössbauer analysis of talc and chlorite. J. Phys. **C141**, 411–412 (1980)

249. M.D. Dyar, R.G. Burns, Mössbauer spectral study of ferriginous one-layer trioctahedral micas. Am. Mineral. **71**, 955–965 (1986)

250. M.D. Dyar, A review of Mössbauer data on trioctahdral micas: Evidence of tetrahedral Fe^{3+} and cation ordering. Am. Mineral. **72**, 102–112 (1987)

251. E. De Grave, J. Vandenbruwaene, M. Van Bockstael, Mössbauer spectroscopic analysis of chlorite. Phys. Chem. Miner. **15**, 173–180 (1987)
252. S.G. Eeckhout, E. De Grave, R. Vochten, N.M. Blaton, Mössbauer effect study of anapaite, $Ca_2Fe^{2+}(PO_4)_2.4H_2O$, and of its oxidation products. Phys. Chem. Minerals **26**, 506–512 (1999)
253. G.M. da Costa, R. Scholz, J. Karfunkel, V. Bermanec, M.L.S.C. Chavez, ^{57}Fe-Mössbauer spectroscopy on natural eosphorite-childrenite-ernstite samples. Phys. Chem. Miner. **31**, 714–720 (2005)
254. R. Vochten, E. De Grave, Mössbauer- and infrared spectroscopic characterization of ferristrunzite from Blaton, Belgium. N. Jb. Miner. Mh. 176–190 (1990)
255. R. Vochten, E. De Grave, K. Van Springel, L. Van Haverbeke, Mineralogical and Mössbauer spectroscopic study of some strunzite varieties of the Silbergrube, Waidhaus, Oberpfalz, Germany. N. Jb. Mineral. Mh. 11–25 (1995)
256. R. Van Tassel, E. De Grave, Ferrostrunzite from Arnsberg, Sauerland, Germany. N. Jb. Miner. Mh. 207–212 (1992)
257. U. Gonser, R.W. Grant, Determination of spin directions and electric field gradient axis in vivianite by polarized recoil-free γ-rays. Phys. Stat. Sol. **21**, 331–342 (1967)
258. J.R. Forsyth, C.E. Johnston, C. Wilkinson, The magnetic structure of vivianite, $Fe_3(PO_4)_2.8H_2O$. J. Phys. C: Solid State Phys. **3**, 1127–1139 (1970)
259. E. De Grave, ^{57}Fe-Mössbauerspectroscopie: fundamentele bijdragen en praktische toepassingen in de fysika, de mineralogie en de technologie. (Thesis Hoger Aggregaat, University of Gent) 1983
260. E. De Grave, R. Vochten, M. Desseyn, D. Chambaere, Analysis of some oxidized vivianites. J. Phys. (Paris) Colloq. **41**, 407–408 (1980)
261. C.A. McCammon, R.G. Burns, The oxidation mechanism of vivianite as studied by Mössbauer spectroscopy. Am. Mineral. **65**, 361–366 (1980)
262. J.L. Dormann, M. Gaspérin, J.F. Poullen, Etude structural de la séquence d'oxydation de la vivianite $Fe_3(PO_4)_2.8H_2O$. Bull. Minér. **105**, 147–160 (1982)
263. L. Aldon, A. Perea, M. Womes, C.M. Ionica-Bousquet, J.-C. Jumas, Determination of the Lamb-Mössbauer fractions of of $LiFePO_4$ and $FePO_4$ for electrochemical in situ and operando measurements in Li-ion batteries. J. Solid State Chem. **183**, 218–222 (2010)
264. A. Van Alboom, E. De Grave, M. Wolfahrt-Mehrens, Temperature dependence of the Fe^{2+} Mössbauer parameters in triphylite ($LiFePO_4$). Am. Mineral. **96**, 408–416 (2011)
265. A. Yamada, S. Chung, Crystal chemistry of the olivine-type $Li(Mn_yFe_{1-y})PO_4$ and $(Mn_yFe_{1-y})PO_4$ as possible 4 V cathode materials for lithium batteries. J. Electrochem. Soc. **148**, A960–A967 (2001)
266. T.H. Fehr, R. Hochleitner, A. Laumann, E. Schmidbauer, J. Schneider, Mineralogy, Mössbauer spectroscopy and electrical conductivity of heterosite $(Fe^{3+},Mn^{3+})PO_4$. Phys. Chem. Miner. **37**, 179–189 (2010)
267. V.G. de Resende, G.M. da Costa, E. De Grave, A. Van Alboom, Mössbauer spectroscopic study of synthetic leucophosphite, $KFe_2(PO_4)_2(OH).2H_2O$. Am. Mineral. **93**, 483–498 (2008)
268. G. Le Caër, J.M. Dubois, H. Fisher, U. Gonser, H.G. Wagner, On the validity of ^{57}Fe hyperfine field distribution calculations from Mössbauer spectra of magnetic amorphous alloys. Nucl. Instrum. Meth. Phys. Res. B **5**, 25–33 (1984)
269. R.E. Vandenberghe, E. Van Ranst, E. De Grave, Mössbauer study of a South African Griffin Farmhill soil profile. ICAME 2005, Montpellier, Book of Abstracts (2005)
270. E. Van Ranst, M. Dumon, A.R. Tolossa, J-Th Cornelis, G. Stoops, R.E. Vandenberghe, R.J. Deckers, Revisiting ferrolysis processes in the formation of Planosols for rationalizing the soils with stagnic properties in WRB. Geoderma **163**, 265–274 (2011)
271. E. Banks, E. Kostiner, G.K. Wertheim, Mössbauer effect in $MnFeO_3$. J. Chem. Phys. **45**, 1189–1191 (1966)
272. S.N. de Medeiros, A. Luciano, L.F. Cótica, I.A. Santos, A. Paesano Jr, J.B.M. da Cunha, Structural and magnetic characterization of the ball-milled α-Fe_2O_3 –Mn_2O_3 and α-Fe–Mn_2O_3 systems. J. Magn. Magn. Mater. **281**, 227–233 (2004)

273. J.M. Han, J.J. Hus, R. Paepe, R.E. Vandenberghe, T.S. Liu, The rock magnetic properties of the Malan and Lishi formations in the loess plateau of China, in *Loess, Environment and Global Change*, ed. by Liu Tungsheng (Science Press, Beijing, 1991), pp. 30–47

274. R.E. Vandenberghe, E. De Grave, J.J. Hus, J. Han, Characterization of Chinese loess and associated palaeosol by Mössbauer spectroscopy. Hyp. Interact. **70**, 977–980 (1992)

275. R.E. Vandenberghe, J.J. Hus, E. De Grave, Evidence from Mössbauer spectroscopy of neo-formation of magnetite/maghemite in the soils of loess/palaeosol sequences in China. Hyp. Interact. **117**, 359–369 (1998)

276. R.C. Mercader, F.R. Sives, P.A. Imbellone, R.E. Vandenberghe, Magnetic and Mössbauer studies of quaternary Argentine loessic soils and paleosols. Hyp. Interact. **161**, 43–53 (2005)

277. C. Algoe, G. Stoops, R.E. Vandenberghe, E. Van Ranst, Selective dissolution of Fe-Ti oxides—Extractable iron as a criterion for andic properties revisited. Catena **92**, 49–54 (2011)

278. E. De Grave, G.M. Da Costa, L.H. Bowen, C.A. Barrero, R.E. Vandenberghe, Characterization of soil-related analogs by applied-field ^{57}Mössbauer spectroscopy. Hyp. Interact. **117**, 245–270 (1998)

279. R.B. Scorzelli, Application of the Mössbauer effect to the study of meteorites—A review. Hyp. Interact. **66**, 249–257 (1991)

280. R.B. Scorzelli, Meteorites: Messengers from the outer space. J. Braz. Chem. Soc. **19**, 226–231 (2008)

281. J. Danon, R.B. Scorzelli, I. Souza-Azevedo, J. Laugier, A. Chamberod, Santa Catharina meteorite and phase composition of irradiated Fe–Ni Invar alloys. Nature **284**, 537–538 (1980)

282. R.B. Scorzelli, I.S. Azevedo, J. Danon, M.A. Meyers, Mössbauer study of shock-induced effects in the ordered alloy $Fe_{50}Ni_{50}$ in meteorites. J. Phys. F. Met. Phys. **17**, 1993–1997 (1987)

283. R.B. Scorzelli, J. Danon, Mössbauer spectroscopy and X-ray diffraction studies of Fe–Ni order-disorder processes in a 35 % Ni meteorite (Santa Catharina). Phys. Scr. **32**, 143–148 (1985)

284. E. De Grave, R.E. Vandenberghe, P.M.A. De Bakker, A. Van Alboom, R. Vochten, R. Van Tassel, Temperature dependence of the Mössbauer parameters of the FeNi phases in the Santha Catharina meteorite. Hyp. Interact. **70**, 1009–1012 (1992)

285. E. De Grave, R.J. Pollard, R.E. Vandenberghe, P.M.A. De Bakker, The effect of high external magnetic fields on the hyperfine interactions in the Fe-Ni phases of the Santa Catharina meteorite. Hyp. Interact. **94**, 2349–2353 (1992)

286. D.G. Rancourt, R.B. Scorzelli, Low spin γ-Fe-Ni (γLS) proposed as a new mineral in Fe-Ni-bearing meteorites: epitaxial intergrowth of γLS and tetrataenite as possible equilibrium state at ~ 20–40 at % Ni. J. Magn. Magn. Mat. **150**, 30–36 (1995)

287. C.L. Herzenberg, D.L. Riley, Mössbauer spectrometry of lunar samples. Science **167**, 683–686 (1970)

288. P. Gay, G.M. Bancroft, M.G. Bown, Diffraction and Mössbauer studies of minerals from lunar soils and rocks. Science **167**, 626–628 (1970)

289. J. Duchesne, J. Depireux, A. Gérard, F. Grandjean, M. Read, Study with Mössbauer spectrometry on iron distribution in mineralogical fractions separated by lunar rocks reported by Apollo-12. Bull. Cl. Sci. Acad. R. Belg. **57**, 1204–1211 (1971)

290. T.C. Gibb, R. Greatrex, N.N. Greenwood, An assessment of results obtained from Mössbauer spectra of lunar samples. Phil. Trans. R. Soc. Lond. A **285**, 235–240 (1977)

291. R.V. Morris, G. Klingelhöfer, R.L. Korotev, T.D. Shelfer, Mössbauer mineralogy on the Moon: The lunar regolith. Hyp. Interact. **117**, 405–432 (1998)

292. G. Klingelhöfer, In situ analysis of planetary surfaces by Mössbauer spectroscopy. Hyp. Interact. **113**, 369–374 (1998)

293. G. Klingelhöfer, R.V. Morris, B. Bernhardt, D. Rodionov, P.A. de Souza, S.W. Squyres, J. Foh, E. Kankeleit, R. Gellert, C. Schröder, S. Linkin, E. Evlanov, B. Zubkov, O. Prilutski, Athena MIMOS II Mössbauer spectrometer investigation. J. Geophys. Res. Planets **108**, 8067 (2003)

294. R.V. Morris, G. Klingelhöfer, B. Bernhardt, C. Schröder, D.S. Rodionov, P.A. De Souza Jr, A. Yen, R. Gellert, E.N. Evlanov, J. Foh, E. Kankeleit, P. Gütlich, D.W. Ming, F. Renz, T. Wdowiak, S.W. Squyres, R.E. Arvidson, Mineralogy at Gusev Crater from the Mössbauer spectrometer on the Spirit Rover. Science 305, 833–836 (2004)
295. R.V. Morris, G. Klingelhöfer, C. Schröder, D.S. Rodionov, A. Yen, D.W. Ming, P.A. De Souza Jr, I. Fleischer, T. Wdowiak, R. Gellert, B. Bernhardt, E.N. Evlanov, B. Zubkov, J. Foh, E. Kankeleit, U. Bonnes, P. Gütlich, F. Renz, S.W. Squyres, R.E. Arvidson, Mössbauer mineralogy of rock, soil, and dust at Gusev Crater, Mars: Spirit's journey through weakly altered olivine basalt on the Plains and pervasively altered basalt in the Columbia Hills. J. Geophys. Res. 111, E02S13 (2006)
296. R.V. Morris, G. Klingelhöfer, C. Schröder, D.S. Rodionov, A. Yen, D.W. Ming, P.A. De Souza Jr, T. Wdowiak, I. Fleischer, R. Gellert, B. Bernhardt, U. Bonnes, B.A. Cohen, E.N. Evlanov, J. Foh, P. Gütlich, E. Kankeleit, T. McCoy, D.W. Mittlefehldt, F. Renz, M.E. Schmidt, B. Zubkov, S.W. Squyres, R.E. Arvidson, Mössbauer mineralogy of rock, soil, and dust at Meridiani Planum, Mars: Opportunity's journey across sulfate-rich outcrop, basaltic sand and dust, and hematite lag deposits. J. Geophys. Res. 111, E12S15 (2006)
297. G. Klingelhöfer, R.V. Morris, B. Bernhardt, C. Schröder, D.S. Rodionov, P.A. de Souza, A. Yen, R. Gellert, E.N. Evlanov, E. Kankeleit, P. Gütlich, D.W. Ming, F. Renz, T. Wdowiak, S.W. Squyres, R.E. Arvidson, Jarosite and hematite at Meridiani Planum from Opportunity's Mössbauer spectrometer. Science 306, 1740–1745 (2004)
298. G. Klingelhöfer, E. De Grave, R.V. Morris, A. Van Alboom, V.G. de Resende, P.A. De Souza, D. Rodionov, C. Schröder, D.W. Ming, A. Yen, Mössbauer spectroscopy on Mars: goethite in the Columbia Hills at Gusev crater. Hyp. Interact. 166, 549–554 (2006)
299. C. Van Cromphaut, V.G. de Resende, E. De Grave, A. van Alboom, R.E. Vandenberghe, G. Klingelhöfer, Characterisation of the magnetic iron phases in Clovis Class rocks in Gusev crater from the MER Spirit Mössbauer spectrometer. Geochim. Cosmochim. Acta 71, 4814–4822 (2007)
300. C. Van Cromphaut, V.G. de Resende, E. De Grave, R.E. Vandenberghe, Temperature dependence of the hyperfine parameters of the iron bearing phases in the Mössbauer spectra collected by the Mars Exploration Rover Spirit. Hyp. Interact. 190, 143–148 (2009)
301. D.G. Agresti, I. Fleischer, G. Klingelhöfer, R.V. Morris, On simfitting MER Mössbauer data to characterize Martian hematite. J. Phys. Conf. Ser. 217, 012063 (2010)
302. C. Schröder, D.S. Rodionov, T.J. McCoy, B.L. Jolliff, R. Gellert, L.R. Nittler, W.H. Farrand, J.R. Johnson, S.W. Ruff, J.W. Ashley, D.W. Mittlefehldt, K.E. Herkenhoff, I. Fleischer, A.F.S. Haldemann, G. Klingelhöfer, D.W. Ming, R.V. Morris, P.A. De Souza Jr, S.W. Squyres, C. Weitz, A.S. Yen, J. Zipfel, T. Economou, Meteorites on Mars observed with the Mars Exploration Rovers. J. Geophys. Res. 113, 06 (2007)
303. D. Rodionov, C. Schröder, G. Klingelhöfer, R.V. Morris, T. Wdowiak, P.A. de Souza Jr, A. Yen, T. Wdowiak, S.W. Squyres, And the Athena Science Team: Mössbauer investigation of "Bounce Rock" at Meridiani Planum on Mars—Indications for the first shergottite on Mars. Meteorit. Planet. Sci. 39, A91 (2004)
304. C. Schröder, R. Gellert, B.L. Jolliff, G. Klingelhöfer, T.J. McCoy, R.V. Morris, D.S. Rodionov, P.A. De Souza Jr, A.S. Yen, J. Zipfel, And the Athena Science team: A stony meteorite discovered by the Mars Exploration Rover Opportunity on Meridiani Planum. Mars. Meteorit. Planet. Sci. 41, 5285 (2006)
305. M. Blumers, B. Bernhardt, P. Lechner, G. Klingelhöfer, C. d'Uston, H. Soltau, L. Strüder, R. Eckhardt, J. Brückner, H. Henkel, J.G. Lopez, J. Maul, The miniaturized Mössbauer spectrometer MIMOS IIA: Increased sensitivity and new capability for elemental analysis. Nucl. Instrum. Methods: Phys. Res. A 624, 277–281 (2010)
306. D. Chambaere, Studie van de strukturele en magnetische eigenschappen van β-FeOOH en van zijn fasetransformatie naar αFe$_2$O$_3$. Ph.D. thesis, (Ghent University, 1983)
307. B.J. Evans, R.G. Johnson, F.E. Senftle, C.B. Cecil, F. Dulong, The ^{57}Fe Mössbauer parameters of pyrite and marcasite with different provenances. Geochim. Cosmochim. Acta 46, 761–775 (1982)

308. S.G. Eeckhout, C. Casteñeda, A.C.M. Ferreira, A. Sabioni, E. De Grave, D.C.L. Vasconcelos, Spectroscopic studies of spessartine from Brazilian pegmatites. Am. Mineral. **87**, 1297–1306 (2002)
309. E. De Grave, S. Geets, [57] Fe Mössbauermetingen aan Belgische Glauconieten. Bull. Soc. Belge Géol. **88**, 237–251 (1979)
310. E. De Grave, A. Van Alboom, Evaluation of ferrous and ferric Mössbauer fractions. Phys. Chem. Miner. **18**, 337–342 (1991)
311. W. Stiers, U. Schwertmann, Evidence for manganese substitution in synthetic goethite. Geochim. Cosmochim. Acta **49**, 1909–1911 (1985)
312. D.J. Vaughan, M.S. Ridout, Mössbauer study of pyrrhotite (Fe_7S_8). Solid State Commun. **8**, 2165–2167 (1970)
313. H.V. Varma, J. Varma, Mössbauer effect study of natural staurolite. Phys. Stat. Solidi (a) **97**, 275–278 (1986)
314. F. Seifert, A note on the Mössbauer spectrum of [57]Fe in ferrocarpholite. Mineral. Mag. **43**, 313–315 (1979)
315. Y. Fuchs, M. Mellini, I. Memmi, Crystal-chemistry of magnesiocarpholite: controversial X-ray diffraction, Mössbauer, FTIR and Raman results. Eur. J. Mineral. **13**, 533–543 (2001)
316. L.G. Dainyak, V.A. Drits, Interpretation of the Mössbauer spectra of nontronite, celadonite and glauconite. Clays Clay Miner. **35**, 363–372 (1987)
317. H. Kodoma, G. Longworth, M.G. Townsend, A Mössbauer investigation of some chlorites and their oxidation products. Can. Mineral. **20**, 585–590 (1982)
318. D.G. Agresti, M.D. Dyar, M.W. Schaefer, Velocity calibration for in situ Mössbauer data from Mars. Hyp. Interact. **167**, 845–850 (2006)
319. C. Van Cromphaut, V.G. de Resende, E. De Grave, R.E. Vandenberghe, Mössbauer study of Meridiani Planum, the first iron-nickel meteorite found on the surface of Mars by the MER Opportunity. Meteorit. Planet. Sci. **42**, 2119–2123 (2007)

Author Biographies

Robert E. Vandenberghe Robert Vandenberghe (°1945) is Professor Emeritus at the Department of Physics and Astronomy (formerly Subatomic and Radiation Physics) of the Ghent University in Belgium. As an experimental physicist, his research started in the field of structural and magnetic properties of magnetic oxides using magnetic measurements and neutron diffraction. At the end of the seventies of the former century, he began to perform Mössbauer spectroscopy on nanoferrite systems. Using this technique, he further investigated a large variety of materials in which his interest was often focused on the methodology of Mössbauer spectral analysis. All the work was based on [57]Fe MS, but, he made some studies of magnetic materials using [61]Ni Mössbauer spectroscopy as well.

During the nineties he continued with fundamental studies on goethite and hematite. His application of Mössbauer spectroscopy went particularly to the characterization of Fe-bearing compounds in soils and sediments. In that respect, he cooperated with many laboratories of geology and soil sciences. His is author or co-author of more than 150 papers and several chapters in books. He was several times invited as plenary speaker at international conferences.

Eddy De Grave Eddy De Grave (°1951) is Professor at the Department of Physics and Astronomy (formerly Subatomic and Radiation Physics) of the Ghent University in Belgium. He was the first researcher doing Mössbauer spectroscopy at Ghent University in 1972. His PhD treated an ^{57}Fe MS study magnesium titano-ferrites for which he stayed for several months in Marburg (Germany) with Prof. S. Hafner in order to perform external-field measurements. Apart from the study of magnetic oxides, he further dedicated a considerable time to the fundamental study soil-related oxides and oxyhydroxides, together with the late Prof. L. Bowen of the North Carolina State University (USA), were he stayed for one year and subsequently several times for a few months. He also cooperated with Dr. Chr. Laurent from the Paul Sabatier University in Toulouse (France) in relation with MS studies on the precursor materials in the synthesis of carbon nanotubes. However, the main connecting thread in his work is undoubtedly the Mössbauer study of a large amount of minerals in which phosphate-based minerals take a great part. In recognition of his significant contributions in the field of mineralogy, he was elected in 2006 "A Fellow" the Mineralogical Society of America. Eddy De Grave is author or co-author of more than 250 papers and several chapters in books.

As a consequence of their experience in magnetic oxides and soil materials, both authors were invited to write a chapter about "Mössbauer effect studies of oxidic spinels" in the series Mössbauer Spectroscopy Applied to Inorganic Chemistry, edited by G. Long and F. Grandjean, and together with H. Bowen about "Mössbauer effect studies of magnetic soils and sediments" in the series Mössbauer Spectroscopy Applied to Magnetism and Materials Science, also edited by G. Long and F. Grandjean.

Our message to the next generation is: *It will still give you great scientific possibilities when you, as a Mössbauer spectroscopist, cooperate with laboratories related to earth sciences, and have an eye for reliable spectral analyses.*

Chapter 4
The Contribution of ^{57}Fe Mössbauer Spectrometry to Investigate Magnetic Nanomaterials

Jean-Marc Greneche

Abstract Fe containing nanomaterials and nanoparticles are quite important because their unusual physical properties make them excellent candidates for different applications. ^{57}Fe Mössbauer spectrometry appears as an excellent tool to provide structural and magnetic data through the hyperfine parameters. After a short definition of nanostructures and their main characteristics originated from confinement effects, we established the relevant features to understand nanoscale magnetism. Some examples have been thus selected to illustrate first how Mössbauer spectrometry contributes to understand the chemical, structural and magnetic nature of nanostructures and the role of surface and grain boundaries. Then, they also demonstrate also how the fitting procedure remains a delicate task to model the hyperfine structure and does require on the one hand large experimental data basis obtained from different techniques including structural, morphological and magnetic parameters and on the other hand materials with high knowledge and control of synthesis conditions.

4.1 Introduction

Nanoscience and nanotechnology are concerned by the understanding of matter and the conception of devices on an ultra-small scale, respectively. It is difficult to establish the early stages as first proposed by Richard Feynman, but the growing interest in nanosciences emerged during the 1980s with the invention of the

J.-M. Greneche (✉)
Institut des Molécules et Matériaux du Mans, UMR CNRS 6283,
Université du Maine, 72085 Le Mans Cedex 9, France
e-mail: jean-marc.greneche@univ-lemans.fr

Y. Yoshida and G. Langouche (eds.), *Mössbauer Spectroscopy*,
DOI: 10.1007/978-3-642-32220-4_4, © Springer-Verlag Berlin Heidelberg 2013

scanning tunnelling microscope (STM) by Binning and Rohrer [1–4] (Nobel Prize in 1986). This new imaging technique led then to the discovery of fullerenes [5, 6] by R. Smalley, R. Curl, J. Heath, S. O'Brien, and H. Kroto in 1985 (Nobel Prize in 1996). In addition the giant magnetoresistance was co-observed on Fe/Cr/Fe tri-layers by P. Grünberg and his group [7] and independently on Fe/Cr multilayers by A. Fert and his group [8] in 1986 (Nobel Prize in 2007), originating the so-called spintronics. Thus, large efforts were essentially devoted in 1990s to the elaboration of well controlled, tunable and reproducible nanostructures including multilayers, and nanostructured powders and the synthesis of nanoparticles by means of different routes with well controlled chemical conditions. Such an approach makes easier the characterization of the nanomaterials and the understanding of their physical properties, increasing their role and the emergence of nanosciences and nanotechnologies. On the contrary, it is obvious but important to emphasize that the non homogeneous structural and morphological properties combined to a lack of reproducibility and time stability prevent definitively from a clear understanding and realistic modelling of the physical properties of these nanomaterials.

During the last decade, the developments in nanotechnology which consists in the studies and processes to manipulate solid matter at the nano and/or molecular scale received a large and explosive debate with social and ethical issues: new policies and regulations on the use of nanotechnologies have to be established. Indeed, nanotechnology aims to design new functional smart materials and devices with a wide range of applications: it is important to emphasize the emergence of new topics such as nanomedicine, nanoelectronics, nanobioengineering, nanofoods, nanoweapons, ... The developments associated to these areas do substantially contribute significant benefits in improving drug delivery, diagnostics and tissue engineering, water and waste treatment, stain-resistant clothes, protective nanopaint, reducing energy consumption, using more environmentally friendly energy systems, increasing information and communication storage, making construction and heavy industry cheaper (weight reduction), faster and safer with nanocomposites, in introducing nanosensors in foods packaging and security devices. However, some aspects concerned by the risks, toxicity and environmental impact of nanomaterials have to be considered. Consequently, both the batch-to-batch reproducibility and the high control of the morphology of nanomaterials have to be systematically checked at the atomic scale, together with their performances, stability and (bio)compatibility, in view of their potential applications, requiring thus the use of a wide set of complementary and specific techniques.

It is first important to classify the different types of nanostructures and their relevant characteristics and morphological features at the nanoscale, according to their dimensionality, as illustrated in Fig. 4.1. One can clearly distinguish 0D nanostructures with nanoparticles, clusters and mesoporous systems as MOFs (Metal Oxide Frameworks), 1D nanostructures with nanotubes and nanowires, 2D nanostructures with multilayers and 3D nanostructures with nanostructured and nanocrystalline materials. In this context, nanocrystalline materials are defined as single- or multi-phase polycrystalline solids with a grain size of a few nanometres

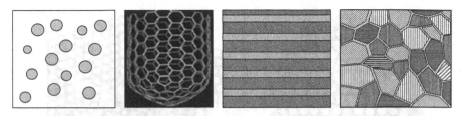

Fig. 4.1 Schematic representation of nanostructures as a function of their dimensionality

up a few tens of nanometres, typically less than 100 nm while nanocrystalline alloys obtained by subsequent annealing of an amorphous precursor consist of nanocrystalline grains embedded in an amorphous remainder. The case of nanostructured powders refer to rather dense packing of nanograins or aggregation of nanoparticles which can be obtained by different physical routes: an example is the high energy ball milling.

Such a description of materials gives unambiguously evidence of the increasing role of the surfaces, interfaces or grain boundaries when the size of the nano-objects is decreasing, particularly below 20 nm as observed in Fig. 4.2. In addition, it does affect some physical properties as the chemical reactivity, mechanical, structural, electronic, vibrational, magnetic, and transport properties, by comparing to those of bulk or massive analogous crystalline materials. The understanding of changes of these physical properties requires a good knowledge of each of the constituents, i.e. the chemical composition and the structure of the particles, grains and layers on the one hand and those of the surfaces, grain boundaries and interfaces on the other hand. The challenge consists thus in the characterization of these different contributions in order to model both the global physical properties. The main questions lye on the elaboration of such nanomaterials and their reproducibility, their chemical and structural homogeneity at different scales, their chemical stability with atmosphere, temperature, ageing, … and then the relevance of their physical properties to make them good candidates for applications.

The investigation of physical properties of microcrystalline solid state materials is usually performed by a complementary set of diffraction techniques, microscopies and different spectroscopic tools, in addition to calorimetric methods and magnetic measurements. Indeed, the modelling of diffraction patterns allow to establish the structural nature and to estimate the lattice parameters while transmission electron microscopy (TEM) can bring an insight on the presence of structural defects, providing the sample has been successfully thinned. In the case of magnetic materials, it is concluded from the usual static magnetic measurements combined to neutron diffraction to the establishment of magnetic structure and the estimation of the magnetic characteristics.

In the case of nanomaterials, the diffraction patterns exhibit a broadening of Bragg peaks: the modelling which requires special methodology included in Rietveld refinement, gives rise to an estimate of the mean lattice parameters and the mean coherent diffraction domain size and its mean morphology, i.e. shape and

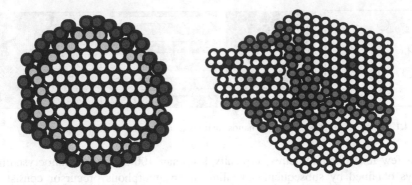

Fig. 4.2 Schematic representation of surface effect in the case of a nanoparticle (*left*) and of grain boundaries in nanostructured powder (*right*)

size of nanoparticles when they are single domain. Example of analysis is given by MAUD software based on the Rietveld method combined with Fourier analysis [9]. Conventional TEM is used to establish the crystalline domain size distribution (nanoparticle size distribution) while High Resolution TEM allows to describe carefully only some nano-objects at the atomic scale. Therefore it is important to note that the sampling remains an important key because the statistics of imaging usually relates to only a few tens or hundreds of nanoparticles, a very small number compared to the total sample which does contain an extremely higher number of nanoparticles. To some extent, local probe techniques do provide a priori relevant and complementary information concerning the atomic structure providing that the confinement effects favour a strong enhancement of surface, interface or grain boundaries. Contrarily to microcrystalline magnetic materials, the magnetic properties of nanostructures require the use of both dc and ac magnetic measurements and the comparison of field-cooled and zero field-cooled (FC–ZFC) magnetization curves, i.e. the thermomagnetic cooling.

Such a view is relevant because the main fundamental questions concern in the case of nanoparticles, the surface and magnetic surface states. Indeed one expects a structural relaxation originating thus some distortions compared to the crystalline lattice: consequently the superficial magnetic structure does result from combined symmetry breaking, surface anisotropy and frustration topological effects arising from the exchange integral, in addition to the reduction or enhancement of the magnetic moment, giving rise a priori to either a dead magnetic layer or a 2 atomic magnetic canted layer shell, as illustrated in Fig. 4.2. In the case of nanostructured powders, the main questions are relative to the chemical composition, the structure, the thickness and the porosity of the grain boundaries, and their influence on the magnetic coupling on neighbouring grains and the bulk magnetic properties.

In this frame, ^{57}Fe Mössbauer spectrometry which is a local atomic probe tool highly sensitive to the atomic neighbouring, appears as an excellent technique to investigate nanomaterials [10–13]: one expects to distinguish atoms belonging to the crystalline zones from those located either at the surfaces of nanoparticles because of symmetry breaking or at the interfaces of multilayers or at the grain boundaries of nanostructured systems because of the topological atomic disorder. In addition, in the case of Fe containing magnetic nanomaterials, ^{57}Fe transmission Mössbauer spectrometry including in-field measurements does contribute to better understand both the intrinsic magnetic structure and the dynamics of magnetic structures, i.e. the superparamagnetic relaxation phenomena which occur in non-interacting single domain nanoparticles. The exploitation of the low temperature magnetic hyperfine structure and the estimation of the blocking temperature provide relevant information to understand the dynamics of magnetic nanostructures. But it is also important to emphasize that the analysis of the Mössbauer hyperfine structure originates large debate because of the difficulties in the modelling leading to various ambiguities.

In the next sections, attention will be paid to the effect of confinement on magnetic properties with different environments but some general features of ^{57}Fe Mössbauer spectrometry will be reported hereafter. Then, some general features are given to introduce nanomagnetism and magnetism of nanoparticles. The last sections are concerned by the review of different situations illustrating the role of ^{57}Fe Mössbauer spectrometry in investigating the structural and physical properties in several types of magnetic nanostructures: nanoparticles, nanocrystalline alloys, nanostructured powders, and mesoporous systems. Most of the examples have been the subject of our own research works developed in collaboration with different groups of chemists. The strategy consists in the elaboration or the synthesis of well controlled nanostructures which have to be well reproduced: thus it requires to well understand the chemical mechanism and the role of all parameters (temperature, pressure, pH, …). It is finally important to mention that the samples reported in the next sections have been widely characterized by means of diffraction techniques, electron microscopies and magnetic measurements.

4.2 General Features on ^{57}Fe Mössbauer Spectrometry

Mössbauer spectrometry is a powerful technique to study solid state materials including magnetic nanostructures and frozen colloids, particularly their magnetic dynamics. It is based on the recoil-free emission of a γ-photon by a nucleus located in a radioactive emitter (source) and the subsequent recoil-free absorption by a similar nucleus located in the absorbing system (sample). The extremely narrow width of the resonance which results from the finite lifetime of the excited state (Heisenberg uncertainty principle) allows the hyperfine interactions between the nuclear and electronic charges to be observed and to be determined. The recoilless nuclear resonance or Mössbauer effect combined to the energy scanning obtained from the periodic movement of the source originate the registration of Mössbauer spectra. Their description gives rise to the hyperfine characteristics of the different Fe species located in the studied sample, which are namely the isomer shift, the quadrupolar splitting, the quadrupolar shift and the hyperfine field (a description of the physical meaning of each hyperfine parameter is given in the chapter "Application of Mössbauer Spectroscopy in Earth Sciences") and their respective proportions.

Additional information is given by the intensities and the shape of lines. In the case of magnetic materials, the relative areas of the six lines sextet are correlated to the Fe moment configuration respect to the γ-beam direction. Indeed, the relative area ratios are given by 3:p:1:1:p:3, where p = 4 $\sin^2\theta/(2 - \sin^2\theta)$ where θ represents the angle between the hyperfine field held by the nucleus probe and the propagation direction of the γ-beam. In the case of a powdered sample, i.e. random distribution of the orientations of the Fe magnetic moments, one expects the following relative ratio 3:2:1:1:2:3. When the magnetic moments are oriented parallel to the γ-beam, one obtains 3:0:1:1:0:3, i.e. the intermediate line disappears: such a situation occurs in a perfect ferromagnetic or ferrimagnetic material submitted to an applied field sufficiently large to saturate the magnetization. One does obviously consider two magnetic components in the case of a ferrimagnetic structure. In addition, it is important to check whether the hyperfine field is either parallel or antiparallel to the magnetic moment, that remains unknown from the zero-field sextet spectrum (one determines only the absolute value of the hyperfine field!). When the effective field is smaller or larger than the hyperfine field, one concludes that the hyperfine field is opposite or parallel to the magnetic moment, respectively, because the magnetic moment is forced to be aligned parallel to the external magnetic field. Consequently, the main contribution to the hyperfine field corresponding to the Fermi or contact term is negative or positive. An other situation occurs when the Fe moments are antiferromagnetically coupled: due to the dominant antiferromagnetic interaction, they are rather oriented perpendicular to the external field and the shape of the Zeeman sextet is 3:4:1:1:4:3 (maximal intensities of intermediate lines) when the magnetic field is applied parallel to the γ-beam. But, the most commonly observed situation gives rise to intermediate intensities values. As the total effective field at the nucleus results from the vectorial sum of the hyperfine field and the applied field, one can establish the following expression

$$B_{hf}^2 = B_{eff}^2 + B_{app}^2 - 2B_{eff}B_{app}\cos\theta \qquad (4.1)$$

which allows the canting angle to be estimated. Figure 4.3 reports theoretical Mössbauer spectra of different typical magnetic structures submitted to an external magnetic field (see reviews [14–18]). In addition, the in-field Mössbauer theoretical spectra characteristics of non collinear static magnetic structures (speromagnetic SP, sperimagnetic SPi and asperomagnetic ASp) earlier evidenced by Coey and Readmann are also illustrated [19]. Finally, one does notice when the external field is applied perpendicular to the γ-beam, the respective relative ratios have to be inversed (3:4:1:1:4:3 becomes 3:0:1:1:0:3 and vice versa).

The natural Mössbauer line profile is lorentzian but the lattice tends to reduce the lifetime of the excited state due to radiation less decay mechanisms originating thus some broadening, in addition to instrumental effects. But some (in) homogeneous

Fig. 4.3 Theoretical Mössbauer spectra on linear and non collinear magnetic structures in presence of external field applied parallel to γ-beam: random powder in zero-field, ferromagnetic with magnetic moments anti-parallel and parallel to the magnetic moment, and antiferromagnetic structures and non collinear magnetic structures (see text)

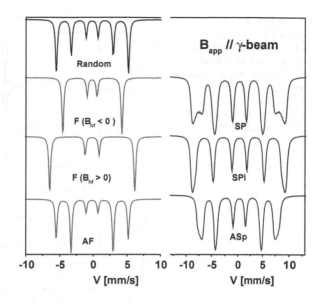

broadening occurs when the local atomic probe environment is disturbed by static effects, i.e. chemical or topological disorder or in presence of electron or magnetic dynamics. In the case of magnetic nanostructures, the magnetization is not systematically static and both its orientation and magnitude may fluctuate. When the Fe moment is submitted to relaxation effects, the hyperfine structure of the Mössbauer spectrum is dramatically modified, particularly its line shape which has widely discussed in the literature. Several theoretical descriptions have been proposed, assuming that the values of $+B_{hf}$ and $-B_{hf}$ and that the relaxation is a stochastic process with an average relaxation time. The calculations become quite complex and some approximation can be done when the superparamagnetic particle exhibits uniaxial symmetry when $KV \gg k_BT$. Figure 4.4 illustrates theoretical ^{57}Fe Mössbauer relaxation spectra using longitudinal relaxation with different relaxation times assuming $B_{hf} = \pm 55$ T and an axial electric-field gradient parallel to B_{hf}. Three regimes can be distinguished by comparing to the characteristic time of the Mössbauer measurement (5.10^{-8} s) which results from the nuclear Larmor precession time in the magnetic hyperfine field. Large ($\tau \geq 5.10^{-8}$ s) and short ($\tau \leq 10^{-10}$ s) relaxation times give rise to broadened lines sextet and quadrupolar doublet (or singlet), respectively, while spectra with very broad lines with a gradual transition from a sextet to a doublet are expected for intermediate range of relaxation times. But it is clear that the physical modelling of the hyperfine structure becomes dramatically complex when the sample consists of an assembly of polydisperse non interacting magnetic nanoparticles because the distribution of time relaxation has to be considered!

Consequently, the interest of Mössbauer spectrometry applied to nanomaterials can be large and relevant in terms of both structural and magnetic characterization. Nevertheless, great attention has to be paid to the modelling of hyperfine structure

Fig. 4.4 Theoretical Mössbauer spectra calculated for various relaxation times using a discrete 2-level relaxation model for uniaxial symmetry: the linewidth is 0.20 mm/s, the quadrupole shift 0.00 mm/s and the hyperfine field ±55T. (From [13])

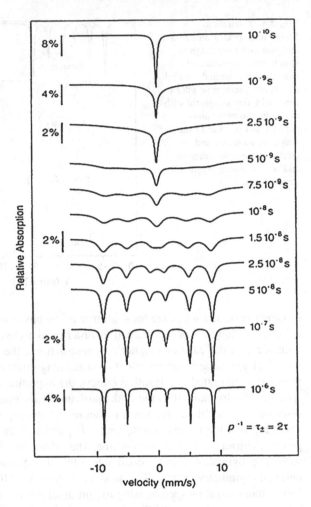

which remains a crucial task. Indeed, the complexity of spectra might give rise to a large number of "mathematical" fitting solutions but probably only a few ones are in agreement with a physically realistic scenario. It is clear that the supporting data established from diffraction techniques, microscopies and magnetic measurements may discard some solutions. But [57]Fe Mössbauer experiments performed versus temperature and/or external magnetic field are suitable to distinguish static from dynamic effects and to see their respective evolution: thus, the physical model does correspond to the fitting solution successfully achieved from this series of spectra.

For those reasons, the content of next sections is constrained to well characterized and homogeneous nanostructured systems to illustrate the relevance but also the limitation of by [57]Fe Mössbauer spectrometry. Consequently, it excludes thus the use of this local spectroscopic technique to study the mechanisms associated to thermal transformations from nanostructures into microstructures, the growth of corrosive layers at the surface of metallic systems submitted to aggressive chemical treatments, the natural minerals which contain nano and microsized particles or inclusions with different cationic species. Indeed, these materials under investigation consist of out-of equilibrium granular structure with a distribution of size and different chemical composition because the processes are a priori non homogeneous. The modelling of corresponding Mössbauer spectra would consequently give an approximate description of physical features resulting from a mixture of static and dynamic phenomena.

4.3 From Magnetism to Nanomagnetism

The characteristics of magnetic materials result from the cooperative contribution of magnetic domains characterized by their individual magnetization which exhibits different orientations (see review and books [10–15, 20–23]). The splitting into magnetic domains separated by domain walls originates from the magneto-static energy acting as a driving force. Coercive field, saturation and remnant magnetization are the three main characteristics obtained from the hysteresis loop which allows thus to qualify the class of magnetic materials as soft, semi-hard or hard magnets according its magnetic stiffness and to understand the mechanism of magnetization reversal. Calculations earlier proposed by Frenkel and Doefman [24] gave rise to establish typical dimensions for magnetic domains in ferro-magnetic crystalline materials with a minimum magnetic domain size, roughly estimated at about 100 nm. The main question arises when the size of the material, typically nanoparticles, becomes similar or lower than the minimum magnetic domain size. Indeed, there exists a critical size of magnetic particles below which the magnetic wall formation becomes unstable: it can be first estimated from a simple approach by comparing the energy of single domain and multidomain particle at equilibrium and at zero K, i.e. the magnetostatic energy and the interfacial wall energy. It can be expressed as

$$NVM_s^2/2 = \gamma V^{2/3} \tag{4.2}$$

where N represents the smallest demagnetizing field factor, V the volume of the particle and γ is the interfacial wall energy per surface unit. One can finally establish the critical length as $l_c \sim \gamma/M_s^2$, assuming the coefficient 2/N related to the particle shape, equal to unity. By substituting the Bloch wall energy expression $\gamma = 4\sqrt{AK}$, one can also write $l_c \sim \delta_e\sqrt{Q}$ where $Q = K/M_s^2$ and $\delta_e \sim \sqrt{A}/M_s$ are the quality factor and the exchange length, respectively, while A, K and M_s^2 represent the exchange, magnetic effective anisotropy and dipolar energies. When the size of the particle does not exceed l_c, it behaves as a single domain magnet.

It is clear that the morphology of the nanoparticles strongly influences its magnetic energy and consequently the critical size which has to be estimated from more rigorous calculations. When considering spherical magnetic particles with uniaxial anisotropy, the magnetic energy is given by $E_a = KV\sin^2\theta$, where θ is the angle between the magnetization vector and an easy direction of magnetization. But, it is important to emphasize that the critical diameter which is typically ranged from 5 to 200 nm, does not depend exclusively on the shape of the particle but also on its magnetic nature: from literature, it is found about 6 and 60 nm for Fe and Fe_3O_4 particles, respectively [25].

It is now important to better understand the magnetic behaviour of an assembly of single domain nanoparticles. The instability of the creation of domain wall was confirmed by the pioneering experiments of geomagnetism performed by Thellier [26]. Indeed, when studying the magnetization of rocks and pottery to be correlated to their cooling history, he found that the magnetization induced by cooling a sample in a magnetic field was unexpectedly decayed with time, that can be considered as the first experimental feature of super paramagnetic relaxation. Then, Louis Néel reported that the magnetization of very small particles may be reversed due to their thermal agitation when the anisotropy energy becomes of the same order of magnitude as the thermal energy, i.e. the initial theory of superparamagnetism [27–29].

Indeed, the theory of Néel consists in considering the precession around the uniaxial anisotropy field of the magnetization characteristics of a small ferromagnetic nanoparticle, possessing a magnetic anisotropy energy given by Eq. 4.1. One observes two energy minima corresponding to two antiparallel easy directions, and separated by an energy barrier (height KV). In this theory, the atomic moments are assumed to be strongly coupled such that their rotation occurs in unison and the precession is perturbed by lattice vibrations. In addition, the average time between magnetization reversals also called the superparamagnetic relaxation time τ, can be calculated using the Boltzmann statistics: for $KV \gg k_BT$, the relaxation time follows approximately an Arrhenius law expressed as $\tau = \tau_0 \exp(KV/k_BT)$. Prefactor τ_0 is of the order of $10^{-13} - 10^{-9}$ s while k_B corresponds to the Boltzmann's constant.

The superparamagnetic relaxation time of a ferromagnetic particle was also derived by Brown [30] from the Langevin equation while the magnetic dynamics of an assembly of non-interacting magnetic nanoparticles has been described by a Fokker–Planck differential equation. Numerous models were also developed assuming that the magnetization reversal follows coherent rotation (unison), buckling or curling modes [31]. It is clear that the improvement of the synthesis routes giving rise to controlled assemblies of magnetic nanoparticles allows suitable experimental data to be obtained: consequently, further theories have been refined including the role of external magnetic field on the magnetization reversal and the symmetry of the anisotropy of the nanoparticles. During the last 15 years, additional approach by means of computer simulations was also devoted to the magnetic structure and the magnetic dynamics of ferromagnetic nanoparticles using Langevin dynamics as a function of size, time at different temperatures. But the theoretical and computer modelling of the magnetic behaviour has not been completely achieved in the case of weakly and strongly interacting nanoparticles.

From the experimental point of view, the timescale characteristic of the applied technique has to be compared to the relaxation time to investigate the dynamics of nanoparticles. Two types of measurement can be distinguished: when the relaxation is fast compared to the timescale, the measurement corresponds to the average value of the magnetization which tends to zero in zero-applied field. On the contrary, the static value of the magnetization observed when the relaxation is slow. Consequently, the decrease of temperature and/or the application of an external magnetic field are two experimental conditions which tend to overcome the barrier energy to cancel thus the relaxation phenomena, i.e. to a blocked magnetic structure. In addition, it is of relevant interest to estimate the blocking temperature values, T_B, which are defined as the temperature at which the superparamagnetic relaxation time is equal to the timescale of the experimental technique, allowing thus the thermal variation of the relaxation time ($\log_{10}\tau_m$ versus $1/T_B$) to be established (see [15] and references therein). Three different regimes can be observed as a function of the interacting nature of the assembly of nanoparticles: (1) in presence of negligible or very weak interactions, the linear behaviour indicates that the properties match the Néel-Brown model; (2) from weak to medium interactions, a weak curvature is observed in agreement with the Néel-Brown model which remains rather valid by taking account of an additional anisotropy temperature dependent contribution, while (3) a critical decrease of the relaxation time in the case of strong interactions suggests a homogeneous dynamical process, i.e. a collective freezing of particle moments, similar to that of the spin freezing in spin glasses which exhibit a phase transition. It is important to note that one can find in the literature pure superparamagnetic, superparamagnetic modified by interactions and collective (glass collective state) regimes.

4.4 Magnetic Nanoparticles

During the last twenty years, the chemical and physical properties of nanoparticles have been widely investigated as well as their potential applications in different topics but one has to emphasize that the elaboration and the manipulation of monodispersed nanoparticles has opened relevant and promising applications in materials science. It is clear that the assemblies of nanoparticles without any control of the size and morphology dispersion and their aggregation due to attractive van der Waals forces prevent from a fundamental knowledge of their physical properties. Indeed, a systematic characterization of monodisperse and monomorphological nanoparticles is required to establish and to control not only the properties of an individual nanoparticle, but also the collective behaviour of an assembly of nanoparticles, favouring thus their structural and magnetic modelling. Consequently, a first crucial point lies on the refinement of the synthesis methods of individual nanoparticles for the last two decades: they are currently based on conventional chemical processes but also on new chemical and physical routes. Then, the ability of nanoparticles to self assemble onto a substrate has been established, in order to elaborate highly ordered monolayers and 3D ordered arrays. Indeed, hexagonal, cubic or random packing of nanoparticles can be achieved with thicknesses ranged from a few tens of nanometres up to a few micrometers.

The main structural characteristics have to be studied by means of diffraction techniques (X-ray, neutrons, EXAFS,…) and transmission electron microscopy to check their crystalline state and the homogeneity in size and aggregation from well established methodology. In this way, because of its local probe sensitivity, ^{57}Fe Mössbauer spectrometry contributes in a better characterization of the local atomic order, i.e. chemical homogeneity intra- and internanoparticles and in the differentiation of surface and core Fe species. In addition, this spectroscopic tool is highly sensitive to the presence of superparamagnetic fluctuations in the case of magnetic nanoparticles, particularly to ferromagnetic, antiferromagnetic and ferrimagnetic species.

Indeed, the intrinsic magnetic properties of nanoparticles are strongly dependent on the size while the extrinsic ones are correlated to the dipolar interactions. By comparing to bulk materials, nanoparticles possess a very high "surface to volume" ratio and their size scale favour thus quantum mechanical effects. Consequently, finite size effects and the large atomic ratio surface/volume originate unusual surface magnetic effects, which are obviously more and more important when the size decreases. The symmetry breaking of the lattice favours thus the reduction of the increase of the magnetic moments, the occurrence of broken exchange bonds leading to the surface anisotropy, which does compete with core–surface interactions and dipole–dipole interactions.

The transmission conventional Mössbauer spectrometry gives rise to hyperfine structures which result from the set of Fe probes located in the sample composed of an assembly of magnetic nanoparticles. The spectra consist of broadened and overlapped lines emerging usually of both magnetic sextets and quadrupolar doublets, the proportions of which are temperature dependent: their interpretation of the spectrum consists in describing by means of a minimum of magnetic and quadrupolar components characterized by physical hyperfine parameters correlated to the chemical nature of the nanoparticles contained in the sample. Consequently, it becomes easier to investigate monodisperse, chemically homogeneous nanoparticles which are well distributed into a matrix or homogeneously aggregated and the Mössbauer study of assemblies of nanoparticles should require first accurate preliminary characterization of structural and microstructural properties. In such a case, the modelling of the hyperfine structure observed by means of ^{57}Fe Mössbauer spectrometry must distinguish surface from bulk Fe species from magnetic and/or structural aspects and bring relevant information to model intrinsic and extrinsic magnetic properties. On the contrary, a non homogeneous assembly of nanoparticles will lead to Mössbauer spectra the description of which cannot be well achieved with physical meaning because of the superimposition of different complex hyperfine structures. Consequently, it remains quite difficult to model accurately the thermal transformations of Fe-bearing materials at the nanoscale because of the non homogeneous process originating distribution of sizes.

Two regimes can be a priori well distinguished: either blocked magnetic states (multi domain particles) or superparamagnetic relaxation phenomena when the thermal energy is prevailing, i.e. fluctuations of the magnetization corresponding to two minima of energy (fine single domain nanoparticles). Those dynamic magnetic states can be well described by the Néel–Brown model [30] and by the Dormann–Bessais–Fiorani model [32] in the case of an assembly of non-interacting and weakly interacting particle. On the contrary, the individual fluctuations are cancelled in the case of strongly interacting nanoparticles, favouring thus a collective magnetic state, as observed in spin glasses [33]. The modelling of the intermediate regime is not yet clearly established.

The main characteristics of a single domain isolated particle are thus its blocking temperature T_B, which corresponds to the progressive transition (not a phase transition!) from the superparamagnetic state to the magnetic state. T_B decreases when the size of the particle decreases and when the distance between particles increases. In addition, T_B increases with decreasing time measurement, indeed its estimation is strongly dependent on the time scale characteristic of the measuring technique. One of the first contribution of zero-field Mössbauer spectrometry to study assembly of non-interacting or interacting single domain oxide nanoparticles is the evaluation of the mean blocking temperature ($T_B^{Möss}$), defined as the magnetically split and un split components representing 50 % each of the spectral area, allowing thus an estimation of the mean anisotropy constant. But it is important to emphasize that the estimate of $T_B^{Möss}$ is not so obvious because it remains quite difficult to distinguish and to estimate accurately the absorption area attributed to paramagnetic and magnetic contributions to the hyperfine structure.

The application of rather intense external field parallel to the easy axis reduces the fluctuations of the magnetization, giving rise to a blocked like magnetic state. The hyperfine structure of in-field Mössbauer spectra have to be compared then to the zero-field Mössbauer spectrum particularly, the comparison between the effective field(s) and the corresponding hyperfine field(s) and the intensity of the intermediate lines: they should allow to better understand the dynamics and the magnetic structure of the nanoparticles, describe the magnetic particle by means of a core shell model. In the case of ferrimagnetic nanoparticles, the magnetization of the core is oriented parallel to the external field while the random distribution of moments occur at the outer shell [34, 35]. In practice, the spectrum is decomposed as the sum of two components corresponding to saturated and random configuration: the thickness of the magnetic shell can be thus estimated, and it is usually found at about 2 atomic layers [34–46]. One does note that such an effect was confirmed from ^{57}Fe NMR experiments [42]. This model is well supported by the surface spin disorder originating from the large surface anisotropy which overcomes the exchange energy contribution [37–39]. Most of studies tend to conclude that the outer layer is preferentially composed of octahedral units, in agreement with chemical considerations. Let us emphasize that it is also supported by computer simulation that predicts different structures according to the surface/magnetocrystalline anisotropy ratio [47–52].

The interpretation of in-field Mössbauer spectra which exhibit rather broadened and overlapped lines, remains rather ambiguous because the presence of structural defects in the core of the particle cannot be clearly distinguished from the surface effects.

In the next section, we illustrate from selected examples how ^{57}Fe Mössbauer spectrometry contributes to characterize structural and magnetic of Fe containing nanoparticles. It is clear that numerous studies have been reported in the literature but currently two different kinds of magnetic nanoparticles, Fe oxides including ferrites and metallic, are mainly being produced, characterized for their potential use in catalysis, data storage, biomedicine, magnetic resonance imaging, magnetic particle imaging and environmental remediation.

Many routes to synthesize magnetic nanoparticles are now well established: they derive from 3 main "bottom up" strategies which are co-precipitation (a facile and convenient method to prepare Fe oxides from aqueous Fe^{2+}/Fe^{3+} salt solutions by the addition of a base), thermal decomposition (endothermic reaction from organometallic compounds in high-boiling organic solvents containing stabilizing surfactants), micro emulsion technique (based on a reaction of thermodynamically stable and isotropic liquid mixtures of oil, water and surfactant with a co-surfactant).

As the Mössbauer spectra of both disordered structures and nanostructures exhibit broadened lines, the first question is concerned by the origin of such a feature: chemical or topological disorder (or both) and dynamics through the presence of relaxation phenomena. Because one cannot a priori establish the origin, the application of an external magnetic field (typically at least 0.5 T up to 10 T) should allow to conclude to the occurrence of dynamic effects by the increase of the magnetic contribution at the expense of the quadrupolar doublet decrease and a better resolution of the magnetic hyperfine structure while the change of the intermediate line intensities is assigned to a magnetic alignment of Fe moments, suggesting rather a disordered structure. The temperature evolution of the hyperfine structure of an assembly of nanoparticles show a progressive collapse from a quadrupolar to a magnetic structure coexisting in varying proportions over a temperature range which is dependent on the particle size, the dispersity in size and the volumetric concentration: the estimation of Mössbauer blocking temperature $T_B^{Möss}$ which consists in describing the spectra by means of quadrupolar splitting and hyperfine field distributions is not obvious because of the higher and lower limits, respectively. Such a value is usually found much higher than that established from magnetic measurements because the characteristic measurement time is much smaller.

Great attention has been devoted to nanoparticles of ferrites: these compounds are spinels with the chemical formula $A^{2+}B_2^{3+}O_4$, where A^{2+} and B^{3+} correspond to metallic 3d cations, respectively. The ferrite spinel structure has a face-centered cubic (fcc) structure in the close packed cubic arrangement of oxide ions. The structure contains two interstitial sites, occupied by oxygen coordinated metallic cations, with tetrahedral (A)-site, and octahedral [B]-site, resulting in a different

Table 4.1 Characteristics of main spinel ferrites with expected distribution of cations in tetrahedral (A) and octahedral (B) sites and theoretical values of magnetization

Ferrite	A-site	B-site	Magnetization (μ_B)
Fe_3O_4	Fe^{3+}	Fe^{2+}, Fe^{3+}	4.0
$CoFe_2O_4$	Fe^{3+}	Co^{2+}, Fe^{3+}	3.5
$NiFe_2O_4$	Fe^{3+}	Ni^{2+}, Fe^{3+}	2.2
$ZnFe_2O_4$	Fe^{3+}	Zn^{2+}, Fe^{3+}	0
$MnFe_2O_4$	Fe^{3+}	Mn^{2+}, Fe^{3+}	4.6
$Mn_{0,5}Zn_{0,5}Fe_2O_4$	Fe^{3+}	$0,5Mn^{2+}$, $0,5Fe^{2+}$, Fe^{3+}	7.0
$CuFe_2O_4$	Fe^{3+}	Cu^{2+}, Fe^{3+}	1.2

local symmetry. As listed in Table 4.1, one can distinguish normal or direct $(A^{2+})[B^{3+}]_2O_4$ spinels where A^{2+} ions occupy the tetrahedral sites and B^{3+} ions the octahedral ones and inverse $(B^{3+})[A^{2+}B^{3+}]_2O_4$ spinels where A^{2+} ions occupy the octahedral sites, half of B^{3+} ions the tetrahedral ones and half of B^{3+} ions the octahedral ones. But spinels with defective structure are usually evidenced: $(A_\lambda B_{1-\lambda})[A_{1-\lambda}B_{1+\lambda}]_2O_4$, where inversion parameter $\lambda = 0$ and 1 stands for the inverse and normal cases, respectively, and 1/3 for random. Spinel ferrites are materials with fascinating magnetic, electronic and transport properties: they can be half metallic such as Fe_3O_4 (magnetite as mixed valence system) or insulating (most of spinel ferrites), ferrimagnetic (most of spinel ferrites) or antiferromagnetic ($ZnFe_2O_4$) as ideal bulk state. Indeed, one has to pay attention to the cationic distribution which strongly influences of the physical properties. Their ferrimagnetic structure was first explained by Néel through two-sublattice model resulting from superexchange interactions between cations (J_{AA}, J_{BB} and J_{AB}) [53]. But cationic inversion and substitution with non-magnetic ions originate non collinear up to spin-glass-like structures. For the last 50 years, numerous experimental, theoretical and numeric studies have been devoted to model the magnetic and electronic structures and to estimate the superexchange constants in microcrystalline ferrites. It is clear that in addition to the role of the chemical homogeneity, parameters as the surface anisotropy related to the surface state and morphology, superparamagnetic relaxation phenomena and dipolar interactions have to be considered to better understand intrinsic and extrinsic magnetic properties in the case of the nanoparticles of ferrites.

Consequently, the first question is to check whether the nanoparticles are chemically homogeneous. Magnetite (Fe_3O_4) appears to be a first excellent illustration: indeed, when cooling, this bulk mixed valent Fe oxide undergoes a charge ordering, i.e. metallic-insulating (Verwey) transition at about 120 K, the nature of which, together with the magnetic properties are related to the metal-to-oxygen stoichiometry [54–57]. It is also important to emphasize that maghemite (γ-Fe_2O_3) has a similar structure but is insulating as it contains only ferric ions. As is illustrated in Fig. 4.5, the 300 K (and also above the Verwey transition) Mössbauer spectrum of microcrystalline magnetite consists in two well resolved magnetic sextets: from the values of hyperfine parameters, the outer sextet is

Fig. 4.5 Typical 300 K Mössbauer spectra of microcrystalline magnetite (*up*) and maghemite (*down*)

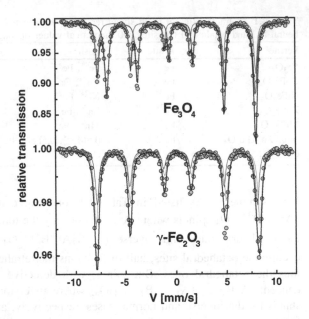

attributed to Fe^{3+} ions located in tetrahedral site while the inner one corresponds to both Fe^{2+} and Fe^{3+} in octahedral sites. The occurrence of a single sextet arises from the electronic hopping phenomenon between Fe^{2+} and Fe^{3+} ions with a characteristic time of $\sim 10^{-9}$ s, slightly smaller than the available time for the Mössbauer measurement. In the case of a stoichiometric magnetite, the hyperfine parameters are well established: isomer shift relative to α-Fe at 300 K: $\delta = 0.26$ and $= 0.67$ mm/s, quadrupolar shift: $2\varepsilon = 0.02$ and 0.00 mm/s, hyperfine field $B_{hf} = 49.0$ and 46.0 T for tetrahedral and octahedral sites [58], respectively, while the relative proportions of each Fe species are derived from their respective absorption area, after correcting the values for the corresponding recoilless factor, f. Below the Verwey temperature, a partial charge ordering gives rise to a complex hyperfine structure which can be usually described by means of 4–5 components attributed to Fe^{3+} in octahedral and tetrahedral sites, Fe^{2+} in octahedral site and Fe with intermediate valency states in octahedral sites [59].

As illustrated in Fig. 4.5, the 300 K Mössbauer spectrum corresponding to maghemite is a sextet resulting from two subcomponents assigned to ferric located in tetrahedral and octahedral sites but the lack of resolution prevents their respective proportions to be estimated. Such a situation occurs at low temperature and an external magnetic field of at least 5 T is necessary to split the hyperfine structure into two well resolved components, as a consequence of the ferrimagnetic structure [60]. An illustration is given in Fig. 4.6. But it is important to mention pioneering studies in 1960s and 1970s carried out on microcrystalline ferrites [61–65]. It can be concluded to the following procedure: (1) the modelling of the in-field Mössbauer spectrum allows the effective field values to be estimated on both tetrahedral and octahedral sites, together with isomer shift and respective

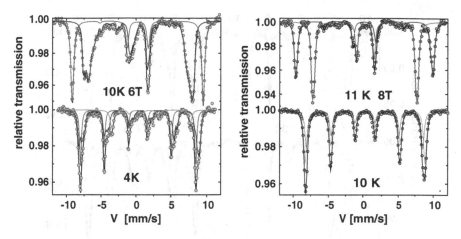

Fig. 4.6 Typical zero-field Mössbauer spectra (*down*) and in presence of an external field applied parallel to the direction of γ-beam on magnetite (*left*) and maghemite (*right*)

Fig. 4.7 Example of Mössbauer spectrum recorded at 300 K on as-prepared assembly of magnetite nanoparticles

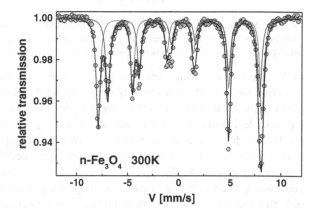

proportions; (2) the respective hyperfine field values can be estimated from Eq. (4.1) and then (3) used to describe the zero-field Mössbauer spectrum. A disagreement with experimental spectrum does give rise to think where it does come from?

Several routes have been used to synthesize nanoparticles of magnetite: it is usually observed for sizes below 40 nm, an increase of the intensity of the left outer line giving rise to a change in the relative proportion of the two sextets as is shown in Fig. 4.7, and to a reduction of the mean value of the isomer shift. Then a progressive collapse of the two sextets into a broadened and asymmetrical lines sextet when the nanoparticles become much finer, preventing thus from an easy modelling based on a discrete number of subcomponents.

The evolution of the relative proportions of the two sextets can be associated to the progressive emergence of a component, the hyperfine characteristics of which are highly close to those of maghemite. Finally, the 300 K Mössbauer spectrum allows to conclude

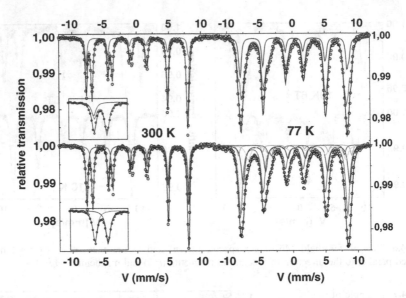

Fig. 4.8 Mösbauer spectra of assemblies of as expected magnetite nanoparticles with two fitting models: assuming different sextets (*low*) or ideal maghemite and magnetite phases (*high*); insert correspond to a zoom of low energy part of the spectrum (from [41])

to the effect of a cationic deficiency in Fe^{2+} which are transformed into Fe^{3+}, i.e. an oxidation process which reasonably occurs rather at the surface of the nanoparticles. Consequently, Mössbauer spectrometry is a priori the most adapted method to evaluate the exact deviation from stoichiometry d in $Fe_{3-d}O_4$ from the isomer shift values which can be used for smaller nanoparticles, i.e. when the superparamagnetic relaxation phenomena occur and prevent from the analysis of broadened lines spectra by means of a 2 components model. As illustrated in Fig. 4.8, a fitting model which consists thus in considering the two ideal maghemite and magnetite phases (including the absorption area of both Fe sites in each component) allows their respective proportions to be estimated [41]. Such a modelling (see Fig. 4.8) can be also applied to 77 K spectra but remains less reasonable, because of the rather complex hyperfine structure of magnetite. By means of a core–shell structural model, one can estimate roughly the thickness of the maghemite layer, assuming stoichiometric Fe phases. Such assumption remains rather valid for magnetite nanoparticles prepared by co-precipitation route, after comparing with magnetic data which reveal the presence of the Verwey transition, i.e. presence of magnetite while the Fe^{2+} content is decreased as observed by XPS: according to its depth sensitivity, it confirms that the Fe^{3+} species are preferentially located within the surface layer.

It turns out that Mössbauer spectrometry is thus able to establish the mean stoichiometry of as-prepared magnetite-like nanoparticles, to follow the charge ordering and to appreciate their stability after ageing. According to the literature, one does conclude that the thickness of the maghemite-like layer is dependent on the morphology and the synthesis conditions: it is generally ranged at about 2–4 nm [41–43, 66, 67], preventing thus the existence of pure ultrafine magnetite nanoparticles, except those highly protected during their synthesis (magnetosome in vesicle and biomineralization in bacterian medium) [68]. It also explains the disappearance of charge ordering by cooling and the Verwey transition, which is not strictly correlated to confinement effects but to the transformation of magnetite into maghemite according to an oxidation process [69, 70].

In the case of other ferrites, the situation might be more complex because the lack of resolution of the hyperfine structure yet observed on bulk systems is worth because of the occurrence of relaxation phenomena: the single method is to apply an external magnetic field to overcome the superparamagnetic relaxation phenomena. In such a case, the Mössbauer spectrum consists in two well resolved components allowing thus the octahedral and tetrahedral Fe proportions, the mean effective fields of each Fe species and the mean direction of the Fe moments under the external field to be accurately estimated [71–76]. When the ferrite nanoparticles contain two cationic species, one can estimate the degree of cationic inversion to be compared with that established from either Extended X-ray Absorption Fine Structure (EXAFS) or X-ray magnetic circular dichroism (XMCD) [71–79]. With more cationic species, EXAFS experiments have to be performed at different edges in order to determine the cationic populations at each site. The magnetization corresponding to the final composition can be thus calculated and compared to that observed by magnetic measurements: the disagreements, if any, have to be correlated to dynamics and confinement effects [71–86]. It is important to note the case of Zn ferrite which behaves as an antiferromagnet below 10 K [87] in the microcrystalline state becomes a ferrimagnet with high magnetic ordering temperature in the nanocrystalline state [72, 83–86].

The profile of the in-field Mössbauer lines has to be carefully analyzed and two main different scenarios can be distinguished. When the lines are quite narrow suggesting an homogeneous Fe environment, the fitting model consists in describing each component with a single effective field and a single angle between Fe moment and γ-beam direction, allowing the hyperfine field to be calculated using Eq. 4.1. It is now possible to simulate the zero-field Mössbauer spectrum (putting together isomer shift, quadrupolar shift, hyperfine field values and proportions of each species) and to compare to the experimental one: a fair agreement would consolidate the homogeneity of the nanoparticles while a disagreement would suggest some remaining relaxation phenomena. On the contrary, components with broadened lines (suggesting different atomic environments of Fe sites) have to be described by means of discrete distribution of effective fields $P(B_{eff})_{A,B}$ assuming the same orientation of Fe magnetic moments respect to the applied field: it is thus easy to evaluate the hyperfine field distributions in order to reconstruct the zero-field Mössbauer spectrum. A disagreement with the experimental spectrum leads to improve the "naive" fitting model into the introduction of joint distribution functions $P(B_{eff}, \theta)_{A,B}$ of the effective field and the canting angle, which is consistent with a non chemical homogeneity within the nanoparticles [64]. Figure 4.9 illustrates the in-field Mössbauer spectrum of 6 nm Co ferrite nanoparticles prepared by thermal decomposition route: one observes clearly broadened lines and a decrease of the intensities of the intermediate lines [73]. The corresponding distributions of effective field and canting angles are reported in Fig. 4.10 for both tetrahedral and octahedral Fe sites and finally the estimated hyperfine field distributions which well describe the zero-field spectrum [73].

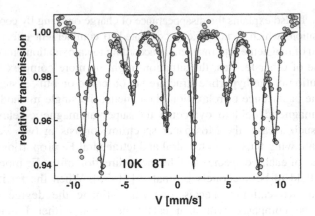

Fig. 4.9 Mössbauer spectrum recorded at 10 K on 6 nm Co ferrite nanoparticles prepared by thermal decomposition route submitted to an external of 8T applied parallel to the γ-beam [73]

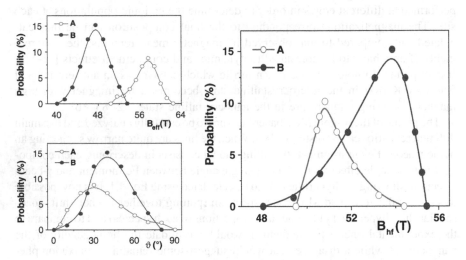

Fig. 4.10 Distributions of the effective magnetic field (*left upper*), of canting angles (*left down*), and of the hyperfine magnetic field (*right*) for both A and B sites [73]

In the case of nanocomposite materials which consist of iron-based alloys mixed with spinel oxide, the main relevant questions are concerned by the exact chemical composition of each of the constituents, i.e. both the metallic and the oxide phases. It has been proposed a strategy combining X-ray diffraction, thermal gravimetric and magnetic measurements, zero-field and in-field ^{57}Fe Mössbauer spectrometry in the case of $\left(Fe_p^0Co_{1-p}^0\right)_a \left(Co_xFe_{3-x-t}Vac_tO_4\right)$ composite materials where a, p, x and t represent the total metal content, p the Fe and Co metallic content, x the Co^{2+} content in the spinel phase and t the vacancies content, respectively [88]. The mean lattice parameters can be estimated by X-ray diffraction, and their values allow thus both the metal composition and the vacancies in the spinel phase to be determined while the elementary analysis gives the Co/Fe ratio in the composites. Then, the thermogravimetric measurements bring an approximate amount of metal in the composite. Finally, the refinement of both zero-field and in-field low temperature Mössbauer spectra provides the amount of metallic Fe, the amount of Fe^{2+} and Fe^{3+} in tetrahedral and octahedral sites of the spinel phase, and consequently the Co content in the metallic part, the Co^{2+} content in the spinel phase and the amount of vacancies [88]. In addition error bars on each of these parameters can be estimated, as detailed in [88].

The study of the functionalization of nanoparticles by molecules requires a good knowledge of both the structure and the chemical nature of the surface state but the chemical procedure may influence on the density of grafting, the chemical and structural evolution of the surface and the chemical bonding at the interface defined by the molecule and the nanoparticulate substrate. It is important to emphasize that the characterization of the chemical bonding is crucial in order to make the functionalized nanoparticles useful and suitable for some biomedical applications as drug delivery. Some relevant information can be obtained by comparing Mössbauer spectra recorded in the same conditions on nanoparticles before and after functionalization, combined to further spectroscopic techniques.

A first illustration is given by the efficient coating of bisphosphonates onto iron oxide nanoparticles as ferrofluids. From transmission electron microscopy, the core size of as-prepared nanoparticles is estimated at about 5 nm while the hyperfine structure is consistent with the presence of superparamagnetic relaxation phenomena, that to confirm the size and the non aggregated assembly of nanoparticles. As is illustrated in Fig. 4.11, the comparison of the 77 K hyperfine structures

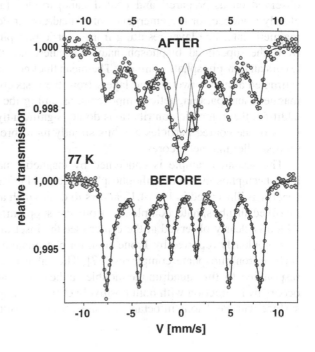

Fig. 4.11 77 K Mössbauer spectra of iron oxide nanoparticles before and after coating of bisphosphonates [89]

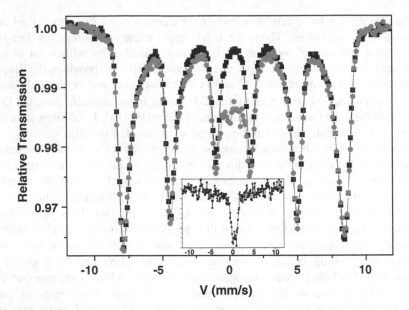

Fig. 4.12 Mössbauer spectra recorded at 77 K obtained on Fe oxide nanoparticle with an average grain size of 12 nm before (in *black*) and after phosphatation (in *red*). The inset corresponds to the spectrum resulting from the difference between the two previous ones, giving clear evidence for a quadrupolar doublet [41]

observed on as-prepared and coated nanoparticles (as frozen ferrofluids) give clearly evidence for the emergence of a quadrupolar doublet attributed to a para-magnetic interface layer resulting from a ferric phosphate oxide phase originated from the superficial bisphosphonate molecules, i.e. the phosphatation reaction favors an iron-phosphate complex. The mean thickness of the paramagnetic layer is estimated at about two atomic layers from the respective absorption areas of the magnetic and paramagnetic components, assuming the same recoilless factors. In addition, the magnetic contributions do not significantly differ, but that character-istics of the coated particles exhibits slightly more broadened lines in agreement with smaller magnetic cores.

The second example is concerned by magnetite nanoparticles in size which have been phosphated in orthophosphoric acid [41]. The difference of Mössbauer spectra as illustrated in Fig. 4.12 allows to distinguish and to identify the iron ions involved in phosphate complexes, from the supplementary quadrupolar doublet which is clearly observed at the center (see the inset of Fig. 4.12). Such a feature was previously reported by Tronc et al. for phosphate maghemite and attributed to surface iron-phosphate complexes [37]. The value of the isomer shift value cor-responding to this quadrupolar doublet indicates clearly that the phosphatation occurs by interaction with both positively charged groups and hydroxyl sites at the surface with Fe^{3+} ions in octahedral sites. Such a result allows to suggest that the main surface species would be a protonated binuclear species and the top layer would be in the (111) plane [41].

At this stage, more complex nanoparticles can be designed, involving several shells with different chemical, structural and magnetic compositions with antiferromagnetic, ferromagnetic, and ferrimagnetic phases, in order to induce some bias exchange phenomena as example. Such a layered spherical structure leads to the occurrence of well defined interfaces and surfaces: this assumes a non atomic diffusive phenomena during the synthesis procedure and a chemical stability versus time. An example of core–shell-shell structure is found in [90] from exotic "onion like" core–shell morphology of γ-Fe nucleus surrounded by a concentric double shell of α-Fe and maghemite-like oxide iron-carbon nanocomposites, synthesized using a commercial activated amorphous and porous carbon.

To go thoroughly into the complete modelling of the magnetic structure, it is necessary to study the effect of size and dispersion of nanoparticles in order to better understand the intrinsic and extrinsic magnetic properties of nanoparticles, providing they are chemically and structurally homogeneous and rather monodisperse. The Zeeman spectrum gives rise to the magnetic effective field at the nucleus probe while the variations of the hyperfine field result from the magnetic and electronic properties. But the thermal fluctuations of the magnetization might originate a reduction of the hyperfine field: consequently, to prevent from the presence of dynamic superparamagnetic relaxation effects, Mössbauer spectra have to be recorded at the low temperature and/or under an intense external field. In such a case, the effective magnetic field characteristic of a Mössbauer atom can be described as

$$H_{eff} = H_{hf} + H_{app} + H_D + H_{dip} + H_L$$

where H_{hf}, H_{app}, H_D, H_{dip} and H_L correspond to the hyperfine field, the applied field, the demagnetizing field, the dipolar field originating essentially from neighbouring particles and Lorentz field, respectively.

Contrary to bulk microcrystalline materials where the demagnetizing field is negligible because of the multidomain structure, the effect of the demagnetizing field in the case of mono domain nanoparticles can be significant, particularly when the hyperfine field is not so large, but dependent on their morphology. Some experiments have been successfully performed to give evidence that the magnetic field at the nuclear probes located in spherical single domain nanoparticles is larger than that characteristic of multi-domain particles (about 0.7 T).

An other important point is related to the surface hyperfine field which may be a priori smaller or larger than the bulk hyperfine field at low temperature in the case of ultrathin films, according to the substrate and the material coating the surface. Mössbauer studies of α-Fe nanoparticles have revealed that the hyperfine field values of inner Fe nuclei are similar to those of bulk crystalline α-Fe at low temperature (i.e. static blocked magnetic regime), but those characteristic of the superficial atomic layer are found either lower or higher [91–96]. Such conclusions are well supported by some examples of the literature with 2 nm α-Fe particles in organic liquids [93],

3–7 nm α-Fe particles on carbon support and 3–4 nm α-Fe particles in alumina [96]. It is also important that some in situ experiments carried out on ^{57}Fe coated nanoparticles to enhance the surface signal: indeed, as natural Fe contains 2.2 at. % ^{57}Fe isotope, the prevailing contribution to the hyperfine structure is due to the surface while that of the bulk becomes negligible [97]. Additional Mössbauer studies versus temperature on 100 nm α-Fe$_2$O$_3$ nanoparticles led to distinguish clearly surface to volume hyperfine structure contributions [97].

4.5 Nanocrystalline Alloys

In 1990s, numerous studies were devoted to nanocrystalline alloys after their discovery by Yoshizawa with that of the so-called FINEMET [98]. These nanocrystalline alloys typically originate from as-quenched amorphous precursor metglasses submitted to a subsequent annealing, when they are characterized by the presence of two crystallization peaks, contrarily to most of metallic glasses for which a single crystallization peak allows to transform directly into a microcrystalline structure. In the first case, the annealing treatment yields to the emergence of nanocrystalline grains embedded within a residual amorphous phase: the proportions of these two components (crystalline giving rise to the volumetric fraction) and the morphology and the size of the crystalline grains can be a priori well controlled from the annealing conditions (time and temperature). But it is also obvious that the final nanocrystalline state is rather dependent on the atomic composition of the amorphous precursor, its amorphicity, i.e. the quenching conditions, and the relative diffusivities of the different atomic constituents. The transformation from the amorphous into the nanocrystalline state, the intergranular amorphous remainder is thus subjected to substantial microstructural and chemical evolutions due to the migration of some atomic species and its enrichment in some elements which are expelled from the first emerged precipitates. But it is important to emphasize that the high energy milling route was also considered to prepare amorphous and/or nanostructured powders with similar chemical composition to compare the physical properties.

The interest of the nanocrystalline alloys is mainly due to their unusual soft magnetic properties combining high magnetic permeability and large saturation magnetization, making them very attractive candidates for applications as inductive devices, loss free transformers [99]. In addition, their two phase structural and magnetic behavior provided them very relevant and fascinating examples from a fundamental point of view [100, 101]. The most prominent nanocrystalline alloys as Finemet (FeCuNbSiB) [98], Nanoperm (FeZr(Cu)B) [102, 103] and Hitperm (FeCoZr(Cu)B) [104] were widely investigated during the 1990s. In those nanocrystalline systems, crystalline grains consist of FeSi with DO$_3$ type structure, bcc-Fe and bcc-FeCo phase, respectively.

In addition, the magnetic properties of nanocrystalline alloys are found to be strongly dependent on the microstructure of nanocrystalline alloys, i.e. the volumetric crystalline fraction and the size and morphology of crystalline grains created

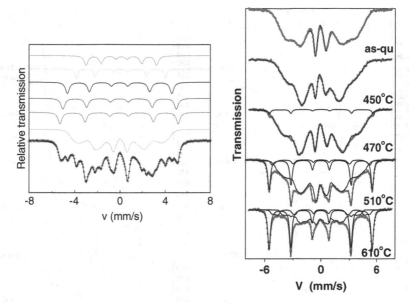

Fig. 4.13   300 K Mössbauer spectrum of annealed $Fe_{73.5}Si_{13.5}B_9Cu_1Nb_3$ (Finemet) alloy and its decomposition (*right*) and a series of 300 K Mössbauer spectra recorded on as-quenched and annealed at given temperatures of Nanoperm $Fe_{80.5}Nb_7B_{12.5}$ alloys (*left*) [110]

during the primary crystallization and of the amorphous remainder. The search of nanocrystalline alloys with well defined physical properties requires to understand the magnetic properties and their evolution versus temperature in 2 phase systems. Consequently, systematical studies were applied to various nanocrystalline alloys where it is first necessary to describe the structure of both the crystalline grains and the amorphous remainder, in order to model the mechanism of magnetic interactions within crystalline grains, amorphous remainder and in-between.

From an experimental point of view, the Fe-containing as-quenched and nano-crystalline samples consist of 20–30 μm thick ribbons perpendicularly oriented to the γ-beam, allowing thus the use of both low and high temperature [57]Fe Mössbauer facilities and in-field [57]Fe Mössbauer measurements (see reviews in [105–109]).

Figure 4.13 illustrates typical Mössbauer spectra obtained at room temperature on nanocrystalline Finemet alloys (left) and on as-quenched amorphous Nanoperm alloys and annealed ones for different times (right) [110]. In the case of the precursor amorphous alloy, the magnetic spectrum consists of broadened and slightly asymmetrical lines attributed to the topological disorder around Fe probes while the hyperfine structure of those characteristics of nanocrystalline substances result from the superimposition of narrow line sextet(s), broad line sextet and broad line quadrupolar feature. These later components are a priori assigned to Fe in the crystalline grains, in the magnetic and paramagnetic amorphous remainder, respectively, with proportions which are temperature dependent. One does note that the complexity of these hyperfine structures prevents from a non ambiguous

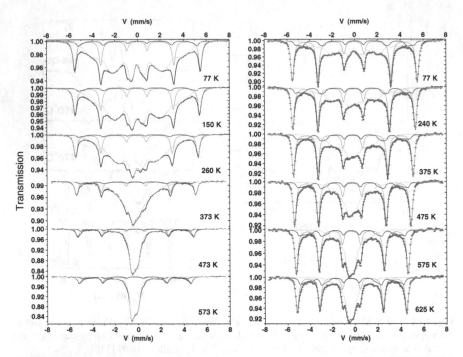

Fig. 4.14 Mössbauer spectra recorded at given temperatures on nanocrystalline NANOPERM Fe$_{80.5}$Nb$_7$B$_{12.5}$ alloys annealed at 783 and 883 K for 1 h, *left-* and *right-*hand column, respectively [110]

modeling, particularly in the case of FINEMET alloys [111–117]. Indeed, the structure of FeSi grains is dependent on the Si content: Si atoms are randomly distributed in bcc Fe lattice up to 10 at. % Si while in the range 12–31 at. % Si, one gets an ordered or disordered DO$_3$ structure, both giving rise to a complex hyperfine structure with several Fe sites [118]. Consequently, the best strategy consists in recording of a series of Mösbauer spectra versus temperature on each nanocrystalline sample and then describing the hyperfine structure of all spectra on the basis of a unique model: an illustration is given in Fig. 4.14 for 2 nanocrys-talline NANOPERM alloys annealed at 2 different temperatures for 1 h. Such a route is a difficult task but it provides a great number of data such as the tem-perature dependencies of isomer shift, quadrupolar shift, quadrupolar splitting and hyperfine field corresponding to Fe probes located in the nanocrystalline grain, the amorphous remainder and the resulting interfacial zone, and finally the tempera-ture independent atomic crystalline fraction (which can be compared to the vol-umetric crystalline fraction). The hyperfine structure is generally described by means of three components ascribed to the nanocrystalline grains (narrow lines magnetic sextet), the residual amorphous phase (broad line sextet which pro-gressively collapses into a quadrupolar doublet) and the interfacial zone (broad line sextet which roughly follows the same temperature dependence as that of the first component) [119–121]. In addition, to check the free-texture behaviour for these different components, one can use the magic-angle configuration which may a priori lead to some simplifications. In addition, f-factors for the resonant atoms located in various structural positions are assumed to be equal, when determining their relative contribution, to the total spectrum area. The fitting procedure of the hyperfine structure has been discussed in various papers [119–125]. Finally, it is obvious to mention that other techniques have to be used in addition to Mössbauer spectrometry but great attention has to be made to compare some results.

Such a strategy allows first to estimate accurately the crystalline fraction in terms of Fe content: its value can be compared to that determined from other techniques, but it requires a careful analysis of the physical meaning of each data. Then, from the fitting model, a third component occurs with a temperature dependence close to that of the nanocrystalline grains: it is attributed to a magnetic interfacial layer occupying approximately 2–3 atomic layers between crystalline grains and amorphous remainder [126–132]. Indeed, this layer resulting from a symmetry restriction consists mainly of Fe atoms which can be structurally ascribed to the periphery of crystalline grains but magnetically influenced by the vicinity of a Zr-rich layer in the amorphous phase which prevents from their growth, and Fe atoms located within the amorphous phase but in close contact with the crystalline grains. Magnetic structures can be described from in-field Mössbauer spectrometry. Figure 4.15 compares zero-field spectrum to in-field ones when external field is applied parallel to the direction of the incident γ-rays and perpendicular to the ribbon plane of the nanocrystalline $Fe_{89}Zr_7B_4$ alloy at 4.2 K [129]. Out of field, one concludes that the magnetic Fe moments of the grains are not randomly distributed but are preferentially oriented within the ribbon plane at about 70° from the normal to the ribbon plane. In presence of an external field of 1 T, one observes a decrease of the in-plane magnetization component giving rise to a roughly random orientation of magnetization in the grains, as the external field slightly overcomes the demagnetizing field [129]. On the contrary, the magnetic Fe moments are preferentially oriented along to the intense applied field but a careful analysis allows to evidence that a non collinear spin arrangement in the topologically disordered region ascribed to the crystalline grain-amorphous matrix interface [129]. The spin glass-like disorder of the interfacial layer is explained by a wide distribution of the Fe–Fe nearest neighbor distances originating competing ferromagnetic and antiferromagnetic interactions and by a large surface/interface anisotropy arising from the magnetocrystalline, magnetoelastic and dipolar shape anisotropies. It is important to emphasize that the estimate of this interface at about 2–3 atomic layers from in-field Mössbauer spectra is consistent with previous one and further computer Monte Carlo based modelling give evidence for a non collinear magnetic layer, whose thickness was found to be with the same order of magnitude [133–135].

The complete description of the hyperfine structure allows to schematize the atomic structure of a nanocrystalline alloy, as illustrated in Fig. 4.16: one notes that small amounts of Zr and B atoms can be included as impurities within the crystalline grains, as discussed in literature [131].

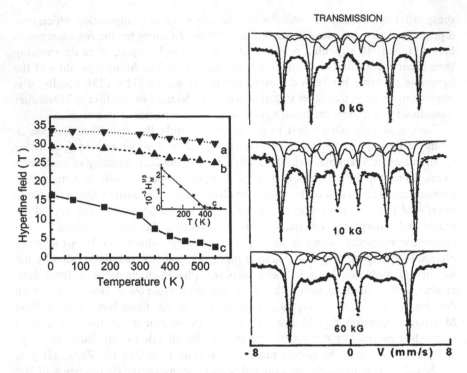

Fig. 4.15 *Left* Temperature dependence of mean hyperfine fields for bcc-Fe crystalline phase (**a**), interfaces (**b**) and amorphous matrix (**c**); the inset shows curve c in reduced coordinates [129]. *Right* Mössbauer spectra of $Fe_{89}Zr_7B_4$ nanocrystalline alloy at 4.2 K in zero field and in parallel field of 1 and 6T [129]

The evolution of the (mean) hyperfine field and other hyperfine parameters versus temperature can be compared to static magnetic data allowing the average composition of the amorphous remainder to be estimated. In the case of nanocrystalline alloys with low crystalline fraction, the temperature dependence of the crystalline phase gives clearly evidence at temperatures exceeding the Curie temperature of the amorphous remainder for the presence of superparamagnetic relaxation phenomena originating thus a degradation of soft magnetic properties. On the contrary, when the volumetric crystalline is high, one observes that the mean value of the hyperfine field does not vanish down to zero: indeed, the amorphous remainder remains partially magnetically ordered beyond the Curie temperature. These small hyperfine field values can be explained by the strong exchange interactions between the grains which penetrate the paramagnetic amorphous matrix. It might be also possible to estimate the strength of the interparticle magnetic coupling [129]. The presence of superparamagnetic relaxation at the crystalline nanograin reduces the intergrain magnetic coupling when the amorphous matrix is paramagnetic, allowing thus to conclude that dipolar interactions are not so small and to understand the degradation of soft magnetic properties [136–142].

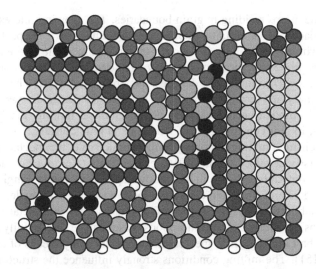

Fig. 4.16 Schematic 2D atomic representation of a nanocrystalline alloy with following caption: *yellow circles* correspond to Fe atoms located in the core of the crystalline grains; *red circles* Fe atoms in the amorphous phase; *blue, dark blue,* and *black* Fe atoms of the interfacial layer as second and first neighbours of the Zr/Nb-rich layer in the crystalline and amorphous phases, respectively; *white* and *green circles* represent B and Zr atoms respectively [131]

The role of the unconventional radio frequency Mössbauer technique [143] has to be emphasized in the case of the nanocrystalline alloys: indeed, it allows to distinguish the magnetically soft amorphous and nanocrystalline phase from the magnetically harder microcrystalline Fe, to determine the anisotropy fields in each phase as a function of the rf field intensity. It was found that the magnetic anisotropy of the amorphous matrix is significantly smaller than that encountered in the nanocrystalline phase. Finally, the rf sidebands effect reveals a strong reduction of magnetostriction related to the formation of the nanocrystalline phase [144, 145].

In conclusion, the Mössbauer studies versus temperature and/or external magnetic field performed on nanocrystalline alloys allow to well estimate the hyperfine characteristics of the different components: this contributes to understand the evolution of magnetic properties with the role of amorphous remainder and the interface (estimated at about 2 atoms thick) in coupling crystalline grains, crucial feature to understand the extremely soft magnetic properties of these alloys.

4.6 Magnetic Nanostructured Powders

Nanocrystalline materials or nanostructured powders are characterized by a microstructure composed with ultrafine grain sizes of 10–100 nm, originated thus high density of defects, interfaces, mainly grain boundaries, i.e. a large volume fraction of the atoms residing in grain boundaries. The growing interest during the last two decades is essentially due to their unusual and unexpected physical properties, which are strongly influenced by the confinement of microcrystalline into nanocrystalline structural domains, giving rise to an increasing contribution of grain boundaries. During the last decades, numerous studies have been essentially devoted to binary and ternary metallic and oxide nanostructured powders prepared by different methods (mechanical alloying, mechanical milling, mechanochemistry, grinding, in situ consolidation of nanoscale atomic clusters,…).

Due to their high mixing efficiency at the atomic scale, the mechanical alloying which was firstly developed in the 1960s by Benjamin to produce oxide dispersion strengthened materials, consists in a top–down approach based on high energy ball milling giving rise to ultra-fine disordered or nanostructured powders [146–150]. Indeed, the elaboration of a large variety of materials such as intermetallics, extended solid solutions, quasicrystals and amorphous phases can be successfully achieved by high energy ball milling, since they cannot be produced by means of conventional procedure [151]. The milling conditions strongly influence the structural and consequently the physical properties of nanostructured powders which need to be characterized in terms of size, shape of crystalline grains, surface area, phase constitution and microstructuration including grain boundaries [151–153].

Consequently, the main questions are essentially concerned by the packing of nanocrystalline grains through the grain boundaries: do the grain boundaries exist? Do their structural and chemical natures differ from that of polycrystalline regime? Do they behave as frozen-gas with lacking both short and long range order? How do they influence the physical properties of nanostructured powders?

Thus, the main goal is to investigate first the structural and microstructural properties involving diffraction techniques and electron microscopies, then atomic scale approach using local probe techniques and finally thermodynamic, electric, magnetic, vibrational properties in order to model the nanostructured state and to understand the confinement effect on physical properties and its relation with the elaboration procedure (milling conditions as example). We do emphasize that the elaboration of nanocrystalline materials and nanostructured powders, and their handling, storage and sampling for experimental studies require special attention because of the surface contamination and/or oxidation processes: these aspects cannot be neglected as they do strongly influence the physical properties and their evolution with ageing.

Below, we report two selected studies which illustrate how the use of ^{57}Fe Mössbauer spectrometry brings relevant information on the structural and magnetic modelling of nanostructured powders. It is clear that the chemical nature of systems under investigation influences remains a fundamental issue: in this context, the case of nanostructured Fe powders which have been widely investigated and debated [154–157], seem quite interesting based on its chemical simplicity while the example of nanostructured ferric fluoride powders illustrates the role of magnetic interactions dependent on the cationic topology, taking into account the combined knowledge of polymorphic crystalline phases and amorphous varieties.

4.7 Fe in Situ Consolidated Clusters

A detailed study was early performed on Fe clusters generated by inert gas condensation and in situ consolidated [158]. A sample pellet of 30–80 µm thick was obtained from compaction of the nanocrystalline powder in high vacuum (5.10^{-6} Pa) applying a unidirectional pressure of 5 GPa. As illustrated in Fig. 4.17 left [158], the Mössbauer spectra clearly exhibit the presence of two magnetic sextets, one consists of narrow lines while the second one is composed of broadened lines. It is clearly observed a magnetic texture from the low intensity intermediate lines which results from the sampling conditions. The evolution of their respective hyperfine field values is strongly temperature dependent with a cross over at about 170 K (as is shown in Fig. 4.17 right [158]).

The narrow line sextet is clearly attributed to the nanocrystalline Fe grain while the second component is assigned to Fe atoms located at the interfaces. Indeed the larger linewidth is consistent with a structural disorder, originating larger Fe–Fe distances. The higher value hyperfine field at low temperature for the grain boundaries can be interpreted in terms of reduced interface density, well supported by the Bethe-Slater curve: indeed the increase of Fe–Fe distance favours that of the Fe magnetic moment, and consequently that of the hyperfine field. In addition, the enhancement of the isomer shift value of Fe atoms located in grain boundaries is consistent with a reduction of the electron density and with a volume expansion. From the temperature dependence of the relative absorption area of the two components where that of the interface decreases faster, one can estimate that the Debye temperature of grain boundaries is lower than that of nanocrystalline grains, supporting that the grain boundaries possess a disordered atomic structure. Finally, from the relative absorption at low temperature (4:1), the mean thickness of interfaces is estimated at about 4 atoms, taking into account the size of nanocrystalline grains.

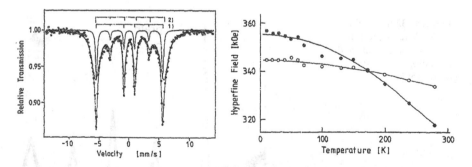

Fig. 4.17 Mössbauer spectrum at 77 K showing the presence of 2 magnetic components (*left*); temperature dependence of the hyperfine field of the 2 magnetic components (*right*) [158]

4.8 Milled Fe Nanostructured Powders

Nanostructured Fe powders were also elaborated from microcrystalline Fe powders by means of high energy ball milling using planetary ball under Ar atmosphere [159]. Their reproducibility is well achieved providing similar milling conditions: ball to powder weight ratio, milling speed, milling time, milling atmosphere, nature of balls and mill vial while some of them have to be considered to prevent from contamination and oxidation of metallic powders.

High statistics X-ray patterns recorded on milled Fe powders for different times allow to model the nanocrystalline structure including the size and morphology of coherent diffraction domain and their relative proportion, providing an additive component is introduced to describe the low part of Bragg peaks. The experimental pattern is compared in Fig. 4.8 to the theoretical one obtained by means of Maud procedure which is based on the Rietveld method combined with Fourier analysis. In addition, the mean morphology of the coherent diffraction domain illustrated in inset of Fig. 4.18 suggests that the nanocrystalline grains behave 10–12 nm pseudo-cubes, favouring a dense packing through thin grain boundaries [159]. In this hypothesis, the thickness of the grain boundary can be roughly estimated at about 0.8 nm, i.e. 2–3 atomic layers, smaller than those observed previously in the case of Fe clusters generated by inert gas condensation and in situ consolidated [158].

On the contrary, Mössbauer spectrometry does not provide any further information except that the presence of sextet typical of bcc-Fe structure remains milling time independent, as observed in Fig. 4.19. Low temperature Mössbauer spectra confirm such a tendency as the hyperfine structure resulting from a single magnetic sextet is not affected. At this stage, one does consider that the grain boundaries which do not contribute to the hyperfine structure are only weakly disordered and quite narrow. High temperature spectra could eventually cause a splitting of the hyperfine structure but they were not recorded because of any

Fig. 4.18 32 h Fe milled powder X-ray pattern fitted with two components and morphology of crystalline grains [159]

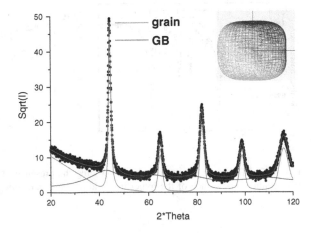

Fig. 4.19 300 K Mössbauer spectra of the samples milled for 16, 32, 48 and 64 h. One notes a quadrupolar component for 64 h milling time attributed to wüstite-like phase originated from reduction effects during milling procedure [159]

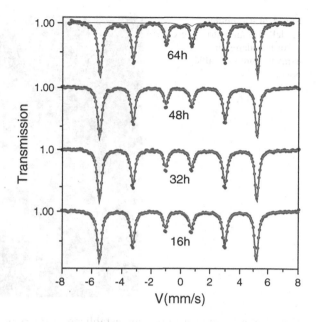

structural transformation of nanocrystalline grains into microcrystalline grains, giving rise to more complex structured powders [159].

In addition to hysteresis loops, ZFC–FC magnetic measurements were performed: the observed irreversibility which gradually reduces with increasing milling time up to 48 h is consistent with the presence of grain boundaries [159].

One concludes from these different techniques that the as-milled nanostructured Fe powders behave as micrometric agglomerates consisting of randomly oriented Fe nanocrystalline pseudo-cubic grains. They are separated by a thin interfacial layer estimated at 2–3 atomic layers where the coherent structural length is small but the nearest neighbour surrounding of Fe atoms remains rather the same as inside the crystalline grains, explaining their non occurrence in the Mössbauer spectra. It can be suggested than the chemical composition of these thin interfacial layer is the same as that inside the crystalline grain, but their topology is slightly different: indeed, the proximity effect of neighbouring grains and strong exchange coupling between Fe atoms from neighbouring particles do cause an apparent increase of the coordination number of surface Fe atoms leading to the hyperfine parameters being very similar to those of the crystalline interior. Consequently, the Fe nanostructured powders behave as strongly correlated ferromagnets and their magnetic behaviour is dominated by interparticle interactions [159].

A further insight in this "rather simple" nanostructure was given by a numeric approach based on the Embedded Atom Method: this new route was developed to model the structure of grain boundaries in metallic nanostructured powders. Numeric aspects are detailed in [160]. As it is illustrated in Fig. 4.20, realistic atomic structures of twisted and tilted double and triple grain boundaries as a function of the relative disorientation of the grains were successfully modelled by

Fig. 4.20 Computer modelled atomic structure of grain boundaries in nanostructured metallic powders

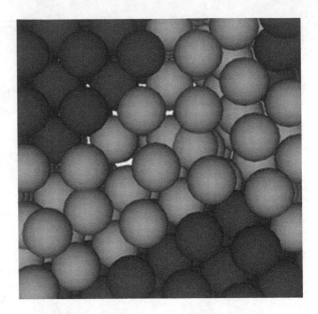

means of a Monte Carlo scheme established from annealing/Metropolis algorithm without using periodic boundaries conditions, with few free parameters [160].

This computer modelling approach allows to determine radial and angular distribution function characteristic of both crystalline grains and grain boundaries: the thickness of grain boundaries is estimated to 2–3 atomic layers, in perfect agreement with experimental features while its atomic configuration which is closer to the bcc structure than the fcc one in the case of Fe nanostructured powders is rather new [160]. Despite no ab initio calculations were yet established, the topology fairly supports the hyperfine structure observed by Mössbauer spectrometry as afore described.

4.9 Ferric Nanostructured Fluoride Powders

The understanding of physical properties of ferric nanostructured fluoride powders requires first a short review of the fundamental interest of ferric fluorides FeF_3 which exhibit three polymorphic crystalline phases and several amorphous varieties. They received great attention during the 1980s because of their simple chemical nature and their non collinear magnetic structures which are governed by the cationic topology and the antiferromagnetic superexchange interactions [161, 162].

The crystalline and magnetic structures of the 3 polymorphic crystalline structures of FeF_3 are illustrated in Fig. 4.21. The more stable crystalline rhombohedral r–FeF_3 phase is a canted antiferromagnet below $T_N = 363$ K, the canting is attributed to the antisymmetric Dzialhozinky-Moriya exchange interaction; the hexagonal tungsten bronze form orders below $T_N = 97$ K with 3 ferromagnetic sublattices in

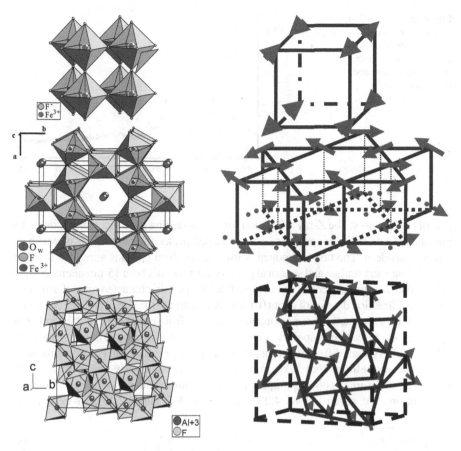

Fig. 4.21 Crystalline (*left*) and magnetic structures (*right*) of rhombohedral r-FeF$_3$ (*top*), hexagonal tungsten bronze HTB-FeF$_3$ (*middle*) and pyrochlore pyr-FeF$_3$ (*bottom*)

the HTB planes which are antiferromagnetically coupled; the pyrochlore form of FeF$_3$ orders below 20 K with 4 ferromagnetic sublattices oriented at 109° from each other. Both the non collinear magnetic structures and the reduction of magnetic ordering temperatures are attributed to the frustration of the antiferromagnetic exchange interactions implied by the cationic topology. The amorphous ferric fluoride varieties can be prepared by either vapour quenching onto a cold substrate or fluorination route [163]. They behave as a speromagnet below the freezing temperature estimated at about 30 K can be thus structurally described from a dense random packing of corner-shared octahedral units [162–164].

The rhombohedral form was ground under Ar atmosphere to prepare nanostructured powders with milling conditions as reported in X and sampling was highly controlled because of the high hygroscopic character of fluoride powder [165]. As illustrated in Fig. 4.22, the X-ray pattern of the ground ferric fluoride powder exhibits high statistics and is expressed with a square root coordinate scale in order to better visualize the low

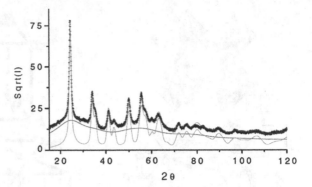

Fig. 4.22 Typical X-ray pattern of milled FeF$_3$ and its decomposition into two components attributed to crystalline grains and grain boundaries [165]

part of Bragg peaks. The X-ray diffraction pattern modelling gave successfully rise to the presence of two main components with broadened peaks, the proportions of which are rather equivalent. The first component with well resolved peaks is attributed to rhombohedral nanocrystalline and spherical grains with a size of about 15 nm diameter while the second component corresponding to the low part is characteristics of structural coherent diffraction domain of about 0.4 nm (equivalent to the size of an FeF$_6$ octahedral unit): it is thus assigned to grain boundaries resulting from a packing of octahedral units.

Series of Mössbauer spectra were recorded for different grinding times and grinding energy at different temperatures and in presence of external magnetic field. The modelling does take into account simultaneously the corresponding hyperfine structures [166]. Examples of spectra and corresponding hyperfine field distributions are given in Fig. 4.23 recorded at 300 and 4.2 K on r-FeF$_3$ in the

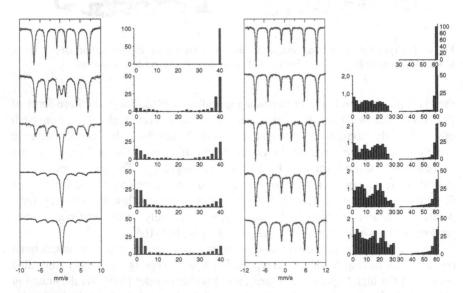

Fig. 4.23 Mössbauer spectra recorded at 300 K (*left*) and 4.2 K (*right*) on as-prepared crystalline FeF$_3$ (*top*) and ground for 0.25, 1, 8 and 16 h (*top to bottom*) [166]

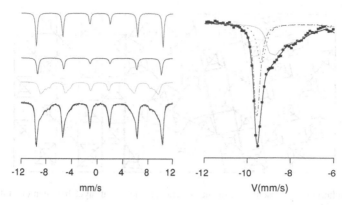

Fig. 4.24 4.2 K Mössbauer spectrum of the 16 h milled ferric fluoride powder and its decomposition into 3 components (*left*) and detailed profile of the refinement focused on the low energy line

as-prepared crystalline state and ground for 0.25, 1, 8 and 16 h. At 300 K, one observes the reduction of the magnetic component with a progressive asymmetrical broadening of lines to the expense of the increase of a single broadened line at the centre. It is important to emphasize that the application of an external magnetic field (even rather small produced by a small magnet) favours an increase of the absorption area of the magnetic component, indicating thus the presence of dynamics originating from superparamagnetic effects.

At low temperature, the hyperfine structures remain a priori rather independent on the grinding time but a more detailed analysis of the baseline reveals also an asymmetrical broadening of sextet lines. These different spectra can be described by means of discrete distribution of static hyperfine fields, since some magnetic dynamics might occur at high temperature. At this stage, first remarks are concerned by the values of isomer shift which are similar to those observed in crystalline and amorphous phases, and consequently consistent with high spin state Fe^{3+} ions (excluding thus the presence of Fe^{2+}) located in FeF_6 octahedral units which are corner-shared.

In the case of the 16 h milled fluoride powder, one can distinguish at 4.2 K from the static hyperfine field distribution 3 components as illustrated in Fig. 4.24 which shows the decomposition and the profile of the low energy external line of the sextet [165]. The narrow line sextet is unambiguously attributed to the crystalline AF FeF_3 grains (similar hyperfine field value) while the broad line low field component fairly compares that of the amorphous FeF_3 phase and is consequently assigned to the grain boundaries. The hyperfine field characteristics of the third component is found intermediate between the two previous ones: thus, it does originate from Fe ions located at the periphery of crystalline grains, giving rise to one Fe atom thick layer facing the grain boundaries, i.e. one octahedral unit layer, as is schematically represented in Fig. 4.25. Such an interfacial layer can be compared to that evidenced in the case of nanocrystalline alloys (see above section).

When the temperature is increasing, one observes a progressive collapse of the magnetic sextet into a single line which can be related to the evolution of the

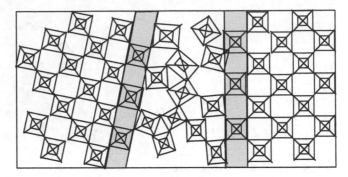

Fig. 4.25 Schematic representation of the interfacial octahedral layer between crystalline grains and grain boundaries

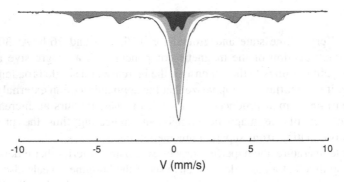

Fig. 4.26 300 K Mössbauer spectrum of the 16 h milled ferric fluoride powder and its decomposition

magnetic structure of the grain boundaries. As is illustrated in Fig. 4.26, the 300 K Mössbauer spectrum of the 16 h milled ferric fluoride powder can be also decomposed into 3 components: a magnetic sextet, a broadened lines quadrupolar doublet and a single line. The magnetic sextet results from Fe moments with slow relaxation dynamics or static Fe moments (the later corresponding to the highest hyperfine field values which well compare that characteristics of r-FeF_3 crystalline phase). However, when the relaxation phenomena are faster (smaller or magnetically uncoupled crystalline grains), the hyperfine structure becomes a broadened single line. Finally, the quadrupolar component which is described by a distribution of quadrupolar splitting the profile of which well compares that of the pure amorphous phase, is assigned to the presence of grain boundaries: indeed the expected topology based on a random packing of corner-sharing octahedral units induces a cancellation of magnetic ordered structure, as compared to the freezing temperature (about 30 K) of the amorphous phase prepared by vapour quenching onto a substrate, corresponding to the transition from the paramagnetic into the speromagnetic states.

Fig. 4.27 4.2 K Mössbauer spectrum under 6 T of FeF$_3$ nanostructured powder ground for 48 h and its decomposition with speromagnetic (*up*) and antiferromagnetic (*middle*) structures attributed to grain boundaries and crystalline grains, respectively [165]

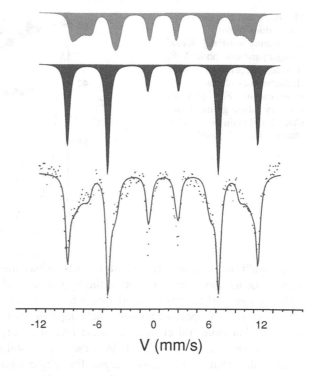

The proportions of each component have to be derived from the fitting model which does describe the Mössbauer spectra recorded at different temperatures, for different grinding times and milling conditions. It is clearly established that the evolution of the crystalline component decreases when the grinding time increases at the expense of the grain boundaries whereas the estimated proportions are found to be rather similar to those estimated from X-ray diffraction patterns, assuming spherical grains.

Furthermore, in-field Mössbauer spectrum was recorded in presence of an external magnetic field of 6T oriented parallel to the γ-beam on the 48 h ground powder. As illustrated in Fig. 4.27, the decomposition into 2 components which gives rise to a good description of the hyperfine structure, is fairly consistent with previous conclusions. Indeed, one component is unambiguously attributed to the antiferromagnetic crystalline grains through the value of the hyperfine field and the significant increase of intermediate line intensities (close to 3:4:1:1:4:3) while the second one results from a speromagnetic structure as in the case of the amorphous varieties, allowing to be assigned to the presence of grain boundaries. The proportions of these two structural components which derive from their respective absorption area, assuming the same values of recoilless factors [167], are consistent with those estimated from series of zero field Mössbauer spectra [165], and those from X-ray diffraction and solid state ^{19}F, ^{69}Ga and ^{71}Ga applied to ball-milled ionic GaF$_3$ [168].

Fig. 4.28 Mean hyperfine field values versus temperature in r-FeF$_3$ (*Black triangle*) and amorphous FeF$_3$ (*Circle*) compared to those characteristics of the nanostructured FeF$_3$ powders with crystalline grains (*Black Circle*) and grain boundaries (*Black diamond*) [165]

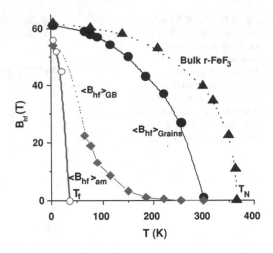

In addition, to magnetic measurements, Mössbauer spectrometry allows to follow the temperature evolution of the hyperfine field characteristic of the crystalline grains and the grain boundaries, which are rather proportional to magnetic moment of the respective phases. Such dependencies are compared to those of massive rhombohedral crystalline phase and of amorphous variety in Fig. 4.28. One observes that the hyperfine field of the nanocrystalline grains remains equal or smaller than that of the massive crystalline phase while that of the grain boundaries is equal or larger than that of the amorphous variety [165].

From Fig. 4.28, one can distinguish 3 temperature regimes: (1) in the low temperature range, the hyperfine field values of nanocrystalline grains and grain boundaries well agree with those of the rhombohedral and of the amorphous phases, respectively: the two phases are magnetically ordered because the superexchange magnetic interactions are strong in both magnetic structures: antiferromagnetic in the crystalline grains and speromagnetic in the grain boundaries, resulting from the different topology involving pseudo-cubic packing and random packing of corner-sharing octahedral units as schematically represented in Fig. 4.29, respectively. (2) when the temperature increases, the grain boundary moments progressively transit to a paramagnetic state, with defrosting temperatures ranged between 100 and 200 K, because the superexchange magnetic interactions are prevailing in the crystalline grains. This large temperature distribution is attributed to both the grain size distribution and the heterogeneous thickness of the grain boundaries. In addition, dynamic phenomena can occur in the smallest grains favoured when grain boundaries are rather thick (2–3 octahedral units which exceed the superexchange antiferromagnetic correlation length between crystalline grains, preventing thus the occurrence of dipolar interactions). In contrast with the topological analogy between amorphous phases and the grain boundaries of milled powders giving rise to same hyperfine characteristics, one does remark that the magnetic freezing temperatures of the grain boundaries are found higher than those of the amorphous phases (100–200 K and 30–40 K,

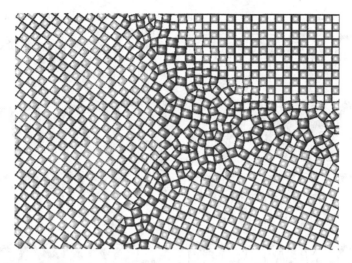

Fig. 4.29 Schematic 2D representation of the microstructure of crystalline and grain boundaries in nanostructured fluoride powder [165]

respectively). Such a disagreement partially originates from the structural disorder of the grain boundaries which can differ from that of the amorphous phases and the symmetry breaking through the presence of neighbouring crystalline grains. Indeed, the polarization of the grain boundaries by the crystalline grain also enhances their magnetic ordering temperature. (3) At high temperature, the grain boundaries behave as a paramagnet and the single domain grains possess thus a superparamagnetic behaviour. Indeed, one expects a progressive magnetic decoupling between grains as effective as the thickness of the grain boundaries is large (exceeding the antiferromagnetic length), leading to an assembly of non-interacting or weak-interacting antiferromagnetic particles. The distributions of both the grain size and of the thickness of grain boundaries give rise to blocking temperature ranged from 200 up to 360 K (corresponding to the Néel temperature of r-FeF$_3$), that is confirmed by static magnetic measurements [166].

Some temperature dependent exchange bias effect was evidenced from the non symmetrical hysteresis loops on some nanostructured ferric fluoride powder recorded after zero field cooling, suggesting the existence of a strong magnetic coupling between grain boundaries and crystalline grains up to the paramagnetic state of grain boundaries.

A computer modelling based on a modified annealing/Metropolis atomistic algorithm was successfully achieved to describe the structural and magnetic properties of the nanostructured ferric fluoride, particularly the topology within the grain boundaries. Long-range interatomic potentials were first adjusted from experimental results of crystalline phases and atomic structures of twisted and tilted GBs were then established as a function of the relative disorientation of the grains without applying any periodic boundary conditions. This approach takes into account the structure of the grains far from the interface in order to constrain

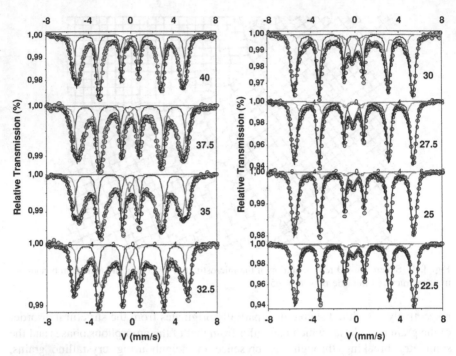

Fig. 4.30 300 K Mössbauer spectra of the MA Fe100-xNix samples milled for 10 h [178]

the relative orientation of the grains. One concludes that a long-range Coulombic fall off of the interatomic potentials is necessary to obtain GB structures presenting a correct local topology but with a smooth transition from crystalline to amorphous states. Indeed, the crystalline grain result from a cubic packing of corner-sharing octahedral units while the grain boundaries consist of corner-sharing octahedral FeF_6 units randomly packed: consequently, the corresponding magnetic structures behave as antiferromagnet and speromagnet, respectively, in agreement with experimental features [169].

The study of nanostructured ferric fluoride powder provides an excellent description of such 3D nanostructures, giving evidence for the important role of grain boundaries and confirms that ferric fluoride appears as an excellent fundamental system to correlate magnetic structure and cationic topology through frustration effects.

High energy ball milling is a route frequently applied to prepare nanoalloys and nanostructured metallic powders (only some selected examples in [170–177] but a lot of studies were reported for the last 25 years). Mössbauer spectrometry contributes in following the alloying process and the characterization of both structural and magnetic properties. But such an approach requires the improvement of the fitting model which does take into account the local atomic disorder, the presence of grain boundaries and the magnetic coupling between nancrystalline grains. An recent example has been developed to describe the Mössbauer spectra of $Fe_{100-x}Ni_x$

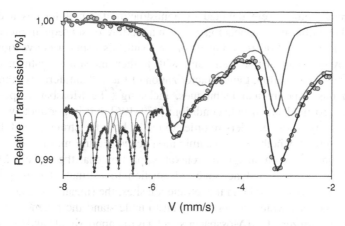

Fig. 4.31 Zoom of peaks 1 and 2 of the Mössbauer spectrum of the Fe$_{67.5}$Ni$_{32.5}$ sample milled for 10 h. The inset represents the whole spectrum. One distinguishes the BCC, FCC magnetic components and the quadrupolar single line component [178]

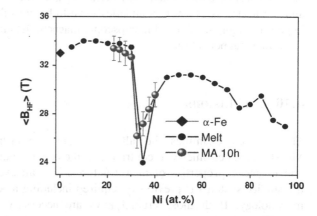

Fig. 4.32 300 K Mean hyperfine field values as a function of Ni content [178]

powders ($22.5 < x < 40$ at.%) mechanically alloyed (MA) for 10 h [178]. It is evidenced from X-ray diffraction that the nanostructured alloys consist of both body-centred cubic and face-centred cubic crystalline phases. Consequently, the hyperfine structures had to be described by a fitting model involving two hyperfine magnetic field distributions and a narrow singlet (see Figs. 4.30 and 4.31). One distribution was attributed to the ferromagnetic BCC grains of tetrataenite) while the second one was assigned to the ferromagnetic grains of taenite, and the narrow singlet to paramagnetic grains of antitaenite (i.e. superparamagnetic grains). This careful description allows to follow the hyperfine field at Fe as a function of the Ni content, giving rise to a jump at about 32.5 at.% Ni as illustrated in Fig. 4.32, attributed to Invar anomaly [178].

Numerous studies were devoted to nanostructured spinel ferrites and garnets elaborated by either mechanochemistry from a mixture of subsequent precursors, or by mechanical activation resulting from the synthesis using a conventional route followed by mechanical treatment, and with further annealing treatments. As discussed in previous section, the characterization of these "nanoferrites" requires the use of various experimental techniques, including ^{57}Fe Mössbauer spectrometry combining zero-field and in-field conditions [179–196]. It is important to well control the homogeneity of the powders in order to model both the structure and the microstructure, to check whether no contamination occurs during milling. In addition to overcome some superparamagnetic relaxation phenomena, the in-field Mössbauer spectra do bring clear and accurate information to estimate the proportions of Fe located in the octahedral and the tetrahedral sites, the (mean) canting angle of Fe magnetic moments, which are both crucial to understand the magnetic properties. The estimate from zero-field Mössbauer spectra is too approximate and the solution is completely dependent on the fitting assumptions, preventing thus a physical approach. It is important to emphasize that the presence of grain boundaries has a relevant role on the magnetic coupling between grains, favoring thus either some spin glass cluster like behavior or superparamagnetic non interacting grains. Consequently, the size of grains and the thickness of grain boundaries have to be considered to explain both local hyperfine magnetic data and macroscopic magnetic features from coercive field and saturation magnetization.

4.10 Conclusions

The selected examples did allow first to demonstrate the ability of ^{57}Fe Mössbauer spectrometry to contribute in the understanding of chemical, structural and magnetic properties of nanostructures. But, an excellent characterization of the studied systems is previously required including a control of an homogeneous morphology. Both those crucial keys are necessary to make thus easier the modelling of the physical properties, particularly local scale structural and magnetic ones, and to understand the role of surfaces and grain boundaries in nanostructures. The fitting of the Mössbauer spectra remains an important step and the results have to be compared to those obtained under different experimental in situ conditions (temperature, external magnetic field). The know-how established from "ideal nanostructures" opens new ways to revisit Mössbauer studies some topics: the phase transformation in some polymorphs [197, 198], the passivation processes and corrosion products and [199], the study of archaeological ceramics [200] and rocks and soils in geology and minerals [201] where non homogeneous nanostructuration does occur.

Acknowledgments This chapter reviews partially studies performed for the last 20 years at the NanoMagnetism and Numeric Modelling Group of the Institut des Molécules et Matériaux du Mans, UMR CNRS 6283 (ex Laboratoire de Physique de l'Etat Condensé UMR CNRS 6287). It is a sincere pleasure to thank first Prof N. Randrianantoandro, Dr N. Yaacoub, Dr Y. Labaye, Prof F. Calvayrac, Dr H. Guérault, Dr M. Grafouté and Dr B. Fongang for their respective significant contributions, and Dr A. Slawska-Waniewska, Dr I. Skorvanek, Prof M. Miglierini, Dr J. Degmova, Dr O. Crisan, Prof. J.M. Borrego, Prof. J. S. Blazquez, Dr D. Peddis, Prof G.A. Perez Alcazar, Dr J.F. Valderruten and Prof J. Restrepo during their respective stays at Le Mans. As mentioned in previous sections, the mutual collaboration with many chemistry groups requires large interactive communication to discuss results in order to optimize (nano)materials and to elaborate more complex (nano)architectures: Dr E. Tronc, Prof J.P. Jolivet, Prof C. Chanéac, Prof S. Ammar (Paris), G. Pourroy and S. Begin (Strasbourg).

References

1. G. Binnig, H. Rohrer, Scanning tunnelling microscopy. IBM J. Res. Dev. **30**, 4 (1986)
2. G. Binnig, C.L. Quate, C. Gerber, Atom force microscope. Phys. Rev. Lett. **56**, 930–933 (1986)
3. G. Binnig, H. Rohrer, Scanning tunnelling microscopy—from birth to adolescence. Rev. Mod. Phys. **59**, 615–625 (1986)
4. G. Binnig, H. Rohrer, Scanning tunnelling microscopy. Surf. Sci. **126**, 236–244 (1983)
5. H.W. Kroto, J.R. Heath, S.C. O'Brien, R.F. Curl, R.E. Smalley, Buckminsterfullerene. Nature **318**, 162–163 (1985). C-60
6. W.W. Adams, R.H. Baughman, Richard E. Smalley (1943–2005)—Retrospective. Science **310**, 5756 (2005)
7. P. Grunberg, R. Schreiber, Y. Pang, M.B. Brodsky, H. Sowers, Layered magnetic-structures—evidence for antiferromagnetic coupling of Fe layers across Cr interlayers. Phys. Rev. Lett. **57**, 2442–2445 (1986)
8. M.N. Baibich, J.M. Broto, A. Fert, F.N. Vandau, F. Petroff, G. Creuzet, A. Friederich, J. Chazelas, Giant magnetoresistance of (001) Fe/(001) Cr Magnetic superlattices. Phys. Rev. Lett. **61**, 2472–2475 (1988)
9. L. Lutterotti, MAUD program, CPD, Newsletter (IUCr) No. 24. (2000), web site: http://www.ing.unitn.it/luttero/maud
10. S. Morup, J. Dumesic, H. Topsoe, in *Applications of Mössbauer Spectroscopy*, ed. by R. Cohen. vol. 2 (Academic, New York, 1980), pp. 1–53
11. J.L. Dormann, Rev. Phys. Appl. **16**, 275 (1981)
12. S Morup, in *Mössbauer Spectroscopy applied to Inorganic Chemistry*, ed. by G.J. Long, vol 2 (Plenum Press, New York, 1987), p. 89
13. J.L. Dormann, D. Fiorani (eds.), *Magnetic Properties of Fine Particles* (North-Holland, Amsterdam, 1992)
14. E. Tronc, Nanoparticles. Nuovo Cimento **D18**, 163–180 (1996)
15. J.L. Dormann, D. Fiorani, E. Tronc, Magnetic relaxation in fine-particle systems. Adv. Chem. Phys. **98**, 293 (1997)
16. J. Chappert, J. de Phys. **C6**(35), 71 (1974)
17. J. Chappert, J. Teillet, F. Varret, J. Magn. Magn. Mater. **11**, 200 (1979)
18. J.M. Greneche, Noncollinear magnetic structures investigated by high-field Mössbauer spectrometry. Acta Physica Slovaca **45**, 45–55 (1995)
19. J.M.D. Coey, P.W. Readman, New spin structure in an amorphous gel. Nature **246**, 476–478 (1973)
20. S. Mørup, M.F. Hansen, Superparamagnetic Particles, in *Handbook of Magnetism and Advanced Magnetic Materials. Novel Materials*, vol. 4, ed. by H. Kronmüller, S. Parkin (Wiley, Chichester, 2007), pp. 2159–2176

21. S. Mørup, C. Frandsen, M.F. Hansen, Magnetic Properties of Nanoparticles, in *The Oxford Handbook of Nanoscience and Technology: Materials, Structures, Properties and Characterization Techniques*, ed. by A.V. Narlikar, Y.Y. Fu (Oxford University Press, Oxford, 2010)

22. S. Mørup, M.F. Hansen, C. Frandsen, Magnetic interactions between nanoparticles. Beilstein J. Nanotechnol. **1**, 182–190 (2010)

23. S. Mørup, M.F. Hansen, C. Frandsen, Magnetic nanoparticles. Compr. Nanosci. Technol. **1**, 437–491 (2011)

24. J. Frenkel, J. Doefman, Spontaneous and induced magnetization in ferromagnetic bodies. Nature **126**, 274–275 (1930)

25. R.C. O'Handley, *Modern Magnetic Materials: Principles and Applications* (Wiley, New York, 2000)

26. E. Thellier, Sur les propriétés de l'aimantation thermorémanente des terres cuites. Comptes Rendus Hebdomadaires des Séances de l'Académie des Sciences **213**, 1019–1022 (1941)

27. L. Néel, Some theoretical aspects of rockmagnetism. Adv. Phys. **4**, 191–243 (1955)

28. L. Néel, Théorie du trainage magnétique des ferromagnétiques en grains fins avec applications aux terres cuites. Annales de Géophysique **5**, 99–136 (1949)

29. L. Néel, Influence des fluctuations thermiques sur l'aimantation de grains ferromagnétiques très fins. Comptes Rendus Hebdomadaires des Séances de l'Académie des Sciences **228**, 664–666 (1949)

30. W.F. Brown Jr, Thermal fluctuations of a single domain particle. Phys. Rev. **130**, 1677–1686 (1963)

31. A. Aharoni, Complete eigen-value spectrum for the nucleation in a ferromagnetic prolate spheroid. Phys. Rev. **131**, 1478–1482 (1963)

32. J.L. Dormann, L. Bessais, D. Fiorani, A dynamic study of small interacting particles-superparamgnetic model and spin-glass laws. J. Phys. C Solid State Phys. **21**, 2015–2034 (1988)

33. D. Fiorani, J.L. Dormann, R. Cherkaoui et al., Collective magnetic state in nanoparticles systems. J. Magn. Magn. Mater. **196–197**, 143–147 (1999)

34. S. Mørup, Superparamagnetism and spin glass ordering in magnetic nanocomposites. Europhys. Lett. **28**, 671–676 (1994)

35. J.M.D. Coey, Noncollinear spin arrangement in ultrafine ferrimagnetic crystallites. Phys. Rev. Lett. **27**, 1140–1142 (1971)

36. E. Tronc, P. Prené, J.P. Jolivet, J.L. Dormann, J.M. Greneche, Spin-canting in γ-Fe$_2$O$_3$ Nanoparticles. Hyperfine Interact. **112**, 97–100 (1997)

37. E. Tronc, A. Ezzir, R. Cherkaoui, C. Chanéac, M. Noguès, H. Kachkachi, D. Fiorani, A.M. Testa, J.M. Greneche, J.P. Jolivet, Surface-related properties of γ-Fe$_2$O$_3$ nanoparticles. J. Magn. Magn. Mater. **221**, 63–79 (2000)

38. E. Tronc, D. Fiorani, M. Noguès, A.M. Testa, F. Lucari, F. D'Orazio, J.M. Greneche, W. Wernsdorfer, N. Galvez, C. Chanéac, D. Mailly, M. Verdaguer, J.P. Jolivet, Surface effects in noninteracting and interacting γ-Fe$_2$O$_3$ nanoparticles. J. Magn. Magn. Mater. **262**, 6–14 (2003)

39. E. Tronc, M. Nogues, C. Chaneac, F. Lucari, F.D. Orazio, J.M. Greneche, J.P. Jolivet, D. Fiorani, A.M. Testa, Magnetic properties of γ-Fe$_2$O$_3$ dispersed particles: size and matrix effects. J. Magn. Magn. Mater. **272–276**, 1474–1475 (2004)

40. A. Slawska-Waniewska, P. Didukh, J.M. Greneche, P.C. Fannin, Mössbauer and magnetisation studies of CoFe$_2$O$_4$ particles in a magnetic fluid. J. Magn. Magn. Mater. **215–216**, 227–230 (2000)

41. T.J. Daou, S. Begin-Colin, J.M. Greneche, F. Thomas, A. Derory, P. Bernhardt, P. Legare, G. Pourroy, Phosphate adsorption properties of magnetite-based nanoparticles. Chem. Mater. **19**, 4494–4505 (2007)

42. T.J. Daou, G. Pourroy, S. Bégin-Colin, J.M. Greneche, C. Ulhaq-Bouillet, P. Legare, P. Bernhardt, C. Leuvrey, G. Rogez, Hydrothermal synthesis of monodisperse magnetite nanoparticles. Chem. Mater. **18**, 4399–4404 (2006)

43. T.J. Daou, J.M. Greneche, S.J. Lee, S. Lee, C. Lefevre, S. Bégin-Colin, G. Pourroy, Spin canting of maghemite studied by NMR and in-field Mössbauer spectrometry. J. Phys. Chem. C **114**, 8794–8799 (2010)

44. E. Lima Jr, E. De Biasi, M. Mansilla Vasquez, M.E. Saleta, F. Effenberg, L.M. Rossi, R. Cohen, H.R. Rechenberg, R.D. Zysler, Surface effects in the magnetic properties of crystalline 3 nm ferrite nanoparticles chemically synthesized. J. Appl. Phys. **108**, 103919 (2010)

45. E.C. Sousa, H.R. Rechenberg, J. Depeyrot, J.A. Gomes, R. Aquino, F.A. Tourinho, V. Dupuis, R. Perzynski, In-field Mössbauer study of disordered surface spins in core/shell ferrite nanoparticles. J. Appl. Phys. **106**, 093901 (2009)

46. A.T. Ngo, P. Bonville, M.P. Pileni, Spin canting and size effects in nanoparticles of nonstoichiometric cobalt ferrite. J. Appl. Phys. **89**, 3370–3376 (2001)

47. R.H. Kodama, A.E. Berkowitz, E.J. McNiff Jr, S. Foner, Surface spin disorder in $NiFe_2O_4$ nanoparticles. Phys. Rev. Lett. **77**, 394 (1996)

48. R.H. Kodama, A.E. Berkowitz, E.J. McNiff Jr, S. Foner, Surface spin disorder in ferrite nanoparticles (invited). J. Appl. Phys. **81**, 5552–5557 (1997)

49. H. Kachkachi, A. Ezzir, M. Noguès, E. Tronc, Surface effects in nanoparticles: application to maghemite γ-Fe_2O_3. Eur. Phys. J. B **14**, 681 (2000)

50. R.H. Kodama, Magnetic nanoparticles. J. Magn. Magn. Mater. **200**, 359 (1999)

51. Y. Labaye, O. Crisan, L. Berger, J.M. Greneche, J.M.D. Coey, Surface anisotropy in ferromagnetic nanoparticles. J. Appl. Phys. **91**, 8715–8717 (2002)

52. J. Restrepo, Y. Labaye, J.M. Greneche, Surface anisotropy in maghemite nanoparticles. Phys. B **384**, 221–223 (2006)

53. L. Néel, Propriétés magnétiques des ferrites-ferrimagnétisme et antiferromagnétisme. Annales de Physique **3**, 137–198 (1948)

54. E.J.W. Verwey, Electronic conduction of magnetite (Fe_3O_4) and its transition point at low temperatures. Nature **144**, 327 (1939)

55. E.J.W. Verwey, E.L. Heilmann, Physical properties and cation arrangement of oxides with spinel structures 1: cation arrangement in spinels. J. Chem. Phys. **15**, 174–180 (1947)

56. E.J.W. Verwey, P.W. Haayman, F.C. Romeijan, Physical properties and cation arrangement of oxides with spinel structures 2: electronic conductivity. J. Chem. Phys. **15**, 181 (1947)

57. I. Leonov, A.N. Yaresko, On the Verwey charge ordering in magnetite. J. Phys. Condens. Matter **19**, 021001 (2007)

58. L. Häggström, H. Annersten, T. Ericsson, R. Wäppling, W. Karner, S. Bjarman, Magnetic dipolar and electric quadrupolar effects on the Mössbauer spectra of magnetite above the Verwey transition. Hyp. Interact. **5**, 201–214 (1978)

59. A.C. Doriguetto, N.G. Fernandes, A.I.C. Persiano, E. Nunes Filho, J.M. Greneche, J.D. Fabris, Characterization of a natural magnetite. Phys. Chem. Miner. **30**, 249–255 (2003)

60. Ö. Helgason, J.-M. Greneche, F.J. Berry, S. Mørup, F. Mosselmans, Tin- and Titanium-doped γ-Fe_2O_3 (Maghemite). J. Phys. Condens. Matter **13**, 10785–10797 (2001)

61. R.J. Armstron, A.H. Morrish, G.A. Sawatzky, Mössbauer study of ferric ions in tetrahedral and octahedral sites of a spinel. Phys. Lett. **23**, 414 (1966)

62. G.A. Sawatzky, F. Van Der Woude, A.H. Morrish, Cation distributions in octahedral and tetrahedral sites of the Ferrimagnetic Spinel $CoFe_2O_4$. J. Appl. Phys. **39**, 1204–1206 (1968)

63. G.A. Sawatzky, F. Van Der Woude, A.H. Morrish, Mössbauer study of several ferrimagnetic spinels. Phys. Rev. **187**, 747–757 (1969)

64. L.K. Leung, B.J. Evans, A.H. Morrish, Low temperature Mössbauer study of a Nickel-Zinc ferrite $Zn_xNi_{1-x}Fe_2O_4$. Phys. Rev. B **8**, 29–43 (1973)

65. A.H. Morrish, P.E. Clark, High-Field Mössbauer study of manganese-zinc ferrites. Phys. Rev. B **11**, 278–286 (1975)

66. J.S. Salazar, L. Perez, O. de Abril, L.T. Phuoc, D. Ihiawakrim, M. Vazquez, J.M. Greneche, S. Begin-Colin, G. Pourroy, Magnetic Iron Oxide nanoparticles in 10–40 nm range: composition in terms of magnetite/maghemite ratio and effect on the magnetic properties. Chem. Mater. **23**, 1379–1386 (2011)

67. T.J. Daou, G. Pourroy, J.M. Greneche, A. Bertin, D. Felder-Flesch, S. Begin-Colin, Water soluble dendronized iron oxide nanoparticles. Dalton Trans. 4442–4449 (2009)
68. D. Faivre, L.H. Böttger, B.F. Matzanke, D. Schüler, Intracellular magnetite biomineralization in Bacteria proceeds by a distinct pathway involving membrane-bound Ferritin and an Iron(II) Species. Angew. Chem. Int. Ed. **46**, 8495–8499 (2007)
69. G.F. Goya, T.S. Berquó, F.C. Fonseca, M.P. Morales, Static and dynamic magnetic properties of spherical magnetite nanoparticles. J. Appl. Phys. **94**, 3520–3528 (2003)
70. I. Dézsi, F. Cs, Á. Gombkötő, I. Szűcs, J. Gubicza, T. Ungár, Phase transition in nanomagnetite. J. Appl. Phys. **103**, 104312 (2008)
71. S. Chkoundali, S. Ammar, N. Jouini, F. Fievet, P. Molinié, M. Danot, F. Villain, J.M. Greneche, Nickel ferrite nanoparticles: elaboration in polyol medium via hydrolysis, and magnetic properties. J. Phys. Condens. Matter **16**, 4357–4372 (2004)
72. S. Ammar, N. Jouini, F. Fiévet, Z. Beji, L. Smiri, P. Molinié, M. Danot, J.M. Greneche, Magnetic properties of zinc ferrite nanoparticles synthesized by hydrolysis in a polyol medium. J. Phys. Condens. Matter **18**, 9055–9069 (2006)
73. D. Peddis, N. Yaacoub, M. Ferretti, A. Martinelli, G. Piccaluga, A. Musinu, C. Cannas, G. Navarra, J.M. Greneche, D. Fiorani, Cationic distribution and spin canting in CoFe$_2$O$_4$ nanoparticles. J. Phys. Condens. Matter **23**, 426004 (2011)
74. M. Artus, L. Ben Tahar, F. Herbst, L. Smiri, F. Villain, N. Yaacoub, J.M. Greneche, S. Ammar, F. Fiévet, Size-dependent magnetic properties of CoFe$_2$O$_4$ nanoparticles prepared in polyol. J. Phys. Condens. Matter **23**, 506001 (2011)
75. S. Burianova, J.P. Vejpravova, P. Holec, J. Plocek, D. Niznansky, Surface spin effects in La-doped CoFe$_2$O$_4$ nanoparticles prepared by microemulsion route. J. Appl. Phys. **110**, 073902 (2011)
76. A. Yang, C.N. Chinnasamy, J.M. Greneche, Y. Chen, S.D. Yoon, K. Hsu, C. Vittoria, V.G. Harris, Large tunability of Néel temperature by growth-rate-induced cation inversion in Mn-ferrite nanoparticles. Appl. Phys. Lett. **94**, 113109 (2009)
77. L. Ben Tahar, M. Artus, S. Ammar, L.S. Smiri, F. Herbst, M.J. Vaulay, V. Richard, J.M. Greneche, F. Villain, F. Fievet, Magnetic properties of CoFe$_{1.9}$RE$_{0.1}$O$_4$ nanoparticles (RE = La, Ce, Nd, Sm, Eu, Gd, Tb, Ho) prepared in polyol. J. Magn. Magn. Mater. **320**, 3242–3250 (2008)
78. Z. Beji, L. Smiri, N. Yaacoub, J.M. Greneche, N. Menguy, S. Ammar, F. Fievet, Annealing effect on the magnetic properties of Polyol-made Ni-Zn Ferrite nanoparticles. Chem. Mater. **22**, 1350–1366 (2010)
79. O. Ersen, S. Bégin, M. Houllé, J. Amadou, I. Janowska, J.M. Grenèche, C. Crucifix, C. Pham-Huu, Microstructural investigation of magnetic CoFe$_2$O$_4$ nanowires inside Carbon nanotubes by electron tomography. Nanoletters **8**, 1033–1040 (2008)
80. Z. Beji, A. Hanini, L.S. Smiri, J. Gavard, K. Kacem, F. Villain, J.M. Greneche, F. Chau, S. Ammar, Magnetic properties of Zn-substituted MnFe$_2$O$_4$ nanoparticles synthesized in polyol as potential heating agents for hyperthermia. Evaluation of their toxicity on Endothelial cells. Chem. Mater. **22**, 5420–5429 (2010)
81. A. Yang, C.N. Chinnasamy, J.M. Greneche, Y. Chen, S.D. Yoon, Z. Chen, K. Hsu, Z. Cai, K. Ziemer, C. Vittoria, V.G. Harris, Enhanced Néel temperature in Mn ferrite nanoparticles linked to growth-rate-induced cation inversion. Nanotechnology **20**, 185704 (2009)
82. B. Antic, A. Kremenovic, N. Jovic, M.B. Pavlovic, C. Jovalekic, A.S. Nikolic, G.F. Goya, C. Weidenthaler, Magnetization enhancement and cation valences in nonstoichiometric (Mn, Fe)$_{3-\delta}$O$_4$ nanoparticles. J. Appl. Phys. **111**, 074309 (2012)
83. V. Blanco-Gutierrez, F. Jimenez-Villacorta, P. Bonville, M.J. Torralvo-Fernandez, R. Saez-Puche, X-ray absorption spectroscopy and Mössbauer spectroscopy studies of super paramagnetic ZnFe$_2$O$_4$ nanoparticles. J. Phys. Chem. C **115**, 1627–1634 (2011)
84. J.F. Hochepied, P. Bonville, M.P. Pileni, Nonstoichiometric Zinc Ferrite nanocrystals: syntheses and unusual magnetic properties. J. Phys. Chem. B **104**, 905–912 (2000)

85. S.J. Stewart, S.J.A. Figueroa, M.B. Sturla, R.B. Scorzelli, F. Garcia, F.G. Requejo, Magnetic ZnFe$_2$O$_4$ nanoferrites studied by X-ray magnetic circular dichroism and Mössbauer spectroscopy. Physica B **389**, 155–158 (2007)

86. G.F. Goya, H.R. Rechenberg, M. Chen, W.B. Yelon, Magnetic irreversibility in ultrafine ZnFe$_2$O$_4$ particles. J. Appl. Phys. **87**, 8005–8007 (2000)

87. J.M. Hastings, L.M. Corliss, An antiferromagnetic transition in Zinc Ferrite. Phys. Rev. **102**, 1460–1463 (1956)

88. N. Viart, G. Pourroy, J.M. Greneche, Study of metal-ferrite composites: complementary use of ^{57}Fe Mössbauer spectrometry, X-ray diffraction and TG analysis. Eur. J. Phys. Appl. Phys. **18**, 33–40 (2002)

89. A. Karimi, B. Denizot, F. Hindre, R. Filmon, J.M. Greneche, S. Laurent, T. Jean Daou, S. Begin Colin, J.J. Le Jeune, Effect of chain length and electrical charge on properties of ammonium-bearing bisphosphonate-coated super paramagnetic iron oxide nanoparticles: formulation and physicochemical studies. J. Nanopart. Res. **12**, 1239–1248 (2010)

90. M.P. Fernández-García, P. Gorria, J.A. Blanco, R. Boada, J. Chaboy, D. Schmool, J.M. Greneche, Microstructure and magnetism of nanoparticles with γ-Fe core surrounded by α-Fe and iron oxide shells. Phys. Rev. B **81**, 094418 (2010)

91. W. Kundig, H. Bommel, G. Constabaris, R.H. Linquist, Some properties of supported small α-Fe$_2$O$_3$ particles determined with Mössbauer effect. Phys. Rev. **142**, 327–333 (1966)

92. F. Bodker, S. Morup, S. Linderoth, Surface effects in metallic iron nanoparticles. Phys. Rev. Lett. **72**, 282 (1994)

93. T. Furubayashi, I. Nakatani, N. Saegusa, Magnetic moment and hyperfine field in colloidal fine particles of iron. J. Phys. Soc. Jpn. **56**, 1855–1858 (1987)

94. B.S. Clausen, S. Mørup, H. Topsøe, Evidence for chemisorption induced changes in the surface electronic and magnetic properties of small iron particles. Surf. Sci. **106**, 143–1438 (1981)

95. R. Birringer, H. Gleiter, H.-P. Klein, P. Marquardt, Nanocrystalline materials an approach to a novel solid structure with gas-like disorder? Phys. Lett. **102A**, 365–369 (1984)

96. F. Bodker, S. Morup, C.A. Oxborros, S. Linderoth, M.B. Madsen, J.W. Niemantsverdriet, Mössbauer studies of ultrafine iron-containing particles on a carbon. J. Phys. Condens. Matter **4**, 6555–6568 (1992)

97. T. Shinjo, N. Hosoito, T. Takada, Magnetism of Fe interfaces studied by Mössbauer spectroscopy. J. Magn. Magn. Mater. **31–34**, 879–880 (1983)

98. Y. Yoshizawa, S. Oguma, K. Yamauchi, New Fe-based soft magnetic-alloys composed of ultrafine grain-structure. J. Appl. Phys. **64**, 6044 (1988)

99. M.E. McHenry, M.A. Willard, D.E. Laughlin, Amorphous and nanocrystalline materials for applications as soft magnets. Prog. Mater. Sci. **44**, 291–433 (1999)

100. G. Herzer, in *Nanocrystalline soft magnetic alloys, Handbook of Magnetic Materials* vol. 10 ed by K.H.J. Buschow (Elsevier Science, 1997), pp. 415–462)

101. G. Herzer, in *Handbook of Magnetism and Advanced Magnetic Materials*, vol. 4, ed. by H. Kronmüller, S. Parkin (Wiley, Hoboken, 2007), p. 1882

102. K. Suzuki, N. Kataoka, A. Inoue, A. Makino, T. Masumoto, High saturation magnetization and soft magnetic properties of bcc Fe Zr B alloys with ultrafine grain structure. Mater. Trans. JIM **32**, 743 (1990)

103. K. Suzuki, A. Makino, A. Inoue, T. Masumoto, Soft magnetic properties of nanocrystalline bcc Fe-Zr-B and Fe-M-B-Cu (M = transition metal) alloys with high saturation magnetization. J. Appl. Phys. **70**, 6232 (1991)

104. M.A. Willard, D.E. Laughlin, M.E. McHenry, D. Thoma, K. Sickafus, J.O. Cross, V.G. Harris, Structure and magnetic properties of (Fe$_{0.5}$Co$_{0.5}$)$_{88}$Zr$_7$B$_4$Cu$_1$ nanocrystalline alloys. J. Appl. Phys. **84**, 6773 (1998)

105. J.M. Greneche, M. Miglierini, Mössbauer spectroscopy materials science, in *Mössbauer Spectrometry Applied to Iron-Based Nanocrystalline Alloys: I High Temperature Studies*, ed. by M. Miglierini, D. Petridis (Kluwer Academic Publishers, Dordrecht, 1999), pp. 243–256

106. M. Miglierini, J.M. Greneche, Mössbauer Spectroscopy in Materials Science, in *Mössbauer Spectrometry Applied to Iron-Based Nanocrystalline Alloys: I Hyperfine Field Distributions*, ed. by M. Miglierini, D. Petridis (Kluwer Academic Publishers, Dordrecht, 1999), pp. 257–272

107. T. Kemény, D. Kaptás, L.F. Kiss, J. Balogh, I. Vincze, S. Szabó, D.L. Beke, Structure and magnetic properties of nanocrystalline soft ferromagnets. Hyperfine Interact. **130**, 181–219 (2000)

108. J.-M. Greneche, Properties and Applications of Nanocrystalline Alloys from Amorphous Precursors, in *Intergranular Phase in Nanocrystalline Alloys: Structural and Magnetic Aspects*, ed. by B. Idzikowski, P. Svec, M. Miglierini, D. Petridis (Kluwer Academic, Dordrecht, 2005), pp. 373–384

109. M. Kopcewicz, Radio-Frequency Mössbauer Spectroscopy in the Investigation of Nanocrystalline Alloys, in *Properties and Applications of Nanocrystalline Alloys from Amorphous Precursors*, ed. by B. Idzikowski, P. Svec, M. Miglierini, D. Petridis (Kluwer Academic, Dordrecht, 2005), pp. 395–407

110. I. Škorvánek, P. Švec, J.M. Greneche, J. Kováč, J. Marcin, R. Gerling, Influence of microstructure on the magnetic and mechanical behaviour of amorphous and nanocrystalline FeNbB alloys. J. Phys. Condens. Matter **14**, 4717–4736 (2002)

111. M. Fujinami, Y. Hashiguchi, T. Yamamoto, Crystalline transformations in amorphous $Fe_{73.5}Cu_1Nb_3Si_{16.5}B_6$ Alloy. Jpn. J. Appl. Phys. **29**, L477–L480 (1990)

112. J.Z. Jiang, F. Aubertin, U. Gonser, H.R. Hilzinger, Mössbauer spectroscopy and X-ray diffraction studies of the crystallization in the amorphous $Fe_{73.5}Cu_1Nb_3Si_{13.5}B_9$ alloy. Zeitschrift fur Metallkunde **82**, 698–702 (1991)

113. J. Jiang, T. Zemcik, F. Aubertin, U. Gonser, Investigation of the phases and magnetization orientation in crystalline $Fe_{73.5}Cu_1Nb_3Si_{16.5}B_6$ alloy. J. Mater. Sci. Lett. **10**, 763–764 (1991)

114. G. Hampel, A. Pundt, J. Hesse, Crystallization of $Fe_{73.5}Cu_1Nb_3Si_{13.5}B_9$ structure and kinetics examined by X-ray diffraction and Mössbauer effect spectroscopy. J. Phys. Condens. Matter **4**, 3195–3214 (1992)

115. A. Pundt, G. Hampel, J. Hesse, Mössbauer effect studies on amorphous and nanocrystalline $Fe_{73.5}Cu_1Nb_3Si_{13.5}B_9$. Zeitschrift fur Physik-Condensed Matter **87**, 65–72 (1992)

116. G. Rixecker, P. Schaaf, U. Gonser, Crystallization behaviour of amorphous $Fe_{73.5}CuINb_3Si_{13.5}B_9$. J. Phys. Condens. Matter **4**, 10295–10310 (1992)

117. G. Rixecker, P. Schaaf, U. Gonser, Depth selective analysis of phases and spin textures in amorphous, nanocrystalline and crystalline ribbons treated with an excimer laser. J. Phys. D Appl. Phys. **26**, 870–879 (1993)

118. G. Rixecker, P. Schaaf, U. Gonser, On the interpretation of the Mössbauer spectra of ordered Fe-Si alloys. Physica Status Solidi A- Appl. Res. **139**, 309–320 (1993)

119. M. Miglierini, Mössbauer-effect study of the hyperfine field distributions in the residual amorphous phase of Fe-Cu-Nb-Si-B nanocrystalline alloys. J. Phys. Condens. Matter **6**, 431–1438 (1994)

120. M. Miglierini, J.M. Greneche, Mössbauer spectrometry of iron-based nanocrystalline alloys: I Fitting model of Mössbauer spectra. J. Phys. Condens. Matter **9**, 2303–2319 (1997)

121. M. Miglierini, J.M. Greneche, Mössbauer spectrometry of iron-based nanocrystalline alloys: II Topography of hyperfine interactions in Fe(Cu)ZrB alloys. J. Phys. Condens. Matter **9**, 2321–2347 (1997)

122. J.M. Greneche, Nanocrystalline iron-based alloys investigated by Mössbauer spectroscopy. Hyperfine Interact. **110**, 81–91 (1997)

123. J.M. Borrego, C.F. Conde, A. Conde, V.A. Peña-Rodríguez, J.M. Greneche, Devitrification process of FeSiBCuNbX nanocrystalline alloys: Mössbauer study of the intergranular phase. J. Phys. Condens. Matter **12**, 8089–8100 (2000)

124. J.M. Borrego, A. Conde, V.A. Peña-Rodríguez, J.M. Greneche, Mössbauer spectrometry of FINEMET-type nanocrystalline alloys: a revisiting fitting procedure. Hyperfine Interac. **131**, 67–82 (2001)

125. O. Hupe, M.A. Chuev, H. Bremers, J. Hesse, A.M. Afanas'ev, Magnetic properties of nanostructured ferromagnetic FeCuNbB alloys revealed by a novel method for evaluating complex Mössbauer spectra. J. Phys. Condens. Matter **11**, 10545–10556 (1999)
126. A. Slawska-Waniewska, A. Roig, E. Molins, J.M. Greneche, R. Zuberek, Surface effects in Fe-based nanocrystalline alloys. J. Appl. Phys. **81**, 4652–4654 (1997)
127. A. Slawska-Waniewska, K. Brzozka, J.M. Greneche, Surface effects in Fe-Based Nanocrystalline alloys. Acta Physica Polonica **91**, 229–232 (1997)
128. J.M. Greneche, A. Slawska-Waniewska, Interface effects in Fe$_{89}$Zr$_7$B$_4$ nanocrystalline alloy followed by Mössbauer spectroscopy. Mater. Sci. Eng. A **226–228**, 526–530 (1997)
129. A. Slawska-Waniewska, J.M. Greneche, Magnetic properties of interface in soft magnetic nanocrystalline alloys. Phys. Rev. B **56**, R8491–R8494 (1997)
130. A. Slawska-Waniewska, Interface magnetism in Fe-based nanocrystalline alloys. J. Phys. IV(8), 11–18 (1998)
131. J.M. Greneche, A. Slawska-Waniewska, About the interfacial zone in nanocrystalline alloys. J. Magn. Magn. Mater. **215–216**, 264–267 (2000)
132. T. Kemeny, L.K. Varga, L.F. Kiss, J. Balogh, T. Pusztai, L. Toth, I. Toth, Magnetic properties and local structure of Fe-Zr-B-Cu nanocrystalline alloys. Mater. Sci. Forum **269–272**, 419–424 (1998)
133. O. Crisan, Y. Labaye, L. Berger, J.M.D. Coey, J.M. Greneche, Exchange coupling effects in nanocrystalline alloys studied by Monte Carlo simulation. J. Appl. Phys. **91**, 8727–8729 (2002)
134. O. Crisan, Y. Labaye, L. Berger, J.M. Greneche, Monte Carlo simulation of magnetic properties in nanocrystalline like Systems. J. Phys. Condens. Matter **15**, 6331–6344 (2003)
135. O. Crisan, J.M. Greneche, Y. Labaye, L. Berger, A.D. Crisan, M. Angelakeris, J.M. Le Breton, N.K. Flevaris, Properties and applications of nanocrystalline alloys from amorphous precursors, in *Magnetic Properties of Nanostructured Materials/Monte Carlo Simulation and Experimental Approach for Nanocrystalline Alloys and Core-Shell Nanoparticles*, ed. by B. Idzikowski, P. Svec, M. Miglierini, D. Petridis (Kluwer Academic, Dordrecht, 2005), pp. 253–266
136. N. Randrianantoandro, A. Slawska-Waniewska, J.M. Greneche, Magnetic interactions of nanocrystallized Fe-Cr amorphous alloys. Phys. Rev. B **56**, 10797–10800 (1997)
137. N. Randrianantoandro, A. Slawska-Waniewska, J.M. Greneche, Magnetic properties of nanocrystallized Fe-Cr amorphous alloys. J. Phys. Condens. Matter **9**, 10485–10500 (1997)
138. M. Miglierini, I. Skorvanek, J.M. Greneche, Microstructure and hyperfine interactions of the Fe$_{73.5}$Nb$_{4.5}$Cr$_5$Cu$_1$B$_{16}$ nanocrystalline alloys: Mössbauer effect temperature measurements. J. Phys. Condens. Matter **10**, 3159–3176 (1998)
139. T. Kemeny, J. Balogh, I. Farkas, D. Kaptas, L.F. Kiss, T. Pusztai, L. Toth, I. Vincze, Intergrain coupling in nanocrystalline soft magnets. J. Phys. Condens. Matter **10**, L221–L227 (1998)
140. T. Kemeny, D. Kaptas, J. Balogh, L.F. Kiss, T. Pusztai, I. Vincze, Microscopic study of the magnetic coupling in a nanocrystalline soft magnet. J. Phys. Condens. Matter **11**, 2841–2847 (1999)
141. D. Kaptas, T. Kemeny, J. Balogh, L. Bujdoso, L.F. Kiss, T. Pusztai, I. Vincze, Anomalous magnetic properties of the nano-size residual amorphous phase in nanocrystals. J. Phys. Condens. Matter **11**, L179–L185 (1999)
142. I. Skorvanek, J. Kovac, J.M. Greneche, Structural and magnetic properties of the intergranular amorphous phase in FeNbB nanocrystalline alloys. J. Phys. Condens. Matter **12**, 9085–9093 (2000)
143. M. Kopcewicz, Mössbauer effect studies of amorphous metals in magnetic radio-frequency fields. Struct. Chem. **2**, 313–342 (1991)
144. M. Miglierini, M. Kopcewicz, B. Idzikowski, Z.E. Horvath, A. Grabias, I. Skorvanek, P. Duzewski, S. Cs Daroczi, Structure, hyperfine interactions, and magnetic behavior of amorphous and nanocrystalline Fe$_{80}$M$_7$B$_{12}$Cu$_1$.(M=Mo, Nb, Ti) alloys J. Appl. Phys. **85**, 1014–1025 (1999)

145. M. Kopcewicz, A. Grabias, I. Skorvanek, J. Marcin, B. Idzikowski, Mössbauer study of the magnetic properties of nanocrystalline $Fe_{80.5}Nb_7B_{12.5}$ alloy. J. Appl. Phys. **85**, 4427–4429 (1999)

146. M. Kopcewicz, A. Grabias, I. Skorvánek, Study of the nanocrystalline $Fe_{73.5}Nb_{4.5}Cr_5Cu_1B_{16}$ alloy by the radio-frequency-Mössbauer technique. J. Appl. Phys. **83**, 935–941 (1998)

147. J.S. Benjamin, Dispersion strengthened superalloys by mechanical alloying. Metall. Trans. **1**, 2943 (1970)

148. J.S. Benjamin, Mechanical alloying. Sci. Am. **234**, 40–49 (1976)

149. J.S. Benjamin, in *New materials by mechanical alloying techniques*, ed, by E. Arzt, L. Schultz, (DGM Informationgesellschaft, Oberursel, Germany, 1989), pp. 3–18

150. J.S. Benjamin, Metal Powder Rep. **45**, 122–127 (1990)

151. C. Suryanarayana, Mechanical alloying and milling. Prog. Mater. Sci. **46**, 1–184 (2001)

152. H. Gleiter, Nanocrystalline materials. Prog. Mater. Sci. **33**, 223–315 (1989)

153. H. Gleiter, Nanostructured materials: basic concepts and microstructure. Acta Mater **48**, 1–29 (2000)

154. L. Del Bianco, A. Hernando, E. Bonetti, E. Navarro, Grain-boundary structure and magnetic behavior in nanocrystalline ball-milled iron, B. Phys. Rev. **56**, 8894–8901 (1997)

155. L. Del Bianco, C. Ballesteros, J.M. Rojo, A. Hernando, Magnetically ordered fcc structure at the relaxed grain boundaries of pure nanocrystalline Fe. Phys. Rev. Lett. **81**, 4500 (1998)

156. E. Bonetti, L. Del Bianco, D. Fiorani, D. Rinaldi, R. Caciuffo, A. Hernando, Disordered magnetism at the grain boundary of pure nanocrystalline iron. Phys. Rev. Lett. **83**, 2829 (1999)

157. L. Del Bianco, A. Hernando, D. Fiorani, Spin-glass-like behaviour in nanocrystalline Fe, (a). Phys. Stat. Sol. **189**, 533 (2002)

158. U. Herr, J. Jing, R. Birringer, U. Gonser, H. Gleiter, Investigation of nanocrystalline iron materials by Mössbauer spectroscopy. Appl. Phys. Lett. **50**, 472 (1987)

159. A. Ślawska-Waniewska, M. Grafoute, J.M. Greneche, Magnetic coupling and spin structure in nanocrystalline iron powders. J. Phys. Condens. Matter **18**, 2235–2248 (2006)

160. M. Grafouté, Y. Labaye, F. Calvayrac, J.M. Greneche, Structure of grain boundaries in nanostructured powders: a Monte-Carlo/EAM numerical investigation. Eur. J. P B **45**, 419–424 (2005)

161. F. Ferey, M. Leblanc, R. De Pape, J Pannetier *Inorganic Solid Fluorides* ed. by P. Hagenmuller (Academic, New York) p. 395

162. J.M. Greneche, F. Varret, in *Mössbauer Spectroscopy Applied to Magnetism and Materials Science*, ed. by G. Long, F. Grandjean (Plenum, New York, 1993), p. 161 and references therein

163. G. Ferey, F. Varret, J.M.D. Coey, Amorphous FeF_3 non crystalline magnet with antiferromagnetic interactions. J. Phys. C Solid State Phys. **12**, L531 (1979)

164. J.M. Greneche, A. Le Bail, M. Leblanc, A. Mosset, F. Varret, J. Galy, G. Ferey, Structural aspects of amorphous fluorides FeF3. J. Phys. C Solid State Phys. **21**, 1351–1361 (1988)

165. H. Guérault, J.-M. Greneche, Microstructural modelling of iron-based nanostructured fluoride powders prepared by mechanical milling. J. Phys. Condens. Matter **12**, 4791–4798 (2000)

166. H. Guérault, M. Tamine, J.M. Greneche, Mössbauer study of nanostructured iron fluoride powders. J. Phys. Condens. Matter **12**, 9497–9508 (2000)

167. H. Guérault, I. Labaye, J.-M. Greneche, Recoilless factors in nanostructured iron-based powders. Hyp. Inter. **136**, 57–63 (2001)

168. B. Bureau, H. Guérault, G. Silly, J.Y. Buzaré, J.M. Greneche, NMR investigation of mechanically milled nanostructured GaF_3 powders. J. Phys. Condens. Matter **11**, L423–L431 (1999)

169. B. Fongang, I. Labaye, F. Calvayrac, J.M. Greneche, S. Zekeng, Coupled structural and magnetic properties of ferric fluoride nanostructures I: a Metropolis atomistic study. J. Magn. Magn. Mater. **322**, 2888–2892 (2010)

170. G. Le Caër, E. Bauer-Grosse, A. Pianelli, E. Bouzy, J. Matteazzi, Mechanically driven syntheses of carbides and silicides. Mater. Sci. **25**, 4726–4731 (1990)

171. G. Le Caer, P. Delcroix, T.D. Shen, B. Malaman, Mössbauer investigation of intermixing during ball milling of $Fe_{0.3}Cr_{0.7}$ and $Fe_{0.5}W_{0.5}$ powder mixtures. Phys. Rev. B **54**, 12775–12786 (1996)

172. R.J. Cooper, N. Randrianantoandro, N. Cowlam, J.M. Greneche, A study of the solid state "amorphisation" reaction in $Fe_{58}Ta_{42}$ by diffraction and Mössbauer spectrometry. J. Phys. Condens. Matter **9**, 1425–1433 (1997)

173. N. Randrianantoandro, R.J. Cooper, J.M. Greneche, N. Cowlam, Study of the solid state "amorphisation" reaction in $Fe_{50}Re_{50}$ by Mössbauer spectrometry and diffraction measurements. J. Phys. Condens. Matter **14**, 9713–9724 (2002)

174. C. González, G.A. Pérez Alcázar, L.E. Zamora, J.A. Tabares, J.M. Greneche, Magnetic properties of $Fe_xMn_{0.600-x}Al_{0.400}$, $0.200 \leq x \leq 0.600$, disordered alloy series. J. Phys. Condens. Matter **14**, 6531–6542 (2002)

175. H. Moumeni, S. Alleg, J.M. Greneche, Structural properties of $Fe_{50}Co_{50}$ nanostructured powder prepared by mechanical alloying. J. Alloys Compd. **386**, 12–19 (2005)

176. S. Azzaza, S. Alleg, H. Moumeni, A.R. Nemamcha, J.L. Rehspringer, J.M. Greneche, Magnetic properties of nanostructured ball-milled Fe and $Fe_{50}Co_{50}$ alloy. J. Phys. Condens. Matter **18**, 7257–7272 (2006)

177. H. Moumeni, S. Alleg, J.M. Greneche, Formation of ball-milled Fe-Mo nanostructured powders. J. Alloys Compd. **419**, 140–144 (2006)

178. J.F. Valderruten, G.A. Perez Alcazar, J.M. Greneche, Mössbauer and x-ray study of mechanically alloyed Fe–Ni alloys around the Invar composition. J. Phys. Condens. Matter **20**, 485204 (2008)

179. G.F. Goya, H.R. Rechenberg, J.Z. Jiang, Structural and magnetic properties of ball milled copper ferrite. J. Appl. Phys. **84**, 1101–1108 (1998)

180. P. Druska, U. Steinike, V. Sepelak, Surface structure of mechanically activated and of Mechanosynthesized Zinc Ferrite. J. Solid State Chem. **146**, 13–21 (1999)

181. V. Sepelak, D. Baabe, K.D. Becker, Mechanically induced cation redistribution and spin canting in nickel ferrite. J. Mater. Synth. Process. **8**, 333–337 (2000)

182. C.N. Chinnasamy, A. Narayanasamy, N. Ponpandian, K. Chattopadhyay, H. Guérault, J.-M. Greneche, Magnetic properties of nanostructured ferrimagnetic zinc ferrite. J. Phys. Condens. Matter **12**, 7795–7805 (2000)

183. C.N. Chinnasamy, A. Narayanasamy, N. Ponpandian, R. Justin Joseyphus, K. Chattopadhyay, K. Shinoda, B. Jeyadevan, K. Tohji, K. Nakatsuka, J.-M. Greneche, Ferrimagnetic ordering in nanostructured $CdFe_2O_4$ spinel. J. Appl. Phys. **90**, 527–529 (2001)

184. C.N. Chinnasamy, A. Narayanasamy, N. Ponpandian, R.J. Joseyphus, K. Chattopadhyay, K. Shinoda, B. Jeyadevan, K. Tohji, K. Nakatsuka, H. Guerault, J.-M. Greneche, Structure and magnetic properties of nanocrystalline ferrimagnetic $CdFe_2O_4$ spinel. Scripta Mater. **44**, 1411–1415 (2001)

185. C.N. Chinnasamy, A. Narayanasamy, N. Ponpandian, K. Chattopadhyay, H. Guerault, J.-M. Greneche, Ferrimagnetic ordering in nanostructured zinc ferrite. Scr. Mater. **44**, 1407–1410 (2001)

186. V. Sepelak, M. Menzel, K.D. Becker, F. Krumeich, Mechanochemical reduction of magnesium ferrite. J. Phys. Chem. **44**, 1411–1415 (2001)

187. N. Ponpandian, A. Narayanasamy, C.N. Chinnasamy, N. Sivakumar, J.-M. Greneche, K. Chattopadhyay, K. Shinoda, B. Jeyadevan, K. Tohji, Néel temperature enhancement in nanostructured nickel zinc ferrite. Appl. Phys. Lett. **86**, 192510 (2005)

188. N. Sivakumar, A. Narayanasamy, N. Ponpandian, J.-M. Greneche, K. Shinoda, B. Jeyadevan, K. Tohji, Effect of mechanical milling on the electrical and magnetic properties of nanostructured $Ni_{0.5}Zn_{0.5}Fe_2O_4$. J. Phys. D Appl. Phys. **39**, 4688–4694 (2006)

189. N. Sivakumar, A. Narayanasamy, K. Shinoda, C.N. Chinnasamy, B. Jeyadevan, J.-M. Greneche, Electrical and magnetic properties of chemically derived nanocrystalline cobalt ferrite. J. Appl. Phys. **102**, 013916 (2007)

190. V. Sepelak, I. Bergmann, A. Feldhoff, P. Heitjans, F. Krumeich, D. Menzel, J. Litterst, S.J. Campbell, K.D.J. Becker, Nanocrystalline nickel ferrite, $NiFe_2O_4$: Mechanosynthesis,

nonequilibrium cation distribution, canted spin arrangement, and magnetic behavior. Phys. Chem. **C**(111), 5026–5033 (2007)

191. A. Mahesh Kumar, K.H. Rao, J.M. Greneche, Mössbauer investigation on Ni–Zn nanoferrite with the highest saturation magnetization. J. Appl. Phys. **105**, 073919 (2009)

192. E.C. Passamani, B.R. Segatto, C. Larica, R. Cohen, J.M. Greneche, Magnetic hysteresis loop shift in $NiFe_2O_4$ nanocrystalline powder with large grain boundary fraction. J. Magn. Magn. Mater. **322**, 3917–3925 (2010)

193. N. Sivakumara, A. Narayanasamya, J.-M. Greneche, R. Murugaraj, Y.S. Lee, Electrical and magnetic behaviour of nanostructured $MgFe_2O_4$ spinel ferrite. J. Alloy. Compd. **504**, 395–402 (2010)

194. C.N. Chinnasamy, J.M. Greneche, M. Guillot, B. Latha, T. Sakai, C. Vittoria, V.G. Harris, Structural and size dependent magnetic properties of single phase nanostructured gadolinium-iron-garnet under high magnetic field of 32 tesla. J. Appl. Phys. **107**, 09A512 (2010)

195. D. Prabhu, A. Narayanasamy, K. Shinoda, B. Jeyadeven, J.-M. Greneche, K. Chattopadhyay, Grain size effect on the phase transformation temperature of nanostructured $CuFe_2O_4$. J. Appl. Phys. **109**, 013532 (2011)

196. M. Guillot, C.N. Chinnasamy, J.M. Greneche, V.G. Harris, Tuning the cation distribution and magnetic properties of single phase nanocrystalline $Dy_3Fe_5O_{12}$ garnet. J. Appl. Phys. **111**, 07A517 (2012)

197. R. Zboril, M. Mashlan, D. Petridis, Iron(III) Oxides from thermal processes synthesis, structural and magnetic properties, Mössbauer spectroscopy characterization, and applications. Chem. Mater. **14**, 969–982 (2002)

198. L. Machala, J. Tucek, R. Zboril, Polymorphous transformations of Nanometric Iron(III) Oxide: a review. Chem. Mater. **23**, 3255–3272 (2011)

199. D.C. Cook, Application of Mössbauer spectroscopy to the study of corrosion. Hyperfine Interact. **153**, 61–82 (2004)

200. F.E. Wagner, A. Kyek, Mössbauer spectroscopy in archaeology: introduction and experimental considerations. Hyperfine Interact. **154**, 5–33 (2004)

201. E. Murad, Mössbauer spectroscopy of clays, soils and their mineral constituents. Clay Miner. **45**, 413–430 (2010)

Author Biography

Jean-Marc Greneche

Jean-Marc Greneche (CNRS Research Director) got his PhD at Le Mans (1987); he was Humboldt fellow at the Kernforschungzentrum Karlsruhe (1988) and guest Professor at Santiago de Compostela (Spain) and at Universidade Federal De Rio de Janeiro (Brazil). He was Director of Laboratoire de Physique de l'Etat Condensé (LPEC UMR CNRS 6087) from 2008 up to 2011 and presently of the Institute of Molecules and Materials at Le Mans (IMMM, UMR CNRS 6283). In addition, after being Secretary of the International Board on Applications of Mössbauer Effect (2001–2007), he is Vice Chairman of the International Board on Applications of Mössbauer Effect since 2007. Now, he is a member of the Editorial Board of Hyperfine Interactions, Editor of Journal of Alloys and

Compounds and of Nanomagnetism. He is co-authored of ca. 410 papers and 10 chapter books, presented ca. 100 invited seminars/plenary lectures.

His current research topics are mainly concerned by magnetic nanoparticles and nanostructures, nanostructured materials and mesoporous structures, with a special emphasis on surfaces, interfaces and grain boundaries, but also magnetic inter-metallics, mineral chemistry with metal organic frameworks, crystalline and glassy systems, physical biogeochemistry and earth sciences. In addition, his strategy was the establishment of numerous national and international collaborations (Algeria, Argentina, Brazil, Colombia, India, Italy, Mexico, Poland, Slovak Republic, and Spain) with chemists and solid state physicists as experts in synthesis and computer modeling, respectively. The present studies of materials have required first collaborative efforts and mutual discussions with chemists, materials scientists and physicists, and then great attention for describing jointly the Mössbauer spectra and other experimental data: such an approach should allow thus to establish structural and magnetic physical models and finally to propose numerical modeling techniques based on *ab initio* or phenomenological calculations.

Chapter 5
Magnetic Multilayers and Interfaces

Teruya Shinjo and Ko Mibu

Abstract Mössbauer spectroscopy is a conventional technique to measure magnetic hyperfine fields. Particularly experiments on ^{57}Fe and ^{119}Sn are easy and observed hyperfine fields furnish us useful information not only for studies on magnetism but also generally to characterize condensed matters. In this article, the usefulness of Mössbauer spectroscopy for studies on magnetic multilayers is introduced. ^{57}Fe spectra of Fe/Mg multilayers are shown as a function of Fe layer thickness and magnetic properties of ultrathin amorphous Fe layers are argued. From the results on Fe/Mn and Fe/Cr multilayers, the features of interfaces are discussed. The results on Fe/RE (rare-earth) indicate that anisotropy of interface plays an important role. Interface magnetic properties of Fe film in contact with MgO and those of Fe oxides have been studied using interface-selectively enriched samples. Finally, several recent studies on magnetic multilayers and also on Heusler alloy films by using ^{119}Sn probes are also described.

T. Shinjo
Kyoto University, Kyoto, Japan
e-mail: shinjoteruya@iris.eonet.ne.jp

K. Mibu (✉)
Department of Engineering Physics, Electronics, and Mechanics,
Graduate School of Engineering, Nagoya Institute of Technology,
Gokiso-cho, Showa-ku, Nagoya 466-8555, Japan
e-mail: k_mibu@nitech.ac.jp

Y. Yoshida and G. Langouche (eds.), *Mössbauer Spectroscopy*,
DOI: 10.1007/978-3-642-32220-4_5, © Springer-Verlag Berlin Heidelberg 2013

5.1 Introduction

In this chapter, the usefulness of the Mössbauer spectroscopy for studies on magnetic multilayers and interfaces of magnetic thin films is described.

Since magnetism of condensed matters is originated from electron spins, magnetic properties are understood by analyzing the behavior of electron spins. Although spins exist also on nuclei of odd numbers, nuclear spins are too small to contribute to the bulk magnetic properties. However, they can serve as very useful probes to investigate fundamental properties of condensed matters, particularly magnetic properties. Experimental techniques to use nuclei as probes for condensed matters are called "nuclear methods" and two methods are very popular: namely NMR (nuclear magnetic resonance) and Mössbauer spectroscopy. As was already introduced in the first chapter of this book, Mössbauer effect can be observed for more than 40 elements but easily measured only for very limited nuclear species. The most convenient is ^{57}Fe and the second is ^{119}Sn. Mössbauer effect measurements are used as a very simple analytical tool if ^{57}Fe or ^{119}Sn is available as a microprobe in a sample. It is very fortunate for investigators in the field of magnetism that Fe nucleus is the most suitable one for Mössbauer effect measurements, since Fe often plays as a leading actor in magnetic materials. ^{57}Fe is a stable isotope found in natural Fe with the abundance of 2.2 %. If a sample includes a certain amount of Fe, Mössbauer absorption (or emission) spectrum is always able to be obtained. If a sample can be enriched in ^{57}Fe, measurements become possible with much less Fe contents. For ^{57}Fe Mössbauer measurements generally, it is conventional to use a radioactive isotope ^{57}Co as a gamma ray source. On the other hand, instead of radioisotope, Mössbauer experiments with synchrotron radiation source are making steady progress in recent years but this chapter is not concerned with synchrotron Mössbauer studies [1].

In our daily life, there can be found a great variety of commercial products containing magnetic materials. Examples are a hard-disk-drive in a personal computer, small permanent magnets on a bulletin board, credit cards with magnetic memory in our pocket, and so on. It is rather usual that the main component of magnetic materials in commercial products is Fe and therefore we can obtain Mössbauer absorption spectra from commercial magnetic goods if they have thin film shapes with such a thickness as gamma-ray can penetrate. In the case that they have bulky shapes and are too thick for gamma-ray penetration, Mössbauer measurements with scattering geometry would be appropriate.

Figure 5.1 shows Mössbauer absorption spectra obtained from very familiar magnetic materials around us; (a) a train ticket and (b) a credit card. Both spectra are well resolved six line patterns and from the values of hyperfine field, we can identify the magnetic materials. The spectrum (a) indicates that the surface of the train ticket is entirely coated by Ba-ferrite and the spectrum (b) shows that the material of magnetic memory part in the credit card is solely γ-Fe$_2$O$_3$. From the intensity ratio of the magnetically split lines, in both cases, it is suggested that the magnetization is preferentially oriented in the plane. These examples may be

Fig. 5.1 ^{57}Fe Mössbauer
absorption spectra for (**a**) a
JR train ticket, and (**b**) a
credit card (magnetic stripe
part)

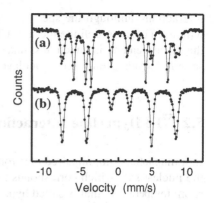

helpful for readers to feel familiar with Mössbauer spectra, even if they have not
yet understood the principles of the Mössbauer effect and to get an image that the
Mössbauer spectroscopy is an easy tool. Indeed, as an analytical method, a merit of
Mössbauer spectroscopy is that we can draw out information intuitively by
glancing over the spectra.

From a viewpoint of fundamental physics, ultrathin magnetic films have been a
very interesting subject and attracted a great attention from theoretical and
experimental viewpoints. A typical example is a monolayer consisting of magnetic
atoms such as Fe or Co. However, even nowadays it is not easy to prepare a
monolayer sample of 3d elements with high crystallographic quality and also to
obtain reliable information from experimental results. A magnetic order in a two-
dimensional lattice becomes spatially unstable at elevated temperatures and
therefore relaxation phenomena are inevitably involved. There has been no conclusive experimental estimation concerning the Curie temperatures of Fe, Co and
Ni monolayers. Although measurements on a sample of single monolayer are
difficult, those for a multilayer sample including hundred monolayers would be
much easier. Multilayers consisting of ferromagnetic and non-magnetic layers
have been actively investigated in 1980s. If a component of multilayer is Fe or Sn,
Mössbauer spectroscopy is a powerful tool for the analysis of the magnetic
properties. If the isotope enrichment in ^{57}Fe is available, for instance, specified
samples for the study of monolayer magnetism are able to be prepared. Interface
properties can be investigated by preparing samples whose interfaces are selectively enriched in a Mössbauer isotope.

There are two methods concerning the geometry of Mössbauer measurements:
transmission and scattering. Most Mössbauer experiments adopt a transmission
mode in which the gamma-ray from a source is counted after passing through an
absorber sample. For magnetic materials, Mössbauer measurements in transmission mode are advantageous to investigate temperature and external field depen-
dences. On the other hand, measurements in scattering mode (Conversion electron
Mössbauer spectroscopy, CEMS) are also possible by collecting the secondary
scattered conversion electrons. From transmission and scattering Mössbauer
measurements, the same information is obtained concerning the hyperfine

interactions, although the spectra in the two modes have the upside-down shapes with each other. A definite merit of CEMS is that the necessary amount of ^{57}Fe for the measurement is much less than absorption measurements. For example, a sample including only one monolayer of ^{57}Fe can be measured.

5.2 ^{57}Fe Hyperfine Interaction in Magnetic Materials

In a magnetic atom, unpaired electron spins produce a very large magnetic field at the nuclear site, which corresponds to 33 T (Tesla) in the case of pure Fe metal at room temperature, that is called hyperfine field. Unpaired electron spins exist in 3d orbitals but 3d electrons have no density at the nuclear site. Such a large field can be produced by the spin polarization of s-orbital electrons. Due to an exchange interaction with 3d electrons, the radial distributions of spin-up and spin-down electrons in core s-orbitals are slightly differentiated. Although the degree of polarization in core electrons is rather small, the induced magnetic field is quite large since the core s-electrons have finite densities at the nuclear site (Fermi contact interaction). If a nucleus experiences a magnetic field, the interaction between the field and nuclear magnetic dipole moment causes a Zeeman splitting of nuclear energy levels. As already introduced in Chap. 1, six transitions are allowed in the case ^{57}Fe, and therefore six peaks are observed in the Mössbauer spectrum. The magnitude of splitting in the Mössbauer spectrum is proportional to the hyperfine field. In the case of pure Fe metal, the hyperfine field at room temperature is 33.0 T and the overall splitting of six lines in the Mössbauer spectrum is 10.657 mm/s in the unit of Doppler velocity. The spin polarization induced at the nuclear site due to the Fermi contact interaction has the opposite direction to the spin polarization of 3d electrons. If a strong external magnetic field is applied, 3d spins are oriented to the field direction and therefore direction of the hyperfine field becomes opposite to that of external field and accordingly the observed hyperfine field is reduced by applying a strong external field. This situation is expressed as the sign of the hyperfine field being negative. The sign of the hyperfine field in ferromagnetic Fe systems is normally negative.

The relation between 3d local magnetic moment and hyperfine field is not straightforwardly proportional but theoretical studies suggest that the hyperfine field delicately depends on the electronic structure. Therefore in general a quantitative argument of hyperfine field is difficult. Nevertheless, magnetic hyperfine field measurements give us valuable information for the understanding of magnetic properties of materials. A rough proportionality between the hyperfine fields and local magnetic moment has been found in many cases of Fe atoms in ferromagnetic alloy systems. The presence of magnetic hyperfine structure in a Mössbauer spectrum is undoubted evidence for the existence of a magnetic order. The temperature dependence of the hyperfine field is proportional to that of the local magnetization, independently of the type of magnetic order, ferromagnetic or antiferromagnetic. The magnetic transition temperature is therefore determined from the collapsing of the magnetic hyperfine structure in the spectrum. The value of hyperfine field corresponding to the full magnetic moment is estimated by extrapolating to zero temperature. In ionic Fe^{3+} compounds, the hyperfine fields range from 50 to 60 T. In the case of Fe^{2+} compounds, hyperfine fields are much smaller, whereas a contribution of orbital angular moment is significant. Mössbauer spectroscopy is very useful for the study of antiferromagnetic materials since similar hyperfine fields are observed also in antiferromagnetic cases, which means the major contribution of hyperfine field is the local spin polarization in the Fe atom. If a sample including two or more phases with different magnetic properties, the observed spectrum becomes a superposition of respective hyperfine structures and the relative amounts of Fe atoms in each phase are estimated from the observed absorption intensities. In the case of ^{57}Fe nucleus, since the gamma ray energy is rather small and the Debye temperature is sufficiently high, the relative amount of observed absorption is almost proportional to the amount of Fe atom. Relative amount of each phase is easily estimated regardless of the spectrum structure, single line, doublet, or magnetically split pattern.

The observed magnetic hyperfine field is an average value in the term of the nuclear Larmor precession time (the order of 10^{-8} s). Therefore in most of magnetically ordered materials, the observed hyperfine field is rigorously proportional to the temperature dependence of local magnetic moment. If a sample includes two valence states, two hyperfine fields are observed as far as each valence state is respectively stable in a much longer time than nuclear Larmor precession time. If two valences are mixed in a faster rate than this characteristic time, an averaged hyperfine field is observed. A typical example is Fe_3O_4 (magnetite). This compound includes Fe^{2+} and Fe^{3+} ions with the ratio of 1:2. The crystal structure (spinel) has 2 crystallographic sites A and B. A site is occupied by Fe^{3+} ions while B site by Fe^{2+} and Fe^{3+} with the ratio of 1:1. This material is known to exhibit a transition (Verwey order) from insulator at low temperature to metal at high temperature and the transition temperature is about 120 K, which depends on the stoichiometry. At the high temperature phase, due to electron hopping, the valence state of Fe in B site becomes the average of Fe^{2+} and Fe^{3+}. Since this compound has a high Curie temperature, the Mössbauer spectrum at room temperature shows two hyperfine splitting: One corresponds to Fe^{3+} at A-site and the other the averaged $Fe^{2.5+}$ at B-site. Since the electron hopping rate is much faster than the nuclear Larmor precession time, the observed spectrum exhibits the averaged hyperfine field of Fe^{2+} and Fe^{3+}.

In nanoscale magnetic systems, the relaxation phenomena may influence the observed Mössbauer spectra. If the particle (or grain) size of a magnetically ordered material becomes very small, on the order of 1 nm, or the film thickness becomes a monolayer region, ordered spin structures are spatially no more stable at room temperature but start to fluctuate by thermal agitation. Such a phenomenon is called superparamagnetism. The Mössbauer spectra with magnetic splitting are only observed if the superparamagnetic relaxation time is longer than the nuclear Larmor precession time. When the particle size is smaller or the temperature is higher, the superparamagnetic relaxation time becomes shorter and accordingly the magnetic structure in the Mössbauer pattern will collapse at a temperature much lower than the Curie temperature. If electron spins oscillate very rapidly, a nucleus does not perceive any magnetic field. In usual magnetic materials at higher than Curie temperature, it is the case. On the other hand, a paramagnetic spin system may exhibit a hyperfine structure if the relaxation time of electron spin is long enough. Such a situation can be realized in insulating systems where the concentration of magnetic spins is low and interactions among them are negligibly weak. A typical example is Al_2O_3 including 1 % Fe_2O_3. The paramagnetic hyperfine structure has different features and is able to be distinguished from that of magnetically ordered state.

The direction of magnetization can be speculated from the intensity ratio of the six lines in a magnetically split Mössbauer pattern. In the case of ideally thin absorbers, the relation between the intensity ratio and the angle between the gamma-ray beam and direction of the magnetic field (i.e. the direction of magnetization) θ is as follows: If the intensity ratio of 6 lines is denoted as $3:X:1:1:X:3$, the dependence of X on θ is expressed as $4\sin^2\theta/(1 + \cos^2\theta)$. For a powder sample where θ is randomly distributed, the ratio of 6 lines becomes $3:2:1:1:2:3$. If the magnetization is oriented in the direction of the gamma ray beam, an intensity ratio of $3:0:1:1:0:3$ would be obtained. If the magnetization is perpendicularly oriented to the gamma ray beam, it becomes $3:4:1:1:4:3$. In case of thin film samples, the direction of spontaneous magnetization against the film plane is estimated by observing the value of X. Examples of spectra in Fig. 5.1 showed that the value X is close to 4, indicating the direction of magnetization is close to the film plane. If a strong magnetic field enough for saturation is applied in the perpendicular direction to the sample plane (parallel to the gamma ray beam) but the observed X is non-zero, the spin structure is suggested to be non-collinear.

5.3 ^{57}Fe Studies on Magnetic Multilayers

5.3.1 Fe/Mg Multilayers

Studies on metallic multilayers with artificially designed superstructures have been actively carried out since 1980s. By alternately depositing two metallic elements with controlling the layer thicknesses in the accuracy of atomic layers, nanoscale superstructures are constructed. In order to control the layer thickness satisfactorily, the deposition rate must be rather slow and therefore the atmospheric vacuum must be sufficiently high. Techniques for ultrahigh vacuum and controlling the layer thickness are necessary conditions to prepare artificial nanoscale superstructures. Concerning the constituents of multilayers, a variety of combinations have been attempted and the establishment of artificial superstructures with the wavelength of a few nm has been confirmed in many cases. Obtained structures are not always epitaxial but can be non-epitaxial and occasionally amorphous layers are also included as components of multilayers. In any case, artificially prepared multilayers are new materials which do not exist in nature and therefore the physical properties, especially magnetic properties are of great interest. Studies on magnetic properties of artificial multilayers have thus been promoted with the expectation to find novel functional materials. Actually the giant magnetoresistance (GMR) was found in the epitaxial multilayers of Fe/Cr by a French group in 1988 and the Nobel prize for physics in 2007 was given to Fert and Grünberg as the discoverers of GMR [2–4].

A typical example of new material is multilayers consisting of Fe and Mg layers since these elements are insoluble even in the liquid phase. Fe has a bcc structure while Mg has an hcp and the difference of their atomic radii reaches to about 30 %. Except amorphous alloys, homogeneously mixed alloys or intermetallic compounds could not be obtained in the phase diagram of Fe–Mg binary system. By depositing Fe and Mg, layer by layer, in ultrahigh vacuum condition, Fe/Mg multilayers are found to be constructed in a rather wide compositional ratio of Fe/Mg and the Fe layer thickness is able to adjust down to a monolayer. Establishment of artificial superstructures as designed are confirmed by the cross-sectional TEM observation and small angle X-ray diffraction measurements. For the study of magnetic properties, ^{57}Fe Mössbauer measurements are very suitable [5].

Figure 5.2 shows Mössbauer absorption spectra at 300 and 4.2 K for Fe/Mg multilayers. Nominal thicknesses of individual layers are noted in the figure. The deposition of Fe/Mg unit was repeated for 30–100 times. From the variation of spectra in Fig. 5.2, we can recognize the magnetic properties as a function of Fe layer thickness. When the thickness is 1.5 nm, the spectra at 4.2 and 300 K have a magnetically split pattern, suggesting that the Curie temperature is very high, and the observed hyperfine field corresponds to that of pure bcc-Fe. Each absorption line is reasonably sharp and the hyperfine field is not much distributed. Indeed, each line has a slightly asymmetric shape (tailing to the inner direction) suggesting

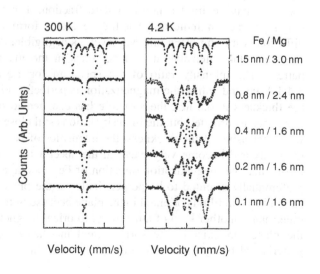

Fig. 5.2 ^{57}Fe Mössbauer absorption spectra for Fe/Mg multilayers at 300 K (room temperature) and 4.2 K [5]. Sample structures (each thickness of Fe and Mg layers) are shown in the figure

the existence of minor fraction with a reduced hyperfine field. This fraction with a reduced hyperfine field is attributed to the interface of Fe layer contacting with Mg layer. In the case of bcc Fe-alloy including a non-magnetic metal impurity of a minor amount, the hyperfine field of Fe, having one non-magnetic neighbor atom (Mg in this case), must be reduced by around 10 %. It is apparent that the inter-mixing of Fe and Mg does not take place and the Fe layers with the thickness of 1.5 nm are regarded as pure bcc Fe sandwiched by Mg layers.

The profiles of spectra for Fe thinner than 0.8 nm are entirely different. At 300 K, spectra have no magnetically split structure but show only a doublet with a small separation due to a quadrupole effect. This result indicates that the Curie temperatures of these very thin Fe layers are lower than room temperature. The reason is the crystallographic structural change from bcc to amorphous. At 4.2 K, all spectra show magnetically split spectra but the each line width is very broad. Such broad six lines are a characteristic for ferromagnetic amorphous Fe alloys, and therefore from the observed line profile, we can judge that the structure of Fe layers thinner than 0.8 nm is amorphous. Amorphous phase of Fe alloys in a bulk form can be obtained by mixing some non-magnetic elements (typically B, N, etc.) but amorphous form of "pure" Fe is known to be unstable and the Curie tem-perature of amorphous pure Fe is not able to estimate experimentally, although it has been suggested to be remarkably low. The spectra for Fe thinner than 0.8 nm therefore furnish us information on the magnetic properties of amorphous "pure" Fe in ultrathin layer forms. According to the results of magnetization measure-ments by using a SQUID magnetometer, as a function of temperature, the Curie temperatures show a variation from about 110 K for 0.8 nm Fe layer to about 35 K for 0.1 nm Fe layer. It is to be noted that the Mössbauer spectrum for 0.1 nm Fe layer is entirely ferromagnetic at 4.2 K and non-magnetic fraction is not visible in the spectrum. The nominal thickness of 0.1 nm corresponds to be less than one perfect monoatomic layer and cannot cover whole area of Mg surface. However, in case that isolated Fe atoms exist in the sample of 0.1 nm film, the Mössbauer spectrum must exhibit a non-magnetic fraction. The fact that there is no non-magnetic fraction indicates that the Fe atoms form fractional monolayers but diffusion or mixing of Fe into Mg layers is negligible. Another interesting feature is seen in the intensity ratio of six lines in the magnetically split Mössbauer pattern. The intensity ratio of the spectrum for 1.5 nm Fe is almost ideally 3:4:1:1:4:3, indicating the magnetization is perfectly oriented in the film plane. If the thickness is 0.8 nm, the structure has changed to be amorphous and the six lines become broader, but the intensity ratio is still close to 3:4:1:1:4:3. In contrast, with decreasing of the thickness, the intensities of No. 2 and 5 lines are reduced gradually and become very weak in the spectra for 0.1 and 0.2 nm. This result means that the magnetization direction of Fe monolayer (perhaps 0.2 nm also) is preferentially oriented to be perpendicular to the plane. It is usual that magneti-zation in a thin film is oriented in the plane because of the shape anisotropy. On the other hand, another kind of anisotropy, to orient magnetization perpendicularly to the plane can exist in the surface (and interface) atomic layers. In magnetic materials of bulky shape, the influence of such interface anisotropy is negligible because the volume ratio of interface fraction to the total is very small. Only in the case that the magnetic layer thickness is extremely thin, the interface anisotropy becomes dominant and the whole magnetization may be oriented perpendicularly. Such a situation can occur, in the present case of Fe/Mg multilayers, if the Fe layer thickness is one or two monoatomic layers. The existence of such interface anisotropy that magnetization is oriented perpendicularly to the plane has been reported for many multilayer systems including very thin Fe or Co layers. An example of magnetic films with perpendicular magnetization caused by the interface anisotropy, being stable even at room temperature, is Co/Pt multilayers including very thin Co layers, which is regarded to have potential as a magnetic recording media material [6].

The spectra shown in Fig. 5.2 were obtained from independent samples including different Fe layer thicknesses but can be interpreted as a sequence during the growing of Fe layer thickness. Namely, Fe atoms deposited on Mg layer surface form fractional monolayers in the initial stage, until the thickness reaches 0.2 nm. The crystal structure is close to amorphous but a stable ferromagnetic order exists in the Fe layer whose direction is perpendicular to the film plane. By depositing more Fe atoms, the thickness of amorphous Fe layer increases up to 0.8 nm, and then the magnetization direction gradually turns to the film plane. A drastic change of the structure occurs in between 0.8 and 1.5 nm, from amor phous to crystalline bcc structure [7]. The Curie temperature jumps up to much higher than room temperature with associating the crystallographic transformation. The spectra for 1.5 nm are perfectly ferromagnetic and the observed hyperfine field indicates that the sample is pure bcc Fe. If the bcc Fe layer includes some percentages of Mg as impurities, the absorption lines should have wider line widths and the average hyperfine field should be smaller than that of pure bcc Fe. If Mg impurities are dissolved in Fe layers, amorphous phase should have been more stabilized and the transformation to the bcc crystalline phase could not occur. Thus, the Fe layer with the thickness of 1.5 nm, sandwiched in Mg layers, is concluded to be pure bcc Fe which does not include Mg impurities. This result is consistent with the phase diagram in the thermal equilibrium, which indicates that Fe and Mg are insoluble with each other.

Concerning the process of film growth, information obtained from the result in Fig. 5.2 is summarized as follows: In the beginning stage of Fe layer growing, the structure is amorphous and if the thickness becomes larger than a critical thickness, which lies between 0.8 and 1.5 nm, the structure as a whole transforms into a crystalline structure (bcc). The composition of ultrathin amorphous Fe layers is supposed to be pure, not including much Mg impurities, since the composition of bcc Fe layer after the transformation from amorphous to crystalline is pure Fe. It has been known that amorphous structure is stable for a mixture of Fe and Mg (i.e. amorphous Fe–Mg alloys). If an amorphous Fe–Mg layer is formed from the initial stage, however, such a transformation from amorphous to crystalline should not occur afterwards since the amorphous alloy phase is always stable. The observed Curie temperature being low is reasonable for pure Fe amorphous layer. This is in contrast to amorphous Fe–Mg alloys which can have rather high Curie

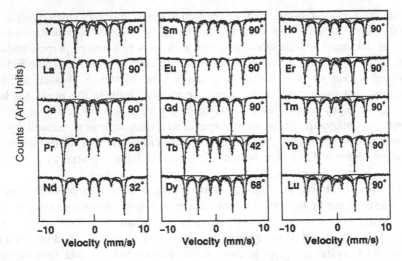

Fig. 5.3 ^{57}Fe Mössbauer spectra at 4.2 K for Fe (4.0 nm)/RE (3.0 nm) multilayers [8]

temperatures. Thus, Fe/Mg multilayers are unique samples to investigate the properties of "pure" amorphous Fe metal, although the thickness is limited to be less than about 1 nm.

5.3.2 Fe/RE (Rare-Earth) Multilayers

Rare-earth (RE) metals have interesting magnetic properties, such as helical magnetic spin structures and large magnetic anisotropies. However, the Curie temperatures of RE metals are lower than room temperature and therefore only intermetallic compounds and amorphous alloys between RE and 3d metals have potentials for applications. Preparation of multilayers by combining 3d and RE metals seems to be very interesting as a novel method to hybridize them. If 3d metal layers (i.e. Fe or Co) and RE metal layers are alternately deposited and the layer thicknesses are not too thin, high Curie temperatures of 3d metals and large anisotropy of RE metals may coexist in the multilayers and novel magnetic features may be resulted. In Fig. 5.3, the Mössbauer absorption spectra of Fe (4.0 nm)/RE (3.0 nm) multilayers measured at 4.2 K are shown [8]. All the spectra have very sharp six lines and hyperfine field is the same as the bulk Fe value. This result means that the Fe layers with the thickness of 4.0 nm in the Fe/RE multilayers have the bulk ferromagnetic properties. In most cases, the intensity ratio of the six lines is 3:4:1:1:4:3, indicating that the magnetization is preferentially oriented in the film plane. In Fe/Pr, Fe/Nd and Fe/Tb multilayers, however, the relative intensities of No. 2 and 5 lines are remarkably smaller, indicating the easy direction of magnetization is oriented toward film perpendicular. The spin direction of Fe layers is thus determined by the species of RE element. It is suggested

Fig. 5.4 Temperature
dependence of the relative
intensity of the No. 2 and 5
lines of the Mössbauer
spectra (X) and the
corresponding direction of
the Fe magnetic moments
relative to the plane normal
(θ) for an Fe (3.9 nm)/Nd
(2.8 nm) multilayer [9]

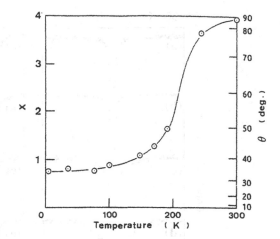

that Fe and RE magnetic moments are coupled through the interface and anisotropy of RE atoms plays a dominant role to determine the easy direction of multilayer magnetization. The key to obtain perpendicular magnetic anisotropy is to
choose RE with "pancake-like" 4f electron distribution. The crystal field at the
interface tends to align the 4f pancakes parallel to the interface and eventually
the magnetic moments of RE, which are parallel to the axis of each pancake, are
aligned perpendicularly to the interface. The magnetic moments of Fe, which
are magnetically coupled with those of RE, also align perpendicularly. When the
temperature increases, the interface magnetic anisotropy caused by the interfacial
RE decreases and the in-plane shape anisotropy of the film becomes dominant, so
that the direction of magnetic moments turns to the film plane. This situation is
observed as an increase of the intensity of No. 2 and 5 lines of Mössbauer spectra
as the temperature increases (Fig. 5.4) [9]. This phenomenon is called "spin
reorientation". The spin reorientation temperature increases when the RE layer
thickness is reduced to 1.0 nm, resulting in a realization of perpendicular magnetization at room temperature [10].

When the Fe layer thickness is reduced to 1.3 nm, the ^{57}Fe Mössbauer spectra
change drastically as shown in Fig. 5.5 [9]. The spectra at 4.2 K show a broad line
profile, and those at 300 K have more paramagnetic-like features. As in the case of
Fe/Mg multilayers already shown in Sect. 5.3.1, thin Fe layers become amorphous
with reduced magnetic transition temperature and wide distribution of hyperfine
fields. The perpendicular magnetic anisotropy appears at 4.2 K in these multilayers
as well, which is clearly evidenced from the intensity ratio of the No. 2 and 5 lines
in the spectra.

Please refer Ref. [2–4] for the results of investigations on the multilayers
with combinations other than Fe/Mg and Fe/RE.

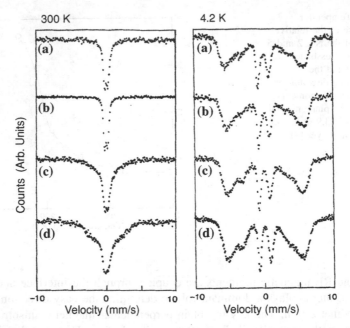

Fig. 5.5 RE layer thickness dependence of the Mössbauer spectra for Fe/Nd multilayers with the Fe layer thickness of 1.3 nm; (**a**) Nd (4.6 nm), (**b**) Nd (2.8 nm), (**c**) Nd (1.4 nm), and (**d**) Nd (0.7 nm) [9]

5.4 [57]Fe Studies on Interfaces of Magnetic Films

5.4.1 Interfaces of Fe/Mn and Fe/Cr

From a viewpoint of fundamental physics, it is very important to investigate the properties of metal surfaces in ultrahigh vacuum atmosphere but such clean vacuum surfaces are just toys which exist only in research laboratories. Real materials are containing various kinds of interfaces, but studies on interfaces are not yet well developed, since experimental means to investigate interfaces from atomic scale are very limited. In this section it is described that the Mössbauer spectroscopy is useful to observe the situations of interface, and particularly if the material is magnetic, hyperfine field observation may give us fruitful information for microscopic interface studies. In case of multilayers consisting of Fe and other metal X, or a sandwich structure, X/Fe/X, there are interfaces of Fe layer in contact with X. Before arguing the magnetic properties of Fe/X interfaces, it is important to clarify the chemical profile of interface. Normally, the degree of chemical mixing at the interface depends on the reactivity of X metal and also the process of layer depositions; Fe-on-X, or X-on-Fe. The Mössbauer spectroscopy is a unique method to study the difference of the two kinds of interface in one Fe layer.

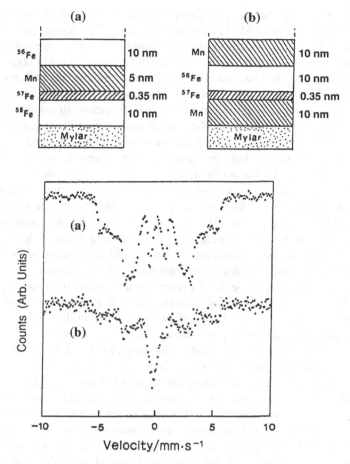

Fig. 5.6 RT ^{57}Fe Mössbauer absorption spectra for Fe/Mn interfaces. Sample structures are illustrated in the figure. (**a**) and (**b**) show the situations of Mn-on-Fe and Fe-on-Mn interfaces, respectively [24]

Figure 5.6 shows the Mössbauer spectra at 300 K to study two kinds of Fe/Mn interfaces. Sample A was prepared by depositing ^{56}Fe, ^{57}Fe and Mn succeedingly, while Sample B, Mn, ^{57}Fe and ^{56}Fe, one after another, on a Mylar substrate and the deposition of the unit structure respectively was repeated for 30 times. Namely Sample A exhibits the situation of Mn-on-Fe, while Sample B, that of Fe-on-Mn. The nominal thickness of 0.35 nm corresponds to about 2 monolayers of Fe in the interface region with Mn. The spectrum for Sample A is a 6-line pattern suggesting that the interface region is entirely ferromagnetic. The line broadening is caused by hyperfine field distribution in a certain degree indicating the existence of a minor fraction with smaller hyperfine field. Each Fe atom in the topmost interface layer has a few Mn neighbor atoms and accordingly the hyperfine field should be reduced. Although the 6-line pattern is fairly broad, the chemical mixing or interdiffusion at the Fe/Mn interface in Mn-on-Fe sample has not very much taken place. In contrast, the spectrum for Sample B, where the ^{57}Fe probing layer was deposited on the surface of Mn layer prior to the thick ^{56}Fe deposition (Fe-on-Mn) was found to have a considerably different profile. A fairly large non-magnetic fraction was observed, which means diffusion of Fe atoms into Mn layers ranging over a significant depth. Thus, interface diffusion (or mixing) depends not only on the combination of the materials but also the procedure of sample preparation. In the process of vapor deposition, an interface is formed with combining a cold substrate and hot vapor atoms. In the present case, a significant reaction occurs between cold Mn and hot Fe, but not so much between cold Fe and hot Mn.

In a multilayer structure built by successive depositions, it can happen that the structures at the top and bottom interfaces of each layer have different chemical profiles. In other words, the compositional modulation may have a unidirectional profile with respect to the film growth direction. Mössbauer spectroscopy is a unique method to clarify the difference of interface chemical profile in two interfaces; top and bottom. In general, if a Mössbauer probe is located at the interface of magnetic layer, observed hyperfine field distribution is useful information for the estimation of degree of intermixing at the interface.

Multilayers consisting of Fe and Cr layers are a particularly interesting system. It is well known that the giant magnetoresistance (GMR) was first found in Fe/Cr multilayers [11] and therefore Fe interfaces in Fe/Cr multilayers have attracted much attention [2]. The GMR phenomenon means the great difference of resistance between the states of spin-parallel and antiparallel and the reason is attributed to the spin-dependent scattering. Since the spin-dependent scattering is considered to occur at interface sites, the relation between spin-dependent scattering probability and the situation of interface is an important issue. Mössbauer spectroscopy is a useful method for the analysis of interface roughness on an atomic scale. Before the detailed argument of interface roughness, however, the difference of interface properties between two kinds of interface, Fe-on-Cr and Cr-on-Fe, has to be clarified, with using a similar procedure as the case of Fe/Mn interfaces. The result of CEMS measurements for the two kinds of Fe layer interface contacting with Cr layer is shown in Fig. 5.7. The nominal structures are Cr 20 ML/natural Fe 12 ML/^{57}Fe 2 ML/Cr 8 ML (upper interface) and Cr 20 ML/^{57}Fe 2 ML/natural Fe 12 ML/Cr 8 ML (lower interface). Both of two obtained spectra are ferromagnetic and the profiles are similar to the previous case of Mn-on-Fe, suggesting that the intermixing at the Fe/Cr interface is being fairly suppressed.

The difference of spectra for the two interfaces can be revealed by analyzing the distribution of hyperfine field as shown in the inserted figure. However, before arguing the degree of mixing between Fe and Cr, an additional problem has to be considered, that is the different degrees of mixing between ^{57}Fe probe and natural Fe matrix at the two interfaces. In the case that hot ^{57}Fe atoms of a minor amount are deposited on a cold Fe substrate, the interdiffusion may be fairly limited, while in the other case, more diffusion might be plausible. Therefore in such a case that the degree of intermixing is small, a quantitative discussion is very difficult. Although there have been several investigations on the relation between interface

Fig. 5.7 RT CEMS (Conversion electron Mössbauer spectroscopy) spectra from epitaxial Fe/Cr (001) superlattices on MgO(001) including 2 ML (monolayers) thick ^{57}Fe-probe layers at upper and lower interfaces (Cr-on-Fe and Fe-on-Cr) respectively. Estimated hyperfine field distributions are also shown in this Figure [12]

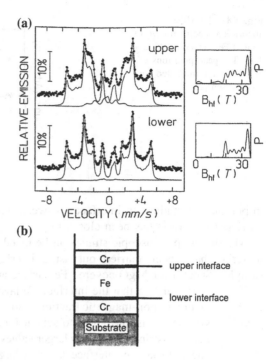

roughness and magnetoresistance, the conclusive answer is not yet presented to the question what is the optimum condition of roughness to obtain the highest probability of spin-dependent scattering [12].

5.4.2 Interfaces of Fe in Contact with MgO

Soon after the discovery of GMR, investigations to apply this phenomenon for technical products have started and a great success was already achieved to use for magnetic recording head sensors, i.e. "GMR head". Thus, the properties of interfaces between magnetic and non-magnetic metals have become a crucial issue to improve the efficiency of magnetoresistance effect (MR ratio). On the other hand, studies on tunneling magnetoresistance (TMR) have made great progress in the 1990s. The TMR effect is caused by the dependence of tunneling current on the spin structures of magnetic layers. The conductivity in spin parallel configuration is much larger than that in spin antiparallel. Nowadays, very big MR ratios have been produced in TMR systems, which are much bigger than those in GMR systems, and already successfully introduced in commercial products of magnetic recording technology, which is called "TMR head". In the initial stage of TMR study, Al_2O_3 has popularly been used as a tunneling barrier material but more recently MgO has been recognized to be a better choice. Because of TMR, the

Fig. 5.8 ^{57}Fe Mössbauer
absorption spectra for MgO-
coated Fe surface at RT and
4 K. The peak positions of
bulk α-Fe are indicated for
comparison [25]

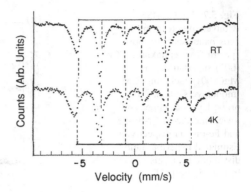

importance of studies on interfaces between ferromagnetic metals and tunneling
barriers (insulators) has been closed up.

Mössbauer spectroscopic studies on Fe metal surfaces covered by non-metallic
materials have been carried out using interface-selectively enriched absorber
samples. Spectra for MgO-covered Fe surface are shown in Fig. 5.8.

The spectra indicate that the interface Fe layers are entirely ferromagnetic and
oxidized fraction or non-magnetic fraction is not observable. The line shapes of each
peak are asymmetric and the tailing to outside is observed. This result means that the
hyperfine field is distributed towards larger values than the bulk, which indicates that
the hyperfine field at the interface layer is larger than the standard bulk value.
Therefore it is suggested that the local magnetic moment of Fe at the topmost
interface layer is enhanced by the interface effect. A large magnetic moment of
interface layer is a favorable condition to hold a tunneling spin current with a high
polarization, which is a key to realize a large TMR effect. This result seems to be
consistent with the recently reported very large TMR effect in Fe/MgO/Fe system.
On the other hand, Mössbauer measurements on very thin Fe layer sandwiched in
Al_2O_3 layers had exhibited a significant amount of non-magnetic fraction, suggesting that Al_2O_3-coated Fe surface tends to loose ferromagnetic moments [13].
The reason why Al_2O_3 is not an appropriate tunneling barrier material seems to be
the interface magnetic effect of Fe layer covered by Al_2O_3. However, both
Mössbauer measurements on MgO-coated and A_2O_3-coated Fe interfaces, introduced here, were old works, which had been carried out before the discovery of large
TMR effect. More refined studies on Fe interfaces contacting with insulating
materials are to be attempted in relation to TMR phenomena.

5.4.3 Interfaces of Fe-Oxides

Magnetic behaviors of surfaces and interfaces in non-metallic materials also are
very interesting subjects as well as those in metallic thin films. In contrast to
metallic cases, local magnetic moments at the top surface layer in non-metallic
systems are not much influenced but surface anomaly should appear in the

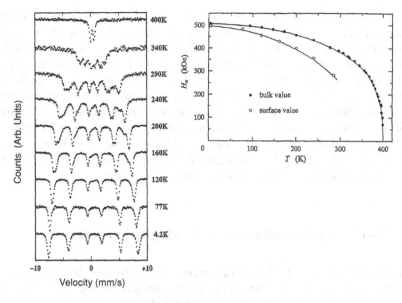

Fig. 5.9 Temperature dependence of surface hyperfine field in α-FeOOH [14]

temperature dependence of local magnetization, since the molecular field must be decreased due to the reduced number of neighboring magnetic atoms. If a surface-selectively enriched sample is available, the Mössbauer spectroscopy can be used to study the surface (interface) of non-metallic materials. The Curie temperature of bcc-Fe is very high (1,043 K) and therefore measurements at elevated temperatures are required for the study of surface (interface) magnetization in metallic cases as a function of temperature. On the other hand, there are several Fe compounds whose T_C or T_N are at moderate temperatures. α-FeOOH (Goethite) is an antiferromagnetic material and the T_N is about 400 K, which is an appropriate value to study the temperature dependence of surface magnetization. By a wet method, precipitates of α-FeOOH were synthesized in an aqueous solution using pure ^{56}Fe isotope. According to a transmission microscope observation (TEM), the shape of each particle was a needle type. By a succeeding chemical process, the surface of the particles was coated with extremely thin ^{57}Fe layers. If the coating is ideally uniform, the nominal thickness of ^{57}Fe layer corresponds to 0.1 nm. The Mössbauer absorption spectra of the surface-enriched α-FeOOH sample were observed at various temperatures below room temperature [14]. In Fig. 5.9, the observed temperature dependence of the hyperfine field at the surface of α-FeOOH is shown in comparison with that of a bulk sample. The hyperfine field at zero temperature is almost the same as the bulk value, but the temperature dependence is considerably different, decreasing much faster with increase of temperature. This behavior is qualitatively accounted for by assuming a primitive molecular field approximation. Here we assume a surface magnetic spin to be $S = 5/2$ and treat as an impurity site with a reduced exchange field. The Néel temperature and the temperature dependence of the exchange field at the normal site (other than surface site) are assumed to have no surface effect. It is known to be a good approximation that the temperature dependence of hyperfine field is proportional to that of local magnetization at each site. As shown in Fig. 5.9, the temperature dependence of surface hyperfine field is well reproduced, if the exchange field at a surface site is assumed to be 70 % of the value at the normal site. The reduction of the exchange field at a surface site, 30 %, appears to be a reasonable size. Similar results were also obtained in the Mössbauer studies for the surfaces of α-Fe_2O_3 and β-FeOOH [15]. These results are examples indicating that Mössbauer studies on surface-selectively enriched samples furnish us unique information on the surface (or interface) magnetic behaviors of non-metallic materials.

5.5 ^{119}Sn Hyperfine Interaction in Magnetic Materials

As mentioned in Sect. 5.1, ^{119}Sn is the second most convenient isotope to measure Mössbauer spectra. The nuclear spin quantum numbers are the same as those of ^{57}Fe, i.e., 1/2 for the ground state and 3/2 for the 1st exited state. Eventually, Mössbauer spectra show six-line patterns when an effective magnetic field is induced at the ^{119}Sn nuclear sites (Fig. 5.10). Sn is essentially a non-magnetic element and the nuclei do not feel a hyperfine field in bulk metallic Sn or in other nonmagnetic compounds. However, in magnetic compounds, the conduction electrons of Sn atoms are spin-polarized due to the hybridization with spin-polarized electrons of the neighboring magnetic elements. This imbalance between the numbers of up-spin conduction electrons and down-spin conduction electrons creates a hyperfine field at the Sn nuclear sites through the Fermi contact interaction. In this way, ^{119}Sn nuclei can also be used as probes to detect electron-spin

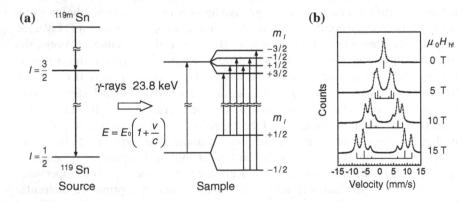

Fig. 5.10 Energy levels of ^{119}Sn nuclei and simulated ^{119}Sn Mössbauer spectra for hyperfine fields of 0, 5, 10, and 15 T

polarization in magnetic materials, which furnish us fruitful information to study magnetic properties of materials.

5.6 ^{119}Sn Studies on Magnetic Multilayers and Interfaces

5.6.1 Spin Polarization in Nonmagnetic Layers

After the discovery of the GMR effect in Fe/Cr multilayers, similar phenomena were found in other ferromagnetic/nonmagnetic multilayer systems. In the systems called "exchange coupling type", antiferromagnetic coupling appears between the ferromagnetic layers through the intervening nonmagnetic layer, resulting in an antiparallel magnetic configuration between the ferromagnetic layers at zero external field. The electric resistance changes drastically when the magnetic configuration turns parallel by applying an external magnetic field. The antiferromagnetic coupling is thought to be mediated by the electron spin polarization in the nonmagnetic layer. By doping the nonmagnetic layers with Sn impurities, and using Mössbauer spectroscopy, such spin polarization can be detected through the induced hyperfine field at the ^{119}Sn nuclear sites. Actually, finite hyperfine fields were observed in the Sn impurity sites in the nonmagnetic layers of Co/Au and other multilayer systems, showing that the conduction electrons in the nonmagnetic layer are really spin-polarized due to the contacting ferromagnetic layers [16]. In this way the ^{119}Sn nuclei can be a sensitive detector of electron spin polarization in nonmagnetic materials.

5.6.2 Magnetism of Cr Layers

Cr, which has a bcc structure as Fe, is an antiferromagnet with the magnetic moments parallel in one (001) atomic plane but antiparallel between the adjacent (001) atomic planes. Besides, the size of the magnetic moments shows a sinusoidal modulation along the <001> direction. This magnetic structure is called "spin density wave". It is also known that the magnetic structure is sensitively influenced by impurities and strains. On the other hand, it is in general not easy to investigate the magnetism of thin antiferromagnetic films experimentally. Therefore, the investigation on magnetism of thin Cr films, including those in Fe/Cr multilayers where the magnetoresistance effect was first discovered, is a challenging subject [17–19].

One attractive method is the use of Sn nuclear probes as a tool to detect electron spin polarization through Mössbauer spectra. When an atomic layer of Cr is replaced by an atomic layer of Sn, the magnetic moments of neighboring Cr atoms tend to be enhanced, and at the same time the Sn nuclei feel a large hyperfine field because of an effect from the magnetic Cr atoms [20]. It was found that the

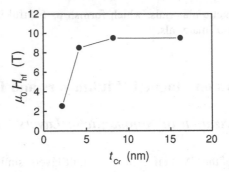

Fig. 5.11 Hyperfine field at the nuclear sites of atomic Sn layers doped at the center of each Cr layer of Fe (1.0 nm)/Cr (t_{Cr}) multilayers as a function of the Cr layer thickness t_{Cr}

magnetic hyperfine field at the Sn sites doped at the center of each Cr layer of Fe/Cr multilayers decreases with decreasing the thickness of the Cr layer (Fig. 5.11). This result indicates that the magnetic moments of Cr in Fe/Cr multilayers reduce as the thickness of the Cr layer decreases [21]. Such kind of information on antiferromagnets cannot be obtained by other experimental methods.

5.6.3 Sn-Containing Ferromagnetic Alloy Films

In some Sn-containing ferromagnetic alloys and compounds, [119]Sn Mössbauer spectra can be an informative tool to investigate the magnetism. A representative intermetallic alloy is Co_2MnSn, which is one of the Heusler compounds having an $L2_1$-type ordered crystallographic structure. Co_2MnSn is expected to have high conduction-electron spin polarization (so-called half metallic), which is a hot topic in the field of "spintronics". This is actually an old material and Mössbauer spectra for bulk Co_2MnSn have been published by several groups. However, it is not easy to prepare alloys with a highly ordered structure and obtain a Mössbauer spectrum with a single Sn environment. Recently, thin films of this alloy have been prepared using atomically controlled alternate deposition to see the possibility of improving the local environments of the constituent atoms [22]. Figure 5.12 shows the Mössbauer spectra and corresponding distribution of hyperfine field for bulk Co_2MnSn prepared by arc-melting and a Co_2MnSn films prepared by atomically controlled alternate deposition. The former spectrum was measured by a transmission method and the latter by a CEMS method. It is clear that the sample prepared by arc-melting has two prominent peaks in the hyperfine distribution suggesting that there are two environments in the Sn sites, whereas the film sample prepared by atomically controlled distribution has almost one [22]. Using such Co_2MnSn films with relatively uniform local environments, studies on interface magnetism of Heusler alloy films, especially its temperature dependence, are now in progress.

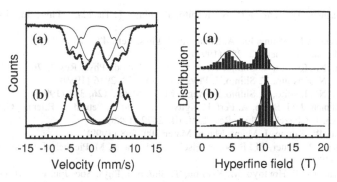

Fig. 5.12 ^{119}Sn Mössbauer spectra and distribution of hyperfine field for (a) bulk Co_2MnSn Heusler alloy prepared by arc-melting, and (b) a Co_2MnSn (40.2 nm) film prepared by atomically controlled alternate deposition

5.7 Remarks

Magnetic hyperfine fields observed by Mössbauer spectroscopy are useful information to characterize condensed matters. If a sample includes a certain amount of Fe or Sn, Mössbauer measurements are easily applied. If the sample is able to be enriched with Mössbauer isotope ^{57}Fe or ^{119}Sn, the limit of concentration required for the measurements can be greatly reduced. Specific samples for interface studies are prepared by doping with isotope selectively at the interface sites. In this article, several examples of Mössbauer studies using interface-selectively enriched samples are introduced.

"Spintronics" is a rapidly growing field where various novel phenomena are being explored [23]. An important concept for spintronics is spin current and spin polarization of conduction electrons is attracting a great attention. Concerning spin polarization of conduction electrons, transferred from a ferromagnetic metal to non-magnetic substance, interface magnetic properties of the ferromagnetic metal may have a significant influence. Thus, the understanding of interface magnetism seems to be crucial for spintronics researches. Application of Mössbauer spectroscopy will be useful for the study of such interface issues.

References

1. P. Gütlich, E. Bill, and A. X. Trautwein (eds.) *Mössbauer Spectroscopy and Transition Metal Chemistry* (Springer, Berlin, 2011). Synchrotron Radiation Experiments have been surveyed in Chap. 9
2. T. Shinjo, T. Takada (eds.) *Metallic Superlattices* (Elsevier, Amsterdam, 1987)
3. T. Shinjo, Surf. Sci. Repts. **12**, 49 (1991)
4. S. Maekawa, T. Shinjo, *Spin Dependent Transport in Magnetic Nanostructures* (Taylor & Francis, London, 2002)

5. K. Kawaguchi, R. Yamamoto, N. Hosoito, T. Shinjo, T. Takada, J. Phys. Soc. Jpn. **55**, 2375 (1986)
6. P.F. Carcia, A.D. Meinhaldt, A. Suna, Appl. Phys. Lett. **47**, 178 (1985)
7. T. Shinjo, Struct. Chem. **2**, 281 (1991)
8. K. Mibu, N. Hosoito, T. Shinjo, Nucl. Instrum. Methods Phys. Res. B **76**, 31 (1993)
9. K. Mibu, N. Hosoito, T. Shinjo, J. Phys. Soc. Jpn. **58**, 2916 (1989)
10. K. Mibu, N. Hosoito, T. Shinjo, J. Magn. Magn. Mater. **126**, 343 (1993)
11. M.N. Baibich, J.M. Broto, A. Fert, F. Nguyen van Dau, F. Petroff, P. Etienne, G. Creuzet, A. Friederich, J. Chazelas, Phys. Rev. Lett. **61**, 2472 (1988)
12. T. Shinjo, W. Keune, J. Magn. Magn. Mater **200**, 598 (1999)
13. O. Lenoble, Ph Bauer, J.F. Bobo, H. Fischer, M.F. Ravet, M. Piecuch, J. Phys, Cond. Matter **6**, 3337 (1994)
14. A. Yamamoto, T. Honmyo, M. Kiyama, T. Shinjo, J. Phys. Soc. Jpn. **63**, 176 (1994)
15. A. Ochi, K. Watanabe, M. Kiyama, T. Shinjo, Y. Bando, T. Takada, J. Phys. Soc. Jpn. **50**, 2777 (1981)
16. T. Emoto, K. Mibu, N. Hosoito, T. Shinjo, J. Phys. Soc. Jpn. **63**, 3226 (1994)
17. H. Zabel, J. Phys. Condens. Matter 11 (1999) 9303
18. D.T. Pierce, J. Unguris, R.J. Celotta, M.D. Stiles, J. Magn. Magn. Mater. **200**, 290 (1999)
19. R.S. Fishman, J. Phys. Condens. Matter **13**, R235 (2001)
20. K. Mibu, T. Shinjo, J. Phys. D Appl. Phys. **35**, 2359 (2002)
21. K. Mibu, M. Almokhtar, S. Tanaka, A. Nakanishi, T. Kobayashi, T. Shinjo, Phys. Rev. Lett. **84**, 2243 (2000)
22. K. Mibu, D. Gondo, T. Hori, Y. Ishikawa, M. A. Tanaka, J. Phys. Conf. Ser. **217**, 012094 (2010)
23. T. Shinjo (ed.), *Nanomagnetism and Spintronics* (Elsevier, Netherlands, 2009)
24. N. Nakayama, T. Katamoto, T. Shinjo, J. Phys. F. Met. Phys. **18**, 935 (1988)
25. S. Hine, T. Shinjo, T. Takada, J. Phys. Soc. Jpn. **47**, 767 (1979)

Author Biographies

Teruya Shinjo

1966–1976	Research Associate, Institute for Chemical Research, Kyoto University
1974–1975	Guest Researcher, Universität Saarbrücken, West Germany
1976–1982	Associate Professor, Institute for Chemical Research, Kyoto University
1982–2002	Professor, Institute for Chemical Research, Kyoto University
1992–2000	Science Advisor, MEXT, Japan
1996–1998	Director, Institute for Chemical Research, Kyoto University

2002   Senior Researcher, International Institute for Advanced Studies

2002   Specially Appointed Professor, Osaka University

Doctor of Science, Kyoto University (1966)
Professor Emeritus, Kyoto University (2002)
Doctor Honoris Causa, Technical University of Ostrava, Czech (2006)

Ko Mibu

1991–1999	Instructor, Institute for Chemical Research, Kyoto University
1999–2002	Associate Professor, Institute for Chemical Research, Kyoto University
2002–2005	Professor, Research Center for Low Temperature and Materials Sciences, Kyoto University
2005	Professor, Graduate School of Engineering, Nagoya Institute of Technology

Doctor of Science, Kyoto University (1993)

Chapter 6
Ion Implantation

G. Langouche and Y. Yoshida

Abstract In this tutorial we describe the basic principles of the ion implantation technique and we demonstrate that emission Mössbauer spectroscopy is an extremely powerful technique to investigate the atomic and electronic configuration around implanted atoms. The physics of dilute atoms in materials, the final lattice sites and their chemical state as well as diffusion phenomena can be studied. We focus on the latest developments of implantation Mössbauer spectroscopy, where three accelerator facilities, i.e., Hahn-Meitner Institute Berlin, ISOLDE-CERN and RIKEN, have intensively been used for materials research in in-beam and on-line Mössbauer experiments immediately after implantation of the nuclear probes.

6.1 Introduction

Among the techniques that can be used to introduce well defined concentrations of impurities into semiconductors, ion implantation turns out to possess particularly attractive properties. It is not dependent on the diffusivity nor the solubility of the dopant atom in the semiconductor host, the substrate does not have to be heated as in a diffusion process, the dosage and the depth distribution of the impurity can be well controlled. On the other hand, in those first years around 1960 when the

G. Langouche (✉)
Department of Physics and Astronomy, University of Leuven, Institute of Nuclear and Radiation Physics, Celestijnenlaan 200D, BE-3001, Leuven, Belgium
e-mail: guido.langouche@kuleuven.be

Y. Yoshida
Shizuoka Institute of Science and Technology, Toyosawa 2200-2, Fukuroi-city, Shizuoka 437-8555, Japan
e-mail: yoshida@ms.sist.ac.jp

Y. Yoshida and G. Langouche (eds.), *Mössbauer Spectroscopy*,
DOI: 10.1007/978-3-642-32220-4_6, © Springer-Verlag Berlin Heidelberg 2013

technique became popular, it soon became clear that the atomic configuration of the implanted atom cannot easily be predicted and that the implantation process can lead to extensive damage. Hence room for scientific research on the ion implantation process itself and on the fate of the implanted ion.

Although ion implantation received its main impetus as a technique to dope semiconductors, its use was not limited to this field. When implanting high enough fluencies in semiconductors, it was found that the technique could be used to synthesise conductive layers at well-defined depth inside the semiconductor host, opening a semiconductor technology field of study of its own.

Also ion implantation into metals and insulators was studied. Just as in the case of semiconductors, at low fluencies this allowed to study the atomic configuration around the implanted ion and its defect association. A special field of study was the investigation of the huge internal magnetic fields that implanted atoms were found to experience in magnetic hosts like Fe, Ni and Co. At high fluencies intermetallic surface layers could be formed, and also phenomena like surface hardening and corrosion resistance upon implantation e.g. steel with nitrogen were intensively studied.

This tutorial will not attempt to deal with all these ion implantation phenomena, although Mössbauer spectroscopy has been used in all these fields. We will give several illustrative examples but we will mainly focus on semiconductors and to rather low implantation fluences where the implanted atoms are still isolated from each other or just start to coalesce and to form precipitates. The phenomena at high fluences and the dynamics of compound layer formation are beyond the scope of this tutorial. The reason for this limitation is that emission Mössbauer spectroscopy on radioactive probe atoms is particularly powerful in this low concentration range and allows to study the more fundamental phenomena of lattice location and defect association at the individual probe level, which is hard to study with other techniques. On the other hand, experience has shown that one has to be extremely careful in drawing conclusions from Mössbauer spectroscopy results only, as the possible interpretation of a particular Mössbauer spectrum is often not unique. Complementary data, e.g. from electron microscopy, X-ray diffraction, transport measurements, channelling experiments, are often more than welcome or even crucial for the interpretation of the hyperfine interaction data.

6.1.1 Probing Local Structures and Their Dynamics

Natural science begins with "seeing". Eyes provide you a first tool to discover wonders of nature. Everyone should have experienced to magnify dragonfly's eyes and crystals in rocks by loupes, or to watch the surface of the moon and the circle of Saturn by telescope. Nowadays, "electron microscope" and "scanning probe microscope" enable us not only to "see" atoms and their arrangements in materials, but also to manipulate them, in order to create new functional materials and biomaterials. Modern "nanoscience" and "nanotechnology" began with the discoveries of these "seeing" technologies, which made it possible for us to perform the intensive research work to realize "single electron transistor", "solar cells with an energy conversion efficiency of higher than 40 %", and "Giant Magneto Resistive (GMR)" and "Tunneling Magneto Resistive (TMR)" sensors to produce Terabyte hard disks and a precise positioning encoder on an atomistic scale [1]. When one starts investigating nature on an atomistic scale, such methodologies lead to create new research fields beyond the old frame of science and technology, such as physics, chemistry, biology, electronics, and mechanics.

The microscopes mentioned above allow to observe atoms on the surface and in thin film with a typical thickness of several 10 nm [2]. But how can we investigate the interior of a material on an atomistic scale without destroying the material? Electric resistivity and magnetic susceptibility provide us information on the characteristics of bulk materials. One would like to distinguish, however, the electric and magnetic state of each atom of which the material consists, the crystal structures, as well as their micro- or even nanostructures. X-ray, electron, and neutron scattering methods are, for this purpose, further combined to obtain such structures, i.e. the atomic arrangements in the material. These methods are based on the interference of waves, such as electromagnetic radiation, electron and neutron "waves" [3]. These methods, however, will have more difficulties to observe "the images of the structure", when the micro- and nano-structures tend to possess a unit of structure of the order of nanometers, because the number of atoms inside the structure becomes too small to produce enough intensity of the interference pattern. Is there any other method to "probe" such small structure?

After the discovery of the Mössbauer effect in 1958 by Rudolf Mössbauer [4–6], an atomic nucleus became such a scientific probe, which opens to study the interior of materials through the study of the "hyperfine interaction", which is the interaction between the nucleus and its surrounding electrons [7]. The energy levels of the nucleus are determined by the hyperfine interaction, causing a shift or a splitting of the nuclear energy levels. This level structure can be observed in a Mössbauer spectrum, as has been explained in the former tutorials. The Mössbauer spectrum can be detected via signals such as γ-rays and/or electrons emitted from the nucleus. These signals bring us information on the electric and magnetic states of the probe atom from the interior of the materials, as is schematically shown in Fig. 6.1. The situation can be compared to a "spy", the nuclear probe, who sends a code signal to deliver us secret information from inside the material.

The probe atoms can be one of the constituents of the material, or they can be impurities introduced into the material from the outside by melting, by diffusion, or by implantation. The former processes make use of the thermal motion of the atoms, while the implantation process injects energetic probe atoms using an accelerator. In this tutorial we will further discuss "implantation Mössbauer Spectroscopy", i.e., the probe atoms will be first implanted into a material, and subsequently Mössbauer spectra will be measured by detecting emitted γ-rays and electrons. The spectra will provide us with atomistic information on the probe atoms through the hyperfine interactions. This situation may be well compared with an analogy of a "spy" which is sent to a place to gather information, and he/she will

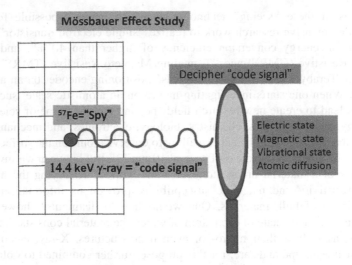

Fig. 6.1 The Mössbauer probe is emitting γ-rays, i.e., a "code signal", which can be deciphered, delivering information on the state of the "spy" in the material

then transmit a code signal to us from this place without disturbing the surroundings. In our case, the code signals are γ-rays and electrons which contain the secret information about the material. The recoil-free γ-ray emission, i.e., Mössbauer effect, will not cause any disturbance at the probe site nor in the surrounding lattice.

6.1.2 Ion Implanters

In this tutorial we will deal with different types of ion implanters. We will first discuss the ion implantation experiments at dedicated "conventional" facilities, meaning that they are of the same type as the ones used in semiconductor industry, but dedicated to the use of radioactive ions. Typical implantation energies in such facilities are in the 50–500 keV range. Figure 6.2 shows a typical ion implanter set-up. It consists of an ion source which usually produces positive ions which are then extracted from this ion source and (pre-)accelerated with a negative voltage of typically about 50–100 kV. This ion beam is then fed into a curved magnet which allows to mass-separate the desired isotope. In the final stage the isotope can still be post-accelerated to a few 100 kV, if desired, and the final stage usually contains magnetic quadrupole lenses to focus the beam and electrodes that allow sweeping the beam over the desired implantation area.

The main difference between a dedicated set-up for hyperfine interaction studies with radioactive probe atoms and a conventional ion implantation set-up is that the dedicated facility is equipped for the handling of radioactive isotopes. As well the ion source stage as the implantation stage need special radioprotection

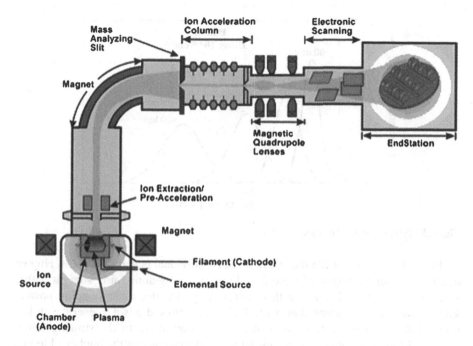

Fig. 6.2 Conventional ion implanter

measures. The inside of the magnet can be covered with a removable lining, which catches the possible radioactive products that are not following the path selected by the magnet setting for the desired isotope. A few laboratories, e.g. at the universities of Groningen, Bonn and Leuven acquired in the 1960s such dedicated experimental facilities to implant relatively long-lived radioactive isotopes. In order to keep the accumulation of radiation in the facility below reasonable limits, the use of radioactive isotopes is these set-ups is normally limited to isotopes with half-lives below one year.

6.1.3 KeV Ion Penetration in Matter

Since the binding energy of atoms in solids is of the order of electron volt, it can be readily expected that ions penetrating in matter with KeV kinetic energy displace many atoms. The phenomena accompanying ions penetrating in matter have been studied in great detail and are well understood. Theoretical treatments can be found in several handbooks [8] and computer codes are available [9] that accurately calculate the implantation profile of particular ions of well-defined energy in solids. Such a profile is characterized by a certain average implantation depth and a certain profile width, called straggling, as shown in Fig. 6.3.

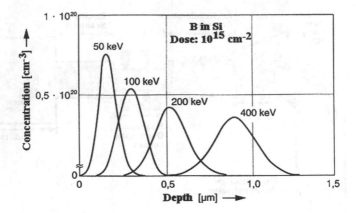

Fig. 6.3 Typical ion implantation profiles

Implantation profiles are the result of many individual penetration paths. Higher energy ions penetrating matter are found to travel first along fairly straight paths, since they are slowed down in this energy range by electronic stopping power, kicking out electrons along their path. Only when slowed down to energies in the tens of KeV range, nuclear collisions start to contribute to the slowing down process, kicking out atoms from their lattice sites and causing the implanted ion to undergo large angle deflection from its original path. These stopping power ranges are illustrated in Fig. 6.4.

Displaced atoms in turn cause damage along their path, so that before finally being stopped in the solid a so-called collision cascade is generated, as shown in Fig. 6.5.

Damage recovery is what happens next. Many displaced atoms finally land up on a regular lattice site again, while others can give rise to permanent defects in the solid. A large variation of possible defect configurations is possible, some of which are more stable than others. The stability of many such defect configurations has been calculated by theorists. The lattice temperature plays a crucial role in the final survival rate of defects and annealing out of defects is a well known procedure after ion implantation in semiconductor industry. Care has to be taken that raising

Fig. 6.4 Stopping power

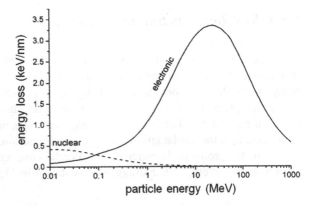

Fig. 6.5 Collision cascade

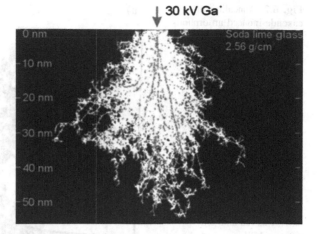

the temperature does not give rise to unwanted diffusion phenomena in the solid. Special annealing tricks such as flash annealing and laser annealing are sometimes used to avoid such unwanted diffusion phenomena.

Implantation damage is known to anneal out completely in metals even below room temperature. Only implantation at very low temperature where vacancies are not mobile results in observable damage. Except for special cases where chemical affinity plays a role or where e.g. oversized atoms are implanted, ion implantation in metals results in a final so-called replacement collision which results in a substitutional lattice location of implanted atoms.

For semiconductors the story is quite different. Many defects created in the collision cascade are not mobile which can result in defect configurations that do not anneal at room temperature. Heating to higher temperatures is often necessary to anneal out most damage. But it is not unlikely that the implanted atom or part of them remains part of a defect structure. Figure 6.6 illustrates defect recovery after implantation and annealing a semiconductor host.

Fig. 6.6 Defect recovery upon annealing

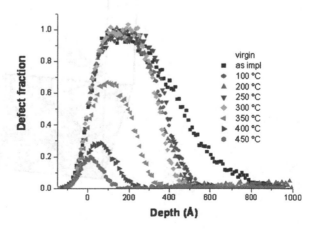

Fig. 6.7 Annealing of cascade-induced amorphous pockets by molecular dynamics. **a** State immediately after the cascade. **b** After 1ns annealing at 1300K the amorphous material has recrystallized and self-interstitials and vacancy clusters are left behind. The vacancies and their clusters induce tensile stress in the neighboring atoms and these are shown as blue spheres. The self-interstitials are shown as red and gray spheres [48]

(a)

(b)

The term "heat spike" has been used to describe the dense mixture of atoms and defects in motion in the core volume (Fig. 6.7) of a collision cascade. Electron microscopy pictures suggests that a heat spike in a semiconductor can lead to the creation of small amorphous zones in semiconductors associated with individual implanted atoms. There is certainly ample evidence that the accumulation of implantation defects can lead to the creation of amorphous layers inside the semiconductor. Implantation at higher temperature can avoid the creation of such amorphous layers and is more effective than post-implantation annealing at this temperature after an amorphous layer has been formed.

6.1.4 Hyperfine Interactions

Hyperfine interaction techniques with implanted radioactive probe atoms turned out to offer a wealth of information to study the atomic configuration and possible defect association of implanted probe atoms. Several hyperfine interaction techniques, e.g. Mössbauer Spectroscopy, Perturbed Angular Correlations, Low Temperature Nuclear Orientation, Muon Spin Rotation, and Beta-NMR are making use of implanted radioactive probe atoms. We refer to the proceedings of the Hyperfine Interaction Conferences, held every two or three years, for reports on many of these studies. Except for the earliest conferences, all these proceedings were published in the journal Hyperfine Interactions.

It is well known to the Mössbauer Spectroscopy community that Mössbauer probes can be studied in absorption spectroscopy, where the Mössbauer probe under investigation is present in its stable ground state, and in emission spectroscopy, where the Mössbauer probe under investigation is formed in the decay of a radioactive parent nucleus. By far the majority of Mössbauer studies are in absorption spectroscopy. Despite the fact that emission spectroscopy studies are more cumbersome since they require the handling of radioactive parent atoms, they have the interesting property that low concentrations of probe atoms can be studied. A very large fraction of Mössbauer ion implantation studies are such emission spectroscopy radioactive probe studies. A large fraction, indeed, but not all: in a few interesting experiments both absorption and emission Mössbauer spectroscopy was used to study the behaviour of some implanted elements in solids. This tutorial focuses on the use of emission Mössbauer spectroscopy to study ion implanted systems.

The largest part of the emission Mössbauer spectroscopy work is devoted to semiconductors since lattice location and defect association are particularly important there and relevant for semiconductor industry. For a more general picture of hyperfine interactions in semiconductors, also including non-nuclear techniques we refer to [10]. An overview of Mössbauer studies on semiconductors, also including diffusion studies and studies where Mössbauer atoms are constituent atoms of the semiconductor matrix (e.g. ZnTe) can be found in [11]. Already in the 1960s implantation studies [12, 13] were performed with radioactive probe atoms and studied using Mössbauer Spectroscopy. The early emission Mössbauer spectroscopy implantation work has been discussed in several review articles [14–17].

In the preceding paragraphs we have described a dedicated conventional ion implantation facility, using fairly long-lived activity so that there is enough time to transport the radioactive source from the ion implanter to a Mössbauer spectrometer for what is generally called *off-line experiments*. A somewhat confusing name maybe in cases where the Mössbauer spectrometer is connected to the implanter in order e.g. not to break the vacuum or to raise the temperature, as described in the Sect. 6.2. The term *on-line* emission Mössbauer experiment is reserved for facilities where the radioactive lifetime of the parent isotope is so short that measurements have to be done in situ. The ISOLDE facility at CERN and the RIKEN facility in Tokyo are such facilities, and are described in Sects. 6.4 and 6.5 respectively. A special type of on-line facilities are those facilities where the production, implantation and Mössbauer measurement occur on the time scale of the lifetime of the involved nuclear Mössbauer state, typically shorter than 10^{-6} s. Implantation generally occurs through the recoil of a Coulomb excited atom. Such experiments are generally called *in-beam* Mössbauer experiments experiments and Sect. 6.3 is devoted to ^{57m}Fe in-beam Mössbauer experiments at such a facility at the Hahn-Meitner Institute in Berlin.

In short, the three different types of emission Mössbauer spectroscopy are distinguished by the lifetime of the parent atom. For the Mössbauer isotope ^{57}Fe these parents are:

- off-line experiments use the ^{57}Co parent with $T_{1/2} = 270$ d
- on-line experiments use the ^{57}Mn parent with $T_{1/2} = 1.5$ min
- in-beam experiments use the ^{57m}Fe parent with $T_{1/2} = 10^{-6}$ s

6.2 Off-line ^{57}Co/^{57}Fe Implantation Mössbauer Studies at the Leuven Ion Implanter

The Ion and Molecular Beam Laboratory at the Institute for Nuclear and Radiation Physics at the University of Leuven has a dedicated conventional ion implanter, as defined in the previous paragraph. Nowadays it is part of a network of coupled facilities, as shown in Fig. 6.8, allowing to move samples from one facility to another, without breaking the vacuum.

The UHV system shown in Fig. 6.8 has different analysis and preparation devices (including two Molecular Beam Epitaxy systems) among which also a facility for Conversion Electron Mössbauer Spectroscopy measurements. It is the lead-wrapped facility on the right of Fig. 6.9. The tube in the middle of Fig. 6.9 is an ultra high vacuum transport tube that can be used to transport samples from one facility to another. The facility on the front left of Fig. 6.9 is a MBE-system. It contains a Knudsen cell allowing to evaporate ^{57}Fe onto surfaces. Other layers can be evaporated on top of this layer so that samples can be prepared for ^{57}Fe Mössbauer studies at surfaces and interfaces.

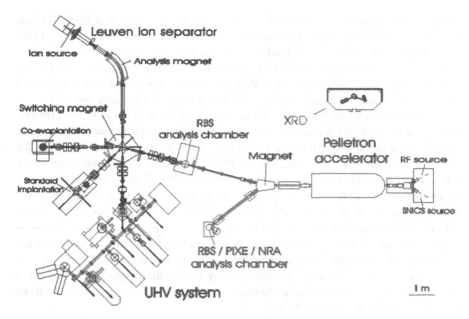

Fig. 6.8 Leuven Ion and Molecular Beam Laboratory

Fig. 6.9 Detailed view of the UHV system of Fig. 6.8

The Pelletron accelerator at the right of the layout shown of Fig. 6.8 is mainly meant to produce 2 MeV He particles for Rutherford Backscattering Spectroscopy studies. Such studies [8] allow a depth selective analysis of the composition of a sample. In channelling geometry it can also be used to determine, with subnanometer precision, the site location of impurity atoms in a single crystal lattice. The minimum impurity concentration needed for such studies is substantially higher than what can be studied in Mössbauer spectroscopy experiments.

Fe in Si shows a complex behaviour which has challenged Mössbauer physicists for almost 50 years and even today still is not fully understood. As will be discussed in Sect. 6.5 of this chapter, it is still a challenge for semiconductor industry.

In a review paper on "The Mössbauer search for Fe in Si" it is stated [18] that among the close to one thousand papers that so far had been published on Mössbauer spectroscopy studies in semiconductors, involving twenty different elements, the Mössbauer work on ^{57}Fe work in Si had been particularly combersome. In early Mössbauer experiments, as early as 1962 [19, 20], shortly after the discovery of the Mössbauer effect in 1958, the role of precipitate formation during diffusion was insufficiently realized, and the interpretation of the data was not very reliable. After ion implantation the Mössbauer spectra of ^{57}Fe in Si, both in emission Mössbauer spectroscopy, after implanting radioactive ^{57}Co [21], as in absorption Mössbauer spectroscopy, after implanting ^{57}Fe [22], were dominated by two single lines. Also in early in-beam experiments by the Stanford group, starting from the ^{57m}Fe parent state, a technique discussed in Sect. 6.3, two lines were observed [13, 23] with asymmetric intensities.

A controversy arose about the interpretation of these two lines. As can be seen in Fig. 6.10, in ^{57}Co/^{57}Fe emission experiments the line intensities varied as a function of implantation fluence. In ^{57}Fe(Si) absorber experiments, after implantation at fluences higher than to the highest fluences used in Fig. 6.10, a fairly symmetric doublet was observed. In the in-beam Mössbauer experiments, at implantation fluences lower than the lowest fluences shown in Fig. 6.10, an asymmetric doublet was observed. Was there a common interpretation possible?

It was thought that the two single lines might be representative of two implantation sites, presumed to be substitutional and interstitial Fe in Si, but the observed dose dependence could not be accounted for. By applying an external magnetic field it could unambiguously be shown [24] that the high fluence ^{57}Fe(Si) spectrum was a quadrupole interaction doublet, with parameters very similar to those of amorphous Fe_xSi_{1-x} films [22] and hence associated with Fe atoms in an amorphized surrounding.

The origin of the asymmetry at lower fluences was first accounted for by new emission Mössbauer experiments whereby ^{57}Co was implanted into Si held at the temperature of 50 K [25]. This cold implantation completely reversed the asymmetry (Fig. 6.11) in the spectrum compared to room temperature implanted samples. The model put forward was that at low temperatures, where vacancies are not mobile, ^{57}Co/^{57}Fe atoms land up in interstitial sites in Si, a site preferred by transition metals in Si. Upon increasing the temperature to room temperature, vacancies become mobile at about 100 K and are trapped by the interstitial Co atoms that become substitutional.

New experiments on Fe in Si with the in-beam Mössbauer spectroscopy technique were performed at the Hahn-Meitner institute in Berlin. The technique is outlined in Sect. 6.3. The results of these Coulomb excitation recoil implantation studies [26] confirmed this picture and led to an unambiguous identification of

Fig. 6.10 $^{57}Co/^{57}Fe$
Mössbauer spectra as a
function of ion implantation
fluence [21]

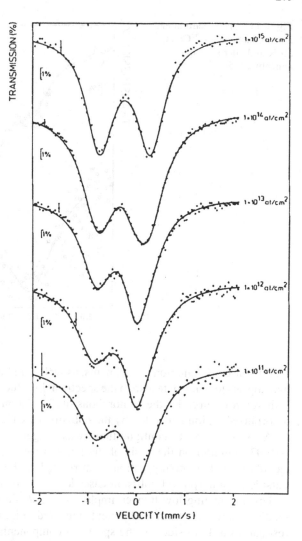

interstitial Fe in Si. The spectra, shown in Fig. 6.12, were interpreted as composed
of the "amorphous" doublet and a superimposed single line (with isomer shift
$\delta = + 0.84(1)$ mm/s with respect to α-Fe).

Due to the improved statistics, an increase in the linewidth of this single line
could be observed for higher target temperatures. This line broadening was analyzed in terms of a diffusional motion of the iron atoms, which are able to make
atomic jumps within the lifetime of the excited nuclear state. Diffusion coefficients
were derived (Fig. 6.13) from the broadening of the Mössbauer lines, and these
were found to be remarkably consistent with the known high- and low-temperature
diffusion coefficients of Fe in Si.

For the "amorphous" site (isomer shift $\delta = + 0.20(3)$ mm/s, quadrupole splitting $\Delta = 0.95(5)$ mm/s no microscopic model exists up to today. It is dominant in

Fig. 6.11 ^{57}Fe spectra after
ion implantation of ^{57}Co at
50 K and after subsequent
annealing [25]

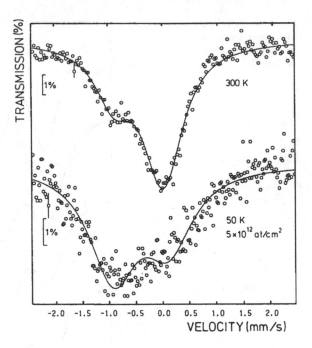

high fluence implantation experiments, as well as in Fe_xSi_{1-x} films for low x values, and appears to contribute also to the spectra at low fluence. This defect configuration is therefore often called the "amorphous site" and thought to be representative of the amorphized region generated by the collision cascade of single implanted ions.

A substitutional configuration was suggested for a third site ($\delta = -0.07(3)$ mm/s) on the basis of the $Fe_i + V \rightarrow Fe_s$ model and from theoretical calculations. This configuration is attributed to Fe atoms that find themselves outside the amorphized collision cascade region, or in a recrystallized part of it.

Upon annealing low fluence implanted ^{57}Co/^{57}Fe a surprising observation was made. A new spectrum component appeared (Fig. 6.14) with extra Mössbauer resonances at both sides of the spectrum components discussed so far.

It would lead us so far to discuss in detail the dependence of this spectrum component on annealing temperature, fluence and substrate doping. Based on these dependencies, it has been assigned to Co_2 pairs in Si [27].

Also other pairs were recognized, formed by normal acceptors in Si and Co donor atoms (CoB, CoAl, CoGa, CoIn) [28]. The different Fe and Co silicides have been characterized in several experimental studies and are observed as small precipitates after diffusion experiments and also in larger (often epitaxial) structures in ion beam mixing experiments, or in experiments to form buried epitaxial layers. A discussion on studies on these larger configurations around Fe in Si is beyond the scope of this tutorial.

We will also not discuss other Mössbauer probes in semiconductors, e.g. the 5sp-elements ^{119}Sn, ^{121}Sb, ^{125}Te and ^{129}I. They were used with success to probe

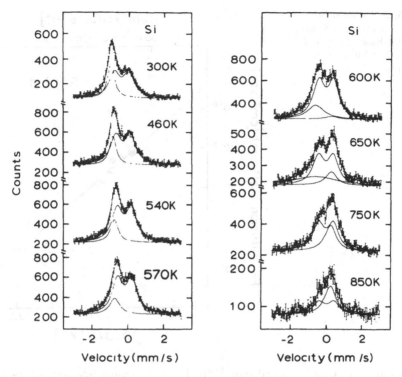

Fig. 6.12 ^{57}Fe spectra after in beam ion implantation of ^{57m}Fe at 50 K and after subsequent annealing [26]

lattice sites and defect configurations after implantation. They offered substantially less difficulties in terms of interpretation compared to Fe in Si. The difference in the chemical nature of these elements is responsible for this less-complicated behavior. On one hand we have the transition metals with their extremely small solubility and extremely high diffusion coefficient in Si, and on the other hand we have an element like Sn, isoelectronic and isostructural with the other group IV semiconductor constituents, and the neighboring elements that can replace host atoms in III-V and II-VI semiconductors, or act as natural substitutional donors or acceptors in these semiconductors.

6.3 In-beam ^{57m}Fe Implantation Mössbauer Studies at the Hahn-Meitner Institute in Berlin

6.3.1 In-beam Mössbauer Spectroscopy Technique

The Stanford group was the first to perform in-beam Mössbauer experiments on ^{57}Fe. The principle of the experiment is to excite stable ^{57}Fe atoms from their ground

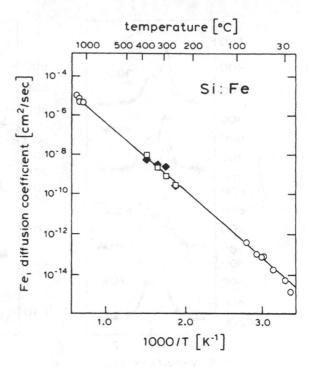

Fig. 6.13 Diffusion coefficients of Fe in Si from the Mössbauer data (*filled squares*) and from other techniques [26]

state into a metastable excited state ^{57m}Fe by the Coulomb interaction between a projectile charged particle and ^{57}Fe nucleus. Subsequently, the excited state ^{57m}Fe decays again to its stable ground state within a lifetime $T_{1/2} = 10^{-6}$ s. The ion implantation and Mössbauer experiment have to be performed within this timespan. If the bombarding energy of such particles is carefully selected to be below the Coulomb barrier, other radioactive nuclei will not be produced, which might otherwise cause a high background for the Mössbauer experiments. As mentioned in the previous section Latshaw [13, 23] applied this technique on Fe in Si.

The technique was further optimized in an experimental set-up at the Hahn-Meitner Institute in Berlin. A sketch of the set-up is shown in Fig. 6.15 [29]. A pulsed beam of 110 MeV ^{40}Ar ions (pulse length ∼ 1 ns, repetition rate ∼ 2.5 MHz) from the VICKSI heavy ion accelerator at the Hahn-Meitner-Institute in Berlin hits an iron-foil target (thickness 3 mg/cm^2, 90 % enriched in ^{57}Fe). More than 90 % of the excited ^{57}Fe recoils ejected from the target have angles between 15° and 75° with respect to the beam direction. These recoiling ions are trapped in a catcher made of the host material under investigation. The Mössbauer γ-radiation is detected in two resonance detectors of the parallel-plate-avalanche type (PPAC) with stainless-steel foil absorbers (53 % enriched in ^{57}Fe). A time resolution of 3 ns allows to measure the 14.4 keV Mössbauer radiation in a time window of about 380 ns between the beam bursts from the accelerator (AT 400 ns). This technique strongly reduces prompt background radiation. This set-up used for the experiments described in this section is shown in Fig. 6.16.

Fig. 6.14 Outer lines are
assigned to Co_2 pairs in
silicon [27]

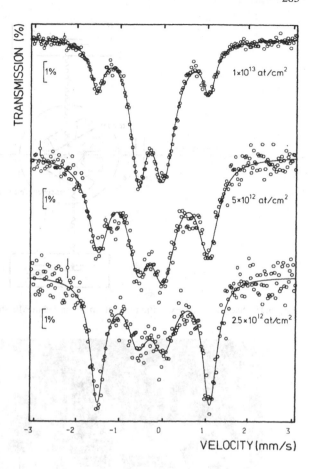

6.3.2 Fast Diffusion in Metals

There are different diffusion mechanisms operating in materials, and their understanding is essential for solid state physics as well as for materials science. In
many crystal systems such as pure iron and metallic alloys, self-diffusion and
impurity diffusion proceed via vacancy mechanism, i.e., atomic jumps by
exchanging an atom with a vacancy (A in Fig. 6.17), while light impurities such as
hydrogen or carbon migrate via an interstitial mechanism, i.e., the impurities
occupy interstitial sites and jump directly between the interstitial sites (B or C in
Fig. 6.17). In the former system, the impurity diffusivities range in the same order
of magnitude as the self-diffusivity, while in the latter systems, the impurity
diffusivities are orders of magnitude higher than that of self-diffusivity.

There are, however, many systems showing anomalous fast diffusion, where
metallic impurities such as Fe, Ni, Co, Cu, and Au in titanium, scandium, alkali,
IVB group metals and semiconductors diffuse several orders of magnitudes faster
than their self-diffusion in these materials, although the atomic sizes of the

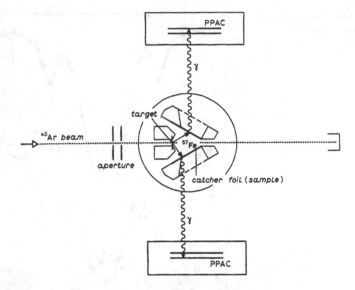

Fig. 6.15 Sketch of the experimental set-up for in-beam Mössbauer spectroscopy [29]

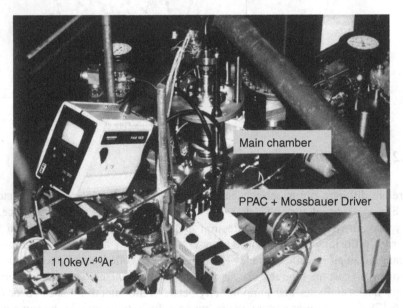

Fig. 6.16 The experimental set-up for in-beam Mössbauer spectroscopy at Hahn-Meitner Institute Berlin

impurities are comparable with those of the host atoms. This is amazing. If one thinks about the following situation, one can better understand how difficult to realize such fast diffusion: in rush-hour time, you, as an impurity, would be squeezed by your neighbors, the host atoms, in the middle of a crowded train, but

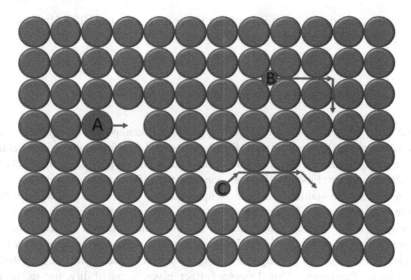

Fig. 6.17 Typical diffusion mechanism in solids

you would like to get out the train much faster than your neighbors. If you knew the mechanism of fast diffusion, you could quickly move through densely packed passengers.

One expects that some impurities occupy interstitial sites, leading to fast diffusion. However, it is rather difficult to understand how some metallic impurities could be incorporated into interstitial sites, when the atomic sizes of the impurities are taken into account. Hydrogen atoms could be easily put on interstitial sites in a crystal, but are not necessarily fast diffusers.

To clarify the origin of such anomalously fast diffusion, one must attempt a direct observation of the fast diffusing impurities by an experimental method. ^{57}Fe Mössbauer spectroscopy appears to be ideal, because it provides the atomistic information on the lattice sites, the charge states, the jump frequency and the jump vector, all of which are inevitably needed to construct an atomistic diffusion model. There were, however, two big problems to overcome experimentally: (1) the reliability of the available atomic jump theories in Mössbauer spectroscopy to deduce the jump parameters mentioned above from experimental spectra, and (2) the lack of Fe solubility which is generally reported in the fast diffusion systems. In order to realize isolated ^{57}Fe impurities in the samples mentioned in the former paragraph, the second problem appears to be impossible to solve as long as conventional absorber experiment are used. This is because neither a simple alloying, nor a low energy implantation of ^{57}Fe into a sample could provide an isolated ^{57}Fe atom without forming clusters of ^{57}Fe, which would mask the fast diffusing components in the Mössbauer spectrum.

To overcome the above mentioned difficulties in the studies of the fast diffusion in metals, several research projects were designed and performed in 1980s for about 10 years using the in-beam Mössbauer spectroscopy technique combining

Coulomb excitation with recoil implantation technique at the VICKSI accelerator facility in the Hahn-Meitner Institute Berlin.

6.3.3 Diffusion Study Using Mössbauer Spectroscopy

In the following sections we will show typical applications of in-beam Mössbauer spectroscopy for the study of fast diffusion in metals. Before getting into details, we are going to explain the principle of how the diffusion phenomenon affects the Mössbauer spectra. First of all, let us consider the following situation in our daily life: a fire engine sounding a siren is approaching you, while you are standing beside the street and are listening to the siren. The acoustic waves with a sound velocity of v_s are emitted continuously during a time interval Δt from the fire engine moving at a velocity of v_f. Consequently, the waves are compressed within a region of $(v_s - v_f) \cdot \Delta t$, leading to an observed sound frequency higher than that of the original frequency. This Doppler effect gives a possibility for us to get information on the motion by measuring the sound frequency even without watching directly the fire engine. More generally, when a matter is emitting a wave, a motion of matter changes the wave form. This will give you a hint to understand the principle of Mössbauer study on atomic jumps.

Now, in a solid matrix we have a Mössbauer probe of ^{57}Fe with a lifetime of 140 ns, as is shown in Fig. 6.18. The 14.4 keV first excited state of the ^{57}Fe nucleus can be fed through different processes, such as electron capture from ^{57}Co, β-decay from ^{57}Mn, Mössbauer absorption of 14.4 keV γ-ray, and Coulomb excitation. Subsequently the 14.4 keV γ-ray will be emitted resonantly without recoil (Mössbauer effect), while in Fig. 6.18, the ^{57}Fe atom is jumping between

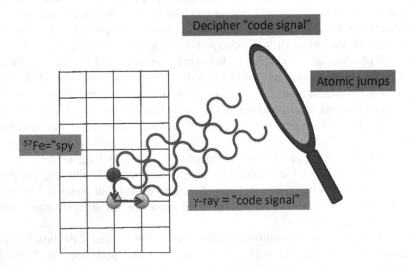

Fig. 6.18 Atomic motion influence on Mössbauer spectrum

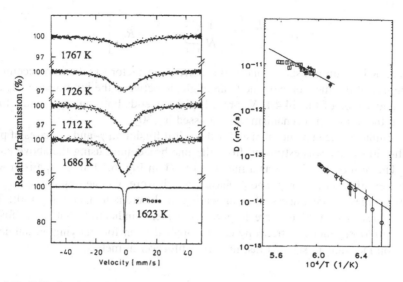

Fig. 6.19 Diffusional broadening of the Mössbauer resonance in Fe self-diffusion [30]

different lattice sites with a typical frequency of 10^7 s^{-1}, which is roughly equal to the inverse of the half-life time. In the crystal lattice, the atom is staying on the lattice positions, and is vibrating with a phonon frequency of 10^{12} s^{-1}. When the atom occasionally obtains thermal energy enough to overcome the potential barrier between the different lattice sites, the atom is able to jump from a lattice site to another, leading to diffusion. The duration of one jump is about 10^{-12} s, which is orders magnitude shorter than that of the atom staying on a lattice site. It turns out that the ^{57}Fe atom performs a few jumps during emitting or absorbing γ-ray, while the coherency of the wave is broken due to the jump process. Consequently, the lifetime of the 14.4 keV γ-ray is practically observed to be shorter than the natural lifetime. Taking into account of the Heisenberg time-energy uncertainty principle, $\Delta E \cdot \tau_M \approx h$, a shorter lifetime provides a broader linewidth of the Mössbauer resonance.

A typical example, that of a Mössbauer study of Fe self-diffusion [30], is shown in Fig. 6.19.

Iron has three different solid phases: the α-phase (bcc) up to 1,184 K, the γ-phase (fcc) up to 1,665 K, and the δ-phase (bcc) up to the melting point of 1809 K. The spectra of the γ- and δ-phases are shown in Fig. 6.19. The linewidths of all the spectra above 1656 K are much broader than that of the γ-phase spectrum at 1,623 K. The line broadenings observed in the δ-phase can be interpreted to be due to the atomic jumps of ^{57}Fe atoms within the lifetime, i.e. self-diffusion in the δ-phase. Since the line broadening, $\Delta \Gamma$, is related to the jump frequency, $1/\tau$, we can deduce the diffusion coefficient, D, using the following formula [30]:

$$D = \frac{R^2}{12\hbar} \cdot f(\theta) \cdot \Delta\Gamma$$

$$\Delta\Gamma = \frac{2\hbar}{\tau} \cdot \left[1 - \frac{1}{N} \sum_{n=1}^{N} \exp\left(i\vec{k} \cdot \vec{R}_n \right) \right]$$

Here R is a jump distance and $f(\theta)$ a correlation factor to take into account of the degree of the incoherency due to an angle, θ, between the jump vector, R_n and the wave vector of the 14.4 keV γ-ray, k. $f(\theta)$ depends both on crystal structures and on the diffusion mechanism, as discussed in [30].

Assuming a vacancy mechanism for the self-diffusion in γ- and δ-phases of pure Fe, the diffusivity was estimated from the line broadening using the above equation. The results are plotted as a function of 1/T in Fig. 6.19, in comparison with the tracer diffusivities for γ- and δ-phases, which were obtained by measuring the depth profiles of Fe tracers after annealing at different temperatures [30]. The diffusivities in both phases are in good agreement, indicating that the diffusion theory in Mössbauer spectroscopy can be applied even for studying an unknown diffusion mechanism such as fast impurity diffusion in metals.

6.3.4 In-beam Mössbauer Study on Fast Diffusion

The in-beam Mössbauer technique combining Coulomb-excitation and recoil-implantation, which was described in the Sect. 6.3.1, provides a unique feature for studying the anomalously fast diffusion, i.e., *one-by-one measurement*: Every γ-ray emission from ^{57m}Fe follows the implantation process. As a consequence, in the lattice the ^{57m}Fe probe always remains fully isolated from other ^{57}Fe atoms implanted before, and therefore, the spectrum obtained with this method is completely free from overlapping cascades as well as from clustering of Fe atoms. Both of them would change completely the diffusion properties of ^{57}Fe atoms. This method, therefore, guarantees an experimental condition under which we can follow a few jumps of ^{57m}Fe atoms immediately after the implantation into anomalously fast diffusion systems, such as α-Zr, Sc, and Pb.

Figure 6.20a–c show a typical spectra obtained from the In-beam Mössbauer experiments on ^{57}Fe in α-Zr [31, 32], Sc, and Pb [33], respectively. The spectra of ^{57}Fe in α-Zr (a) and Sc (b) consist mainly of a singlet and a doublet at the left and the right hand side, respectively, while the spectra of ^{57}Fe in Pb (c) appear to be only a singlet. First of all we discuss about the spectra of ^{57}Fe in α-Zr, which can be fitted with two different doublets. Their resonance areas, the quadrupole splittings of ΔEq, and the center shifts of δ are plotted as a function of temperature in Fig. 6.21a–c, respectively: (1) the electron density at ^{57}Fe nucleus, i.e., the isomer shift, is higher for the doublet than that of the singlet, and (2) the doublet shows dynamical behavior at around 50 K, which is observed as a quadrupole relaxation accompanied by a sudden decrease in the resonance area (Fig. 6.21a–c). These observations can be explained by the fast local jumps between equivalent

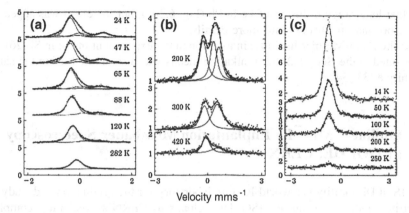

Fig. 6.20 In-beam Mössbauer spectra of ^{57}Fe in **a** α-Zr, **b** Sc, and **c** Pb

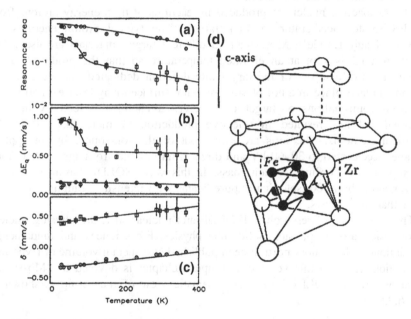

Fig. 6.21 Cage motion of Fe at interstitial equivalent positions. Fitting parameters for Mössbauer spectra of Fe in α-Zr as functions of temperature, and a possible model to explain the temperature dependence: **a** resonance areas, **b** quadrupole splitting ΔE_q, and **c** center shifts δ [31]

"cage sites" as is shown in Fig. 6.21d. This cage motion is considered as a precursor to the long-range interstitial diffusion.

In the case of Sc in Fig. 6.20b, there seems to be a similar cage motion accompanied by both a quadrupole relaxation and a strong decrease of the area on the doublet at the right hand side at 300 and 450 K. In the case of Pb, on the other hand, the singlet shows a slight line broadening above 250 K, which could be due to a fast long-range diffusion of Fe in Pb (Fig. 6.20c), but the effect was appeared to be too small to discuss into more details.

At the Hahn-Meitner Institute, in addition to the experiment on Fe in Si [26], as mentioned in the Sect. 6.2, Fe in alkaline metals was also studied using the same technique [33, 34].

6.4 On-line ^{57}Mn/^{57}Fe Implantation Mössbauer Spectroscopy at ISOLDE, CERN

The ISOLDE facility is a world-leading laboratory for the production and study of shortlived radioactive nuclei. ISOLDE belongs to CERN's accelerator complex situated on the border between Switzerland and France. The ISOLDE facility has been in operation since its start in 1967.

The radioactive nuclei are produced in reactions of high-energy protons from the PS-Booster accelerator in thick targets. The typical proton energies are between 1 and 1.4 GeV. More than 25 different target materials are used. The target material is kept at an elevated temperature so that the produced radio-active atoms diffuse out of the target into different dedicated ion sources. Ion-isation can take place in a hot plasma, on a hot surface or by laser excitation. By judicious combinations of target-ions sources a chemical selectivity may be obtained and has resulted in selective production of more than 70 of the chemical elements. The ions are swept out of the ion-source by an applied voltage, accelerated to 30–60 kV and directed into an electro-magnet where they are separated according to their mass. In this way ISOLDE has been able to deliver more than 700 isotopically pure beams with intensities ranging from 1 to more than 10^{10} ions/s.

The main lines of research at ISOLDE are: nuclear structure physics, nuclear astrophysics, atomic physics, solid state physics, life sciences and fundamental interactions. A laboratory portrait been published as a special volume of Hyperfine Interactions [35]. In this volume a complete chapter is devoted to "Mössbauer Spectroscopy at ISOLDE" [36]. A schematic view of the set-up is shown in Fig. 6.22.

6.4.1 Site Selective Doping of III–V Semiconductors

The ^{119}Sn Mössbauer isotope can be studied in off-line emission Mössbauer experiments starting from the longlived ^{119m}Sn isomeric parent. The ISOLDE facility offers the interesting possibility to access the Mössbauer transition from the shortlived radioactive ^{119}In ($T_{1/2} = 2,3$ min) and ^{119}Sb ($T_{1/2} = 38$ h) precursor parents.

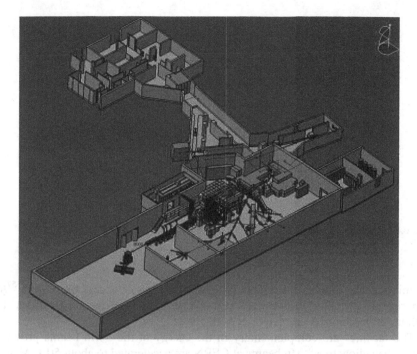

Fig. 6.22 ISOLDE lay-out

A very nice example of the power of Mössbauer spectroscopy was the demonstration [37] of site selective doping of compound III-V semiconductors such as GaAs, where it was concluded from the measured ^{119}Sn isomer shift that implanted In and Sb radioactive parent ions selectively populate III and V sites respectively.

6.4.2 Charge-State Dependent Diffusion of Fe in Si

As stated in the beginning of this tutorial, there are three different types of emission Mössbauer spectroscopy methods to study Fe in Si. They differ by the lifetime of the parent atom. We have first discussed in Sect. 6.2, the *off-line experiments* using the ^{57}Co radioactive parent with $T_{1/2} = 270$ d. We then mentioned in the Sect. 6.3 the *in-beam experiments* using the ^{57m}Fe parent with $T_{1/2} = 10^{-6}$ s. The third emission Mössbauer spectroscopy method makes use of a much shorter lived parent than ^{57}Co namely ^{57}Mn with a lifetime of $T_{1/2} = 1.5$ min. Two experimental facilities are nowadays making use of this method in *on-line experiments*: the ISOLDE facility at CERN and the RIKEN facility in Tokyo. These are huge experimental facilities to which access is granted through

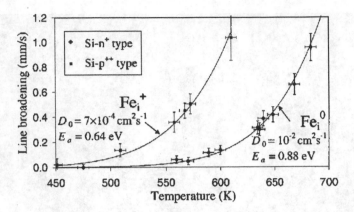

Fig. 6.23 Charge state dependent diffusivity of Fe in Si [38]

submission and approval of project proposals. In practice access for on-line ^{57}Mn experiments is limited to one week per year, at both experimental facilities. The authors of this tutorial are active at both sites.

The facilities are complementary in the energy of the ^{57}Mn beams provided. While the radioactive ^{57}Mn beams at CERN are accelerated to about 50 keV, as in the ^{57}Co off-line experiments discussed in Sect. 6.2, and result in shallow implantation depths of the order of 10 nm, the radioactive ^{57}Mn beams at RIKEN are accelerated to several 100 MeV resulting in implantation depths of hundreds of micrometers. As will be demonstrated in the Sect. 6.5, this allows to study materials deep below surface layers, as e.g. in electrode-covered solar cells.

Both on-line experimental facilities have been and are still involved in studies on Fe in Si. In the framework of this tutorial we show one such example [38]. The authors compared the Mössbauer spectra of ^{57}Fe in the decay of ^{57}Mn in two silicon samples with very different doping levels, extremely p-type (p^{++}) and highly n-type (n^{+}). They found that interstitial Fe atoms were created in silicon at 400–800 K as a result of the recoil imparted on these daughter atoms in the β-decay of ion-implanted, substitutional ^{57}Mn. Then they observed that diffusional jumps of the interstitial ^{57m}Fe cause a line broadening in their Mössbauer spectra, which is directly proportional to their diffusivity, as discussed in the preceding chapter. Thus, the charge-state-dependent diffusivity has been determined in differently doped material as shown in Fig. 6.23 [38].

6.5 On-line ^{57}Mn/^{57}Fe Implantation Mössbauer Studies at RIKEN

6.5.1 RIKEN RI Beam facility for On-line Mössbauer Spectroscopy

The Radioactive Ion Beam Factory (RIBF) accelerators will allow us to provide the world's most intense RI beams at energies of several hundreds MeV/nucleon over the whole range of atomic masses (Fig. 6.24). Recently, this facility has been upgraded by installing several heavy-ion accelerators on the bases of the former complex consisting of a K540 MeV ring cyclotron (RRC) and a couple of different types of the injectors: a variable-frequency heavy-ion linac (RILAC) and a K70 MeV AVF cyclotron (AVF). In the facility, a projectile-fragment separator (RIPS) provides the world's most intense light-atomic-mass (less than nearly 60)

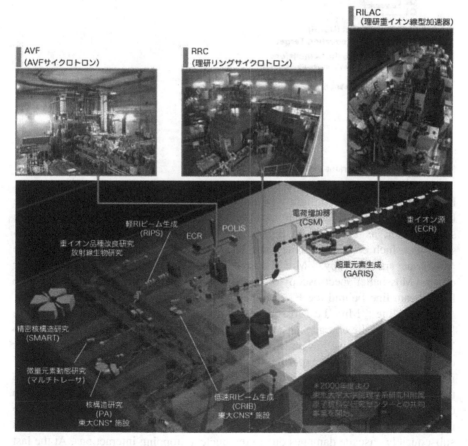

Fig. 6.24 RIKEN-RI beam facility

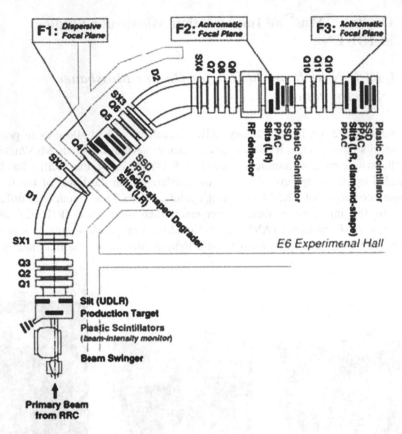

Fig. 6.25 RIKEN-separator configuration

RI beams. RIPS is an in-flight type radioactive isotope (RI) separator to produce RI beams using the fragmentation of the heavy ion beams. RIPS is an achromatic spectrometer, as is shown in Fig. 6.25, consisting of two magnetic separation sections, which enables us to separate RI beams with high resolution. For on-line Mössbauer spectroscopy, ^{57}Mn/^{57}Fe isotopes are separated by RIPS. The set-up for on-line Mössbauer spectroscopy of ^{57}Mn/^{57}Fe isotopes is installed at the end of the RIPS beam line behind the F3 chamber, as is shown in Fig. 6.26.

The on-line ^{57}Mn/^{57}Fe Mössbauer experiments were performed to clarify the lattice sites and the charge states of ^{57}Fe in Si materials at high temperatures up to 1,200 K [39–42] as well as under light illumination [43]. The implantation energy of ^{57}Mn/^{57}Fe used in this experiment is an order of GeV, which is several orders of magnitude higher than that used in other implantation experiments, and therefore the highest energy applied so far for solid state physics. At CERN, for instance, the energy is 60 keV. Generally speaking, a GeV-nuclear probe implanted into material is known to create a columnar defect due to the electrical stopping, and subsequently cascade damages due to the nuclear stopping interactions. At the last

Fig. 6.26 Mössbauer set-up scheme at RIKEN

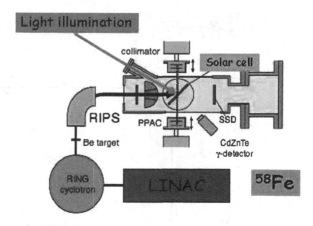

period of the landing processes, the nuclear probe will find different lattice sites close to this damage region either by landing directly on interstitial sites, or by substituting host atoms. In the case of ^{57}Mn implantation, the β-decay from ^{57}Mn to ^{57}Fe causes an additional small turbulence on the lattice sites of the probe, i.e., a recoil process of ^{57}Mn probe when emitting a highly energetic electron. The experimental results will tell us about the lattice sites as well as the charge states, as will be given in the followings.

Starting with a primary beam of ^{58}Fe, ^{57}Mn beam will be produced through the projectile fragmentation following a reaction with a Be target. Subsequently, the fragment of ^{57}Mn nuclei will be implanted into a sample, and will be stopped deeply in the matrix with an order of hundreds μm from the surface. The stopping range and their struggling, which are evaluated by TRIM code [9], are shown in Fig. 6.27 for the case of a Si crystal, and the result is compared with that of 100 keV-^{57}Fe implantation. Since the ion range of ^{57}Mn is 4 orders of magnitude higher than that of conventional low energy implantation, as described in the sections before, yielding us unique experimental conditions such as the low concentrations of ^{57}Mn as well as defects. One would speculate that GeV-implantation would produce only "amorphous", and therefore, no atomistic information could be studied through Mössbauer spectroscopy. The experimental results have, however, showed that we can send our "spy", i.e., ^{57}Mn/^{57}Fe deeply into the sample, and the spy send us, indeed, unique atomistic information on the nuclear probes and their surrounding lattice in Si materials.

6.5.2 ^{57}Mn/^{57}Fe Experiments on Si

As stated before, iron impurities in Si have been intensively investigated for more than 50 years by different experimental techniques [44] including ^{57}Fe Mössbauer spectroscopy. This is because Fe impurities can be easily incorporated into Si

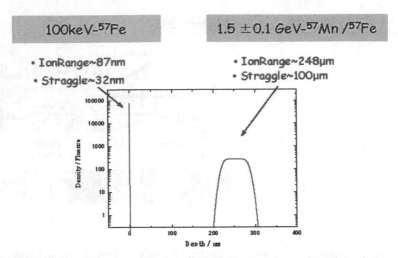

Fig. 6.27 Ion ranges of ^{57}Mn/^{57}Fe at RIKEN compared with conventional separator

matrix during the industrial processes, causing serious degradation in the electronic properties of silicon-based devices and also solar cells on one hand. Atomistic information on Fe impurities in Si, on the other hand, is essential to understand dynamical properties such as atomic diffusion as well as carrier transport. Fe impurities are known to form the deep levels in the Si band gap, producing strong trapping centres for the carriers in the devices: Interstitial Fe_i are well known to form a donor level at 0.39 eV from the valence band edge, while substitutional Fe_s is expected to form an accepter level of 0.69 eV from the first principle calculation [45]. Although interstitial and substitutional Fe have been detected as spectral components in ^{57}Fe Mössbauer experiments, the charge states could not be well distinguished experimentally so far in terms of isomer shifts.

Typical Mössbauer spectra of ^{57}Fe in n-type FZ-Si in the temperature region between 330 K and 1,200 K are shown in Fig. 6.28. The spectra from 300 to 700 K can be fitted by two singlets of Lorentzians, while the spectra from 800 up to 1,200 K can be analyzed only by a broad singlet. The two singlets at 330 K at the left and the right hand site have been assigned to ^{57}Fe atoms on interstitial and substitutional sites in Si matrix, respectively. This assignment is based on a theoretical calculation of the isomer shifts [46], which relate with the charge density at ^{57}Fe nucleus. The spectra and their fitting parameters are changing anomalously with elevating temperature. These behaviors are in good agreement with the former experiments on p-type FZ-Si wafers up to 800 K [39–41]. The resonance area of the interstitial ^{57}Fe component at the left hand side decreases strongly above 500 K.

A "peak position of a component" in the Doppler velocity scale provides us information both on the different lattice sites of ^{57}Fe atoms and their charge states. This shift is called "center shift", depending on the isomer shift and the second-order Doppler (SOD) shift. As a reference, the broken line in Fig. 6.28 shows the

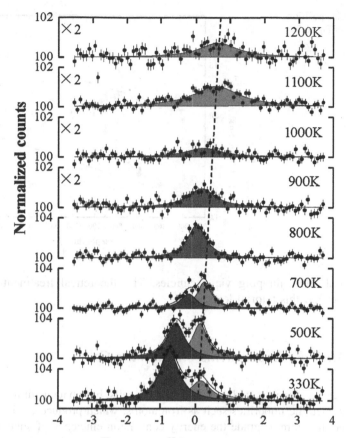

Fig. 6.28 Mössbauer spectra of ^{57}Fe in n-type FZ-Si measured between 330 and 1,200 K. Notice that the vertical scale for the region above 900 K is blown up with a factor of two, in order to show the small and broad resonance effects more clearly [42]

usual temperature dependence of a spectrum component due to the SOD shift. Both the interstitial and the substitutional components, however, do not simply follow the SOD shift. The center shift of the interstitial component deviates from the SOD shift, moving continuously to the substitutional position with increasing temperature. This anomaly is accompanied by the strong decrease of the resonance area. Furthermore, the center shifts of the substitutional singlet between 800 and 1,000 K are different from the extrapolation of the center shift, but above 1,100 K the shift goes back to the position of the substitutional again. This is essential observation to interpret the whole dynamical phenomena (Fig. 6.29).

Above 500 K interstitial Fe atoms are supposed to start migrating within the lifetime, subsequently finding vacancies. Consequently, the singlet between 800 and 1,000 K is thought to be a "motional-averaged component" due to substitutional ^{57}Fe formation. Above 1,100 K the isomer shift of the broad singlet coincides with that of substitutional Fe, indicating that Fe atoms stay dominantly on

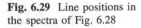

Fig. 6.29 Line positions in
the spectra of Fig. 6.28

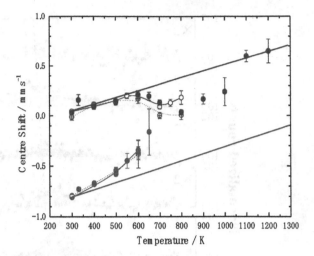

substitutional sites, jumping via vacancies. The theoretical treatment [42] is, however, beyond this tutorial level.

6.5.3 ^{57}Mn/^{57}Fe Experiments on Si Solar-Cells

Multi-crystalline silicon is widely used for solar cells, but contains different lattice defects and metallic impurities such as iron atoms, which produce carrier trapping centers, and therefore, degrade the energy conversion efficiency of solar cells. No direct observation on the charge states of Fe atoms has been achieved in multi-crystalline Si solar cells during operation, i.e. under light illumination. The Fermi Level will be shifted by injecting the excess carriers, and consequently different charge states of interstitial and substitutional Fe atoms are expected to appear in the on-line Mössbauer spectra of ^{57}Mn/^{57}Fe in mc-Si under light illumination [43]. Figure 6.30 shows the top and the back surfaces of mc-Si solar cell. Ag electrodes on an anti-reflection Si–N layer can be seen and the ^{57}Mn implantation was performed through this top surface. During a Mössbauer spectral measurement under dark condition, I–V characteristic of the p-n junction was measured every one hour, in order to control the defect accumulation due to ^{57}Mn implantation. The I–V curve, however, did not change with increasing the implantation dose of ^{57}Mn.

The spectrum of ^{57}Mn/^{57}Fe in the p-region of the p-n junction, i.e., solar cell, was measured at 400 K under light illumination. In Fig. 6.31a this spectrum (red points) was compared with that in p-type multi-crystalline (mc)-Si (Black points). The "black spectrum" consists of two components, as is shown with red and green components in Fig. 6.31a. They are assigned to interstitial and substitutional Fe in Si matrix with the isomer shifts of 0.8 and −0.06 mms^{-1}, respectively. On the other hand, the "red spectrum" of the solar cell is very broad, and therefore, is

Fig. 6.30 Top and back surfaces of a silicon solar cell (30 × 30 mm in square) which was specially provided by Kyocera

Fig. 6.31 a Mössbauer spectrum of ^{57}Mn/^{57}Fe in p-type multi-crystalline-Si (*black points*) is compared with that in Si-solar cells (*red points*) at 400 K under Xe lamp illumination. The *black spectrum* is fitted with *red* and *green* components. **b** The difference of the *red* and *black spectra* is shown with blue points. The difference spectrum can be analyzed by the appearing *three* singlets (*pink, yellow*, and *light blue* components and the *two* disappearing singlets (*red* and *green* components)

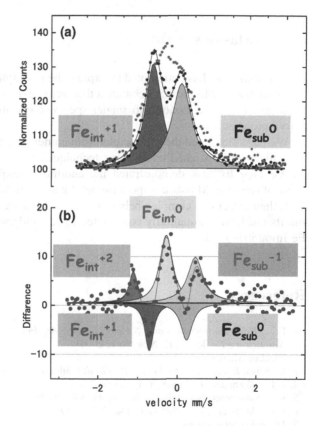

difficult to analyze by a superposition of singlets. To interpret such broad spectrum, we subtract the "black spectrum" from the "red spectrum". The difference spectrum (blue points) in Fig. 6.31b can be fitted by three appearing components (pink, yellow, and light blue) and two disappearing ones (red and green).

These singlets are considered to correspond to different charge states of interstitial Fe_i and substitutional Fe_s: Fe_i^{2+}, Fe_i^{1+}, Fe_i^0, Fe_s^0, and Fe_s^{-1} are colored by pink, red, yellow, green and light blue, respectively in Fig. 6.31b. The isomer shift values are in good agreement with those of previous absorber experiments obtained in ^{57}Fe-diffused mc-Si [47]. Notice that the ^{57}Mn probes were implanted into the p-type region of mc-Si solar cell, which is the same material as the p-type mc-Si wafer. The present results indicate that the light illumination changes not only in the Fermi level (quasi-Fermi level) by the excess carrier injection, but also in the carrier trapping processes. The latter must be due to the directional excess carrier flow through the p-n junction, which affects the carrier trapping kinetics with electrons and holes at the Fe impurities, leading to the different charge states on both Fe substitutional and interstitial sites in the p-region in mc-Si solar cell. This is, in fact, the first in situ observation of the carrier trapping processes at Fe impurities in mc-Si solar cell, which degrades the energy conversion efficiency.

6.6 Conclusions

In this tutorial we have attempted to explain the principles of ion implantation and of the dedicated Mössbauer techniques that were developed to perform off-line, in-beam and on-line emission Mössbauer spectroscopy after implantation of radioactive probe atoms.

We have illustrated these principles with numerous examples, focusing on Fe in Si, which can be studied by all three techniques.

We hope to have demonstrated the enormous resolving power on atomistic scale of emission Mössbauer spectroscopy for such studies. For Fe impurities in Si, with their extremely complex behavior, these techniques have clearly shown their merits and have substantially contributed to our understanding of the behavior of Fe impurities in Si.

References

1. E.L. Wolf, *Nanophysics and Nanotechnology: An Introduction to Modern Concepts in Nanoscience* (Wiley-VCH, New York, 2006)
2. Electron Microscope
3. C. Kittel, *Introduction to Solid State Physics*, 8th edn. (Wiley, New York, 2005
4. R.L. Mössbauer, Z. Physik, **151**, 124 (1958)
5. R.L. Mössbauer, Naturwissenschaften, **45**, 538 (1958)
6. R.L. Mössbauer, Z. Naturforsch. **14a**, 211 (1959)
7. Hyperfine Interactions
8. L.C. Feldman, J.W. Mayer (eds.), *Fundamentals of Surface and Thin Film Analysis.* (Appleton and Lange, New York, 1986)
9. J.F. Ziegler, J.P. Biersack, U. Littmark, *The Stopping and Range of Ions in Solids* ed. (Pergamon Press, New York, 1985); http://www.srim.org/

10. G. Langouche, *Hyperfine Interaction of Defects in Semiconductors* (Elsevier, Amsterdam, 1992)
11. G. Langouche, in *Mössbauer Spectroscopy Applied to Inorganic Chemistry*, G. Long, F. Grandjean (eds.), Vol. 3, (Plenum Press, New York and London, 1989), pp. 445–512
12. H. de Waard, S.A. Drentje, Phys. Lett. **20**, 38 (1966)
13. G.L. Latshaw, Stanford University, PhD Thesis, 1971
14. L. Niesen, Hyperfine interact. **13**, 65–88 (1983)
15. G. Weyer, Hyperfine Interact. **27**, 249–262 (1986)
16. H. de Waard, Hyperfine Interact. **40**, 31–48 (1988)
17. G. Langouche, Hyperfine Interact. **45**, 199–216 (1989)
18. G. Langouche, Hyperfine Interact. **72**, 217–228 (1992)
19. M. de Coster, H. Pollak, S. Amelinckx, in *Proceedings of the 2nd International Conference on the Mössbauer Effect,* D.M.J. Compton, A.H. Schoen (eds.) (Wiley, New York, 1962), p. 289
20. P.C. Norem, G.K. Wertheim, J. Phys. Chem. Solids **23**, 1111 (1962)
21. G. Langouche, M. de Potter, I. Dézsi, M. Van Rossum, Radiat. Effect Lett. **67**, 404 (1982)
22. J.A. Sawicki, B.D. Sawicka, Phys. Stat. Sol. b **86**, K159 (1978)
23. G.L. Latshaw, P.B. Russell, S.S. Hanna, Hyperfine Interact. **8**, 105–127 (1980)
24. J.A. Sawicka, B.D., Sawicki, J.A. Phys. Lett. A **64**, 311 (1977)
25. G. Langouche, M. de Potter, Nucl. Instrum. Methods B **19/20**, 322 (1987)
26. P. Schwalbach, S. Laubach, M. Hartick, E. Kankeleit, B. Keck, M. Menningen, R. Sielemann, Phys. Rev. Lett. **64**, 1274 (1990)
27. G. Langouche, M. de Potter, D. Schroyen, Phys. Rev. Lett. **53**, 1364 (1984)
28. W. Bergholz, Physica B **16**, 312 (1983)
29. M. Menningen, R. Sieleman, G. Vogl, Y. Yoshida, K. Bonde-Nielsen, G. Weyer, Europhys. Lett. **3**, 927–933 (1987)
30. A. Heiming, K.H. Steinmetzt, G. Vogl, Y. Yoshida, J. Phys. F: Met. Phys. **18**, 1491–1503 (1988)
31. Y. Yoshida, M. Menningen, R. Sielemann, G. Vogl, G. Weyer, K. Schroeder, Phys. Rev. Lett. **61**, 195 (1988)
32. Y. Yoshida, Hyperfine Interact. **47**, 95–113 (1989)
33. R. Sielemann, Y. Yoshida, Hyperfine Interact. **68**, 119–130 (1991)
34. B. Keck, R.Sielemann, Y. Yoshida, Phys. Rev.Lett.
35. D. Forkel-Wirth, ISOLDE laboratory portrait. Hyperfine Interact. **129** (2000)
36. G. Weyer, Hyperfine Interact. **129**, 371–390 (2000)
37. G. Weyer, J.W. Petersen, S. Damgaard, H.L. Nielsen, Phys. Rev. Lett. **44**, 155–157 (1980)
38. H.P. Gunnlaugsson, G. Weyer, M. Dietrich and the ISOLDE collaboration, M. Fanciulli, K. Bharuth-Ram, R. Sielemann, Appl. Phys. Lett. **80**, 2657–2659 (2002)
39. Y. Kobayashi, Y. Yoshida et al., Hyperfine Interact. **126**, 417 (2000)
40. Y. Yoshida, K. Kobayashi et al., Defect Diffus. Forum **194–199**, 611 (2001)
41. Y. Yoshida; ALTECH 2003 Analytical and Diagnostic Techniques for Semiconductor Materials, Devices, and Processes, 479 (2003)
42. Y. Yoshida, Y. Kobayashi, K. Hayakawa, K. Yukihira, A. Yoshida, H. Ueno, F. Shimura, F. Ambe; Physica B, **376-377**, 69 (2006)
43. Y. Yoshida, K. Suzuki, Y. Kobayashi, T. Nagatomo, Y. Akiyama, K. Yukihira, K. Hayakawa, H. Ueno, A. Yoshimi, D. Nagae, K. Asahi, G. Langouche, Hyperfine Interact. **204**, 133–137 (2012)
44. A.A. Istratov, H. Hieslmair and E. R. Weber; Appl. Phys. A **69**, 13 (1999)
45. S. K. Estreicher, M. Sanati, N. Gonzalez Szawacki, Phys. Rev. B, **77**, 125214 (2008)
46. J. Kübler, A. E. Kumm, H. Overhof, P. Schwalbach, M. Hartick, E. Kankeleit, B. Keck, L.Wende, R.Sielemann, Z. Phys., B **92**, 155 (1993)
47. Y. Yoshida, S. Horie, K. Niira, K. Fukui and K. Shirasawa; Physica B, **376–377**, 227 (2006)
48. T. Diaz de la Rubia and G. H. ilmer, Phys. Rev.Lett., **74**, 2507-2510 (1995)

Author Biographies

Guido Langouche

Since 2010 Guido Langouche is emeritus professor in nuclear solid state physics at the University of Leuven. After obtaining his doctoral and habilitation degrees from K.U.Leuven, he was post-doc at the universities of Stanford and Groningen and guest professor at the universities of Osaka, Lyon and Kinshasa.
From 1995 till 2005 he was vice-rector of K.U.Leuven.
From 2005 till 2010 he was chairman of the Coimbra Group, an academic collaboration network of 40 of Europe's longest-established research-intensive universities.
He is presently vice-president of NVAO, the *Accreditation Agency for the Netherlands and Flanders*, residing in The Hague, where he was appointed in 2007 jointly by the Dutch and Flemish Ministers of Education.
Since 2011 he is also Secretary of INQAAHE, the *International Network for Quality Assurance Agencies in Higher Education*.
He is editor-in-chief of the *Hyperfine Interactions* journal.

Representative publication

1. G. Langouche "Hyperfine interactions of defects in semiconductors" Elsevier 1992, 489 pages.
2. G. Langouche "Characterization of semiconductors by Mössbauer Spectroscopy" in "Mössbauer Spectroscopy applied to inorganic chemistry – Volume 2" p 445-512, editors G. Long and F. Grandjean, Plenum Press, New York 1989.
3. G. Langouche "Ion implantation in semiconductors studied by Mössbauer Spectroscopy", Hyperfine Interactions 45 (1989) 199–216.
4. G. Langouche "Ion implantation" Hyperfine Interactions 68 (1991) 95–106.

Yutaka Yoshida

Since 2004 Yutaka Yoshida is professor in Materials and Life Science at Shizuoka Institute of Science and Technology, Japan, following associate professor at the same institute between 1991 and 2003. After obtaining his doctoral degree from Osaka University under the guidance of emeritus Professor F.E. Fujita, and he stayed in the group of Professor Gero Vogl as a guest scientist at the Hahn-Meitner Institute Berlin, Germany, in the period between 1983 and 1985, and 1990, and also as a research assistant at the Institute of solid state physics, Universität Wien, Austria, between 1986 and 1989.

Since 1993 he is a guest Scientist at the RIKEN, Japan. He is the chairperson of ICAME 2011 at Kobe.

Representative publication:

1. Y. Yoshida; "In-Beam Mössbauer Study of Atomic Jump Processes in Metals", Hyperfine Interactions, 47, 95-113 (1989).
2. Y. Yoshida and F. Shimura, "In-Situ Observation of Diffusion and Segregation of Fe atoms in Si Crystals at High Temperature by Mössbauer spectroscopy", in Electrochemical Society Proceedings, 98-1, 984-996 (1998).
3. Y. Yoshida, "Mössbauer spectroscopy to investigate atomistic jump processes on an atomistic scale", Hyperfine Interactions, 113, 183-198 (1998).
4. Y. Yoshida, "Direct Observation of Substitutional and Interstitial Fe atoms in Si by high-temperature and In-beam Mossbauer Spectroscopy", in ALTECH 2003, ECS, Salt Lake City, US, 479-482 (2003).

Index

Y. Yoshida and G. Langouche (eds.), *Mössbauer Spectroscopy*,
DOI: 10.1007/978-3-642-32220-4, © Springer-Verlag Berlin Heidelberg 2013